DATE			

BAKER & TAYLOR

CELLULOSICS:
CHEMICAL, BIOCHEMICAL AND MATERIAL ASPECTS

CELLULOSICS: CHEMICAL, BIOCHEMICAL AND MATERIAL ASPECTS

Editors:

J. F. KENNEDY B.Sc., Ph.D., D.Sc., C.Biol., C.Chem., F.I.Biol, F.I.F.S.T., F.I. Mgt., F.R.S.C.
Director of the Research Laboratory for the Chemistry
of Bioactive Carbohydrates and Proteins,
School of Chemistry, University of Birmingham, England,
and Professor of Applied Chemistry North East Wales
Institute of Higher Education, Wrexham, Wales

G. O. PHILLIPS B.Sc., Ph.D., D.Sc., F.R.S.C.
Chairman of Newtech Ltd, Deeside, Wales

P.A. WILLIAMS B.Sc., Ph.D., C.Chem., F.R.S.C.
Head of the Polymer and Colloid Chemistry Group
North East Wales Institute of Higher Education, Wrexham,
Wales

ELLIS HORWOOD
NEW YORK   LONDON   TORONTO   SYDNEY   TOKYO   SINGAPORE

First published 1993 by
Ellis Horwood Limited
Market Cross House, Cooper Street
Chichester
West Sussex, PO19 1EB
A division of
Simon & Schuster International Group

© Ellis Horwood Limited   1993

Printed and bound in Great Britain
by Bookcraft, Midsomer Norton

British Library Cataloguing in Publication Data

A catalogue record for this book is available from the British Library

ISBN 0–13–053042–5

Library of Congress Cataloging-in-Publication Data

Available from the publisher

1 2 3 4 5   97 96 95 94 93

ACKNOWLEDGEMENTS

This volume arises from the Internatonal Conference on cellulose etc., - Cellulose 91 - which was a joint meeting of the Cellucon Conferences (operated by the Cellucon Trust), the ACS Cellulose, Paper and Textile Division, and the Syracuse Cellulose Conferences.

MEMBERS OF THE ORGANISING COMMITTEE – CELLULOSE 91

This meeting owed its success to the invaluable assistance of the Organising Committee.

Dr. N.R. Bertoniere (Co-Chair)	ACS Cellulose, Paper and Textile Division and Southern Regional Research Center, USDA, USA.
Prof. G.O. Phillips (Co-Chair)	The Cellucon Trust and Newtech Ltd. Clwyd, UK.
Prof. T.E. Timell (Co-Chair)	Syracuse Cellulose Conferences and SUNY College of Environmental Science and Forest, New York, USA.
Dr. H.L. Chum (Programme Chair)	Solar Energy Research Institute, Colorado, USA.
Dr. E.J. Blanchard	Solar Energy Research Institute, Colorado, USA.
Prof. W.G. Glasser	Virginia Tech., Virginia, USA.
Dr. H. Hatakeyama	Industrial Products Research Institute, Tsukuba, Japan.
Prof. J.F. Kennedy	The North East Wales Institute and The University of Birmingham, Birmingham, UK.
Prof. H.L. Needles	University of California, USA.
Dr. R.P. Overend	Solar Energy Research Institute, Colorado, USA.
Prof. A. Sarko	SUNY College of Environmental Science and Forestry, New York, USA.
Dr. P.A. Williams	The North East Wales Institute, Clwyd, UK.
Prof. W.T. Winter	SUNY College of Environmental Science and Forestry, New York, USA.

INDUSTRIAL SPONSORSHIP

The organisers express their grateful thanks to the following for their financial support of Cellulose 91.

U.S. Department of Energy
  Conservation and Renewable Energy
    Office of Industrial Technologies
    Office of Transportation Technologies
    Office of Utility Technologies
National Renewal Energy Laboratory
  (formerly Solar Energy Research Institute)
U.S. Department of Agriculture
  Southern Regional Research Center
Aqualon Co., USA
Asahi Chemical Co., Japan
Daicel Chemical Industries Ltd., Japan
ITT Rayonier, Inc., USA
Kodak/Tennessee Eastman, USA
Monsanto, USA
NOVO Nordisk, Denmark
The Proctor & Gamble Cellulose Co., USA
Westvac, USA

THE CELLUCON TRUST

incorporating

CELLUCON CONFERENCES

International Educational Scientific Meetings
on Cellulose, Cellulosics and Wood

CELLUCON CONFERENCES AND THE CELLUCON TRUST

Cellucon Conferences as an organisation was initiated in 1982, Cellucon '84 was the original Conference, which set out to establish the strength of British expertise in the field of cellulose and its derivatives. This laid the foundation for subsequent conferences in Wales (1986), Japan (1988), Wales (1989), Czecho-Slovakia (1990) and USA (1991). They have had truly international audiences drawn from the major industries involved in the production and use of cellulose pulp and derivatives of cellulose, plus representatives of academic institutions and government research centres. This diverse audience has allowed the cross fertilization of many ideas which has done much to give the cellulose field the higher profile that it rightfully deserves.

Cellucon Conferences are organised by The Cellucon Trust, an official UK charitable trust with worldwide objectives in education in wood and cellulosics. The Cellucon Trust is continuing to extend the knowledge of all aspects of cellulose worldwide. At least one book has been published from each Cellucon Conference as the proceedings thereof. This volume is one of two volumes arising from the 1991 conference held in New Orleans, USA, and the conferences in Wales 1992, Sweden 1993 etc., will generate further useful books in the area.

Mr. T. Greenway	Berol Nobel Ltd., UK
Dr. J. Guthrie	University of Leeds, UK
Dr. H. Hatakeyama	Industrial Products Research Institute, Japan
Dr. A. Henderson	DOW Chemicals Europe, Switzerland
Dr. M.B. Huglin	University of Salford, UK
Dr. P. Levison	Whatman Specialty Products Division, UK
Dr. J. Meadows	The North East Wales Institute, UK
Prof. Y. Nakamura	Gunma University, Japan
Mr. W.B. Painting	Hoechst (UK) Ltd., UK
Mr. A. Poyner	Aqualon (UK) Ltd., UK
Mr. R. Price	Shotton Paper Co. Ltd., UK
Prof. J. Roberts	Institute of Science and Technology, University of Manchester, UK **CUT**
Mr. J.F. Webber	The Forestry Commission, UK
Dr. C.A. White	Fisons Plc, Scientific Equipment Division, UK

Cellucon Conferences are sponsored by:

The Biochemical Society - Chembiotech Ltd., - Hoechst (UK) Ltd., - The North East Wales Institute - USAF European Office of Aerospace Research and Development - US Army Research, Development and Standardisation Group, UK - Welsh Development Agency - Whatman Specialty Products Division.

Cellucon Conferences are supported by:

The American Chemical Society (Cellulose, Paper and Textile Divison) - Aqualon (UK) Ltd., - Berol Nobel Ltd., - Courtaulds Chemicals and Plastics - Ministry of Defence - Crown Berger Europe Ltd., - DOW Chemicals - The Forestry Commission - Shotton Paper Company Ltd., - Syracuse Cellulose Conferences.

Table of contents

Contents

viii   Contents

Contents

PREFACE

Cellulose '91, by any standards, was a monumental International meeting. Three major organisations, with a proven track record in organising cellulose conferences joined forces to celebrate the 50th Anniversary of the Southern Regional Research Center of the United States Department of Agriculture. These were:

- ACS Cellulose, Paper & Textile Division
- Cellucon Conferences
- Syracuse Cellulose Conferences

What could be more appropriate? Celebrating the fifty years of research achievement by the famed Southern Regional in the historic home of cotton - New Orleans. The venue alone ensured a massive turn-out of 500 scientists and camp followers. The meetings were held in the old-style Monteleone Hotel in the French Quarter; dinner was served on the Cajun Queen on the Mississippi River; the daily creole cuisine all provided the stage for a memorable Conference.

The science more than matched the occasion, as this volume testifies. Europe, Japan and the USA combined to provide a high-level cross-section of the world's best cellulose research. The output, set out in this book, confirms the resurgence in cellulose research, development and application. Lignocellulosic sources now offer the most versatile new functional, structural and varied industrial products which are environmentally friendly.

The areas covered in this book support this generalisation:

- Cellulose blends and composites
- Cellulosic enzymes
- New cellulose fibers
- Cellulose fiber-reinforced plastics

- New pulping processes
- High performance derivatives
- New fabrics and finishes
- Biogenesis of cellulose and related polymers
- Cellulosic liquid crystals
- Innovations in cellulose fiber technology
- High performance polymers from lignocellulosics
- Biological degradation

Some years ago I observed the headline in the famous New Orleans daily, the Times Picayune: *KING COTTON RIDES AGAIN*. This would be an appropriate title to this volume, if were not for the fact that a wide variety of cellulosic and lignocellulosic sources are encompassed in the papers. The series of volumes, on cellulosic research, of which this book forms the latest addition, have provided a great stimulus for cellulosic research on the five continents. We all came together in New Orleans in a spirit of friendship and cooperation in the interests of cellulose research. May we thank all who made this memorable meeting possible.

NOELIE R. BERTONIERE, CO-CHAIR CELLULOSE '91

GLYN O.PHILLIPS, CO-CHAIR, CELLULOSE '91

TORE E. TIMELL, CO-CHAIR, CELLULOSE '91

PART 1

1

Molecular genetics of cellulose synthesis in *Agrobacterium tumefaciens*

Ann G. Matthysse - Department of Biology, CB # 3280, University of North Carolina, Chapel Hill, NC 27599, USA.

ABSTRACT

Agrobacterium tumefaciens is a plant pathogenic bacterium which elaborates cellulose fibrils when it is in contact with host cells. The role of these fibrils appears to be to anchor the bacteria firmly to the host cell surface. Cellulose synthesis is induced by low molecular weight compounds from the plant cells. The genes required for cellulose synthesis are clustered on the bacterial chromosome in a 15 kb region.

INTRODUCTION

Infections of dicotyledonous plants with *Agrobacterium tumefaciens* result in the formation of crown gall tumors. This gram-negative bacterium is found in soil with a world-wide distribution. It generally infects plants through a wound. Wounds at the region where the stem of the plant enters the soil (the crown of the plant) represent the most common site of infection; hence the name of the disease, crown gall. The mechanism of pathogenesis involves the transfer of a segment of plasmid DNA (the T or transferred region of the tumor-inducing plasmid, pTi) from the bacterium to the plant host cell. The T DNA becomes integrated into the host cell chromosomes and is expressed; it encodes enzymes which are involved in the synthesis of two plant growth hormones, auxin and cytokinin. The genes encoding these enzymes have constitutive plant promoters and it is their unregulated expression which gives rise to the formation of a tumor at the infection site. The T DNA also encodes enzymes involved in the synthesis and secretion of compounds referred to as opines (some of these opines are modified amino acids, for example, octopine made by the conjugation of pyruvic acid and the alpha amino group of arginine). These opines are not used by the plant cells which make them. Instead they are secreted and the bacteria use them as carbon and nitrogen sources. This allows the bacterium to tap into the energy and carbon compounds available to the plant through photosynthesis [1].

In order to transfer DNA to the plant cell the bacteria must attach to the plant cell surface. In my laboratory we have examined the attachment of the bacteria to plant cells. Because bacterial infections of wound sites in whole plants are very difficult to study we have examined the interaction of bacteria with suspension culture cells derived from a variety of plants including tobacco and carrot. The bacteria can transfer T DNA to such plant suspension culture cells and the system appears to be a reasonable model for the bacterial-plant interaction in the wound site.

ATTACHMENT OF *A. TUMEFACIENS* TO TISSUE CULTURE CELLS

The binding of bacteria to plant cells was examined by adding bacteria to a suspension culture of plant cells, incubating the two together for varying times and then filtering the bacteria-plant cell mixture through Miracloth which allows the passage of free bacteria but retains plant cells along with any bacteria bound on the plant cell surface. The numbers of free and bound bacteria are then determined. When *A. tumefaciens* was added to carrot suspension culture cells, the bacteria bound to the surface of the plant cells in increasing numbers with increasing times of incubation (Figure 1). In the light microscope bacteria were visible on the surface

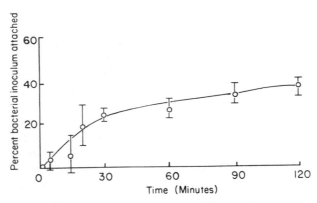

Figure 1. Time course of attachment of *A. tumefaciens* to carrot cells [2].

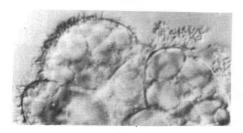

Figure 2. Bacteria bound to the surface of live carrot cells as seen in the light microscope with Nomarski optics [4].

of the carrot cells; both individually attached bacteria and large clusters of attached bacteria could be seen (Figure 2). When bacterial attachment was examined in the scanning electron microscope (SEM) a net of fibrillar material was observed covering the surface of the plant cells after about 12 hours incubation with bacteria. Only a minority of the bacteria were directly attached to the plant cell surface. The majority of the bacteria were found in large aggregates attached to the plant cell surface only at the base of the aggregate. The bacteria appeared to be trapped in these aggregates by a network of thin fibrillar material. These fibrils were made by the bacteria and not by the plant cells since they were observed when live bacteria bound to the surface of heat-killed carrot cells. The bacteria could be induced to elaborate fibrils with a similar appearance when grown in pure culture by the addition of plant extracts such as soytone to the culture medium (Figure 3A & B). The fibrils were purified from bacteria grown with soytone and shown by several techniques including infrared spectroscopy to be composed of cellulose [2,3].

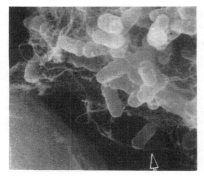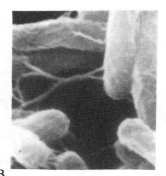

A.   B.

Figure 3. A. Bacteria bound to carrot cells as seen in the SEM B. Bacteria grown in pure culture with soytone [2].

The production of cellulose represents an interesting pathogenic strategy on the part of the bacteria. The plant may not recognize this bacterial cellulose as foreign; even if the plant does recognize the bacterial cellulose fibrils as foreign it can not digest the fibrils without at the same time digesting its own cell wall. A model for the attachment of *A. tumefaciens* to plant cells was proposed (Figure 4). In this model bacterial attachment to carrot cells is a two step process. The first step involves bacterial attachment to the plant surface using bacterial binding sites located in the outer membrane and plant receptors located on the surface of the plant cells. This initial binding is loose and reversible. Substances coming from the plant cell induce the bacteria to make cellulose fibrils which anchor the bacteria tightly to the plant cell surface and entrap additional bacteria forming aggregates. In order to test this model and to examine the role of cellulose in bacterial pathogenesis bacterial mutants which were altered in their ability to attach to plant cells were obtained [3].

CELLULOSE SYNTHESIS MUTANTS OF *A. TUMEFACIENS*

Transposon mutagenesis with Tn*5* was used to obtain mutants of *A. tumefaciens* which were altered in their ability to bind to plant cells. Nonattaching mutants were obtained by actually screening individual mutants for their ability to bind to carrot cells. Bacteria were incubated with plant cells overnight and the results observed in the light microscope. All mutants which failed to bind to carrot cells were avirulent suggesting that the initial binding of the bacteria to host cells is required for DNA transfer and virulence. Despite their inability to bind to plant cells these nonattaching mutants were all able to synthesize cellulose when incubated with plant extracts [4].

Cellulose-minus mutants were obtained by plating bacteria mutagenized with Tn*5* on plates containing neomycin (to select for those bacteria carrying Tn*5*) and the dye cellufluor which fluoresces under ultraviolet light with many b-linked polysaccharides including cellulose. Those

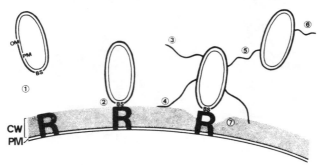

Figure 4. A model of the binding of *A. tumefaciens* to plant cells [2].

colonies which failed to fluoresce were screened further for failure to produce material which fluoresced with cellufluor under a variety of growth conditions. The resulting cellulose-minus mutants failed to make fibrillar material when incubated with plant cells [5]. These *cel* mutants retained virulence, but the virulence of many of these mutants was found to be markedly reduced (10 to 100-fold) using a semi-quantitative virulence assay [6].

The location of the Tn*5* insertion in the cellulose-minus mutants was mapped on the bacterial chromosome. All of the *cel* mutants obtained mapped to one location. Two mutants were obtained in the original mutagenesis which overproduced cellulose. One of these mutants mapped very near to the *cel* mutants on the chromosome; the other mapped about 1/3 of the total chromosome away from the location of the *cel* mutants [7]. These mutants (Cel-B1 and Cel-B4) which overproduce cellulose have not been characterized further at the present time.

ROLE OF BACTERIAL CELLULOSE IN CROWN GALL DISEASE

All wild type strains of Agrobacteria of all three biotypes which we have examined produce cellulose. Cellulose production was not required for virulence in the laboratory. This raises the question of the role of bacterial cellulose production in nature. Why do the bacteria produce what must be a rather expensive material in response to the presence of plant cells? To examine the answer to this question leaves of *Bryophyllum daigremontiana* were inoculated with a wild type bacterial strain and *cel* mutants of the same strain using tooth pick wounds. If the plants were left undisturbed after inoculation, then all inoculated sites formed tumors. However, if the leaves were washed slowly with 50 ml of water 2 hours after inoculation, then only the sites inoculated with the wild type strain formed tumors. The *cel* mutants were removed by the gentle water wash. Thus the role of cellulose production by *Agrobacterium tumefaciens* appears to be to anchor the bacteria tightly to the surface of the plant [5].

IN VITRO CELLULOSE SYNTHESIS

We have obtained the incorporation of ^{14}C-UDP-glucose into material which appears to be cellulose using membrane vesicles prepared from *A. tumefaciens*. We have found it helpful to use mutants of *A. tumefaciens* which fail to synthesize b-1,2-glucan (*chvB* mutants) since the precursor for both this polysaccharide and for cellulose is UDP-glucose. Incorporation of UDP-glucose required divalent cations. We were unable to detect a requirement for cyclic diguanylic acid in this system, although this compound, which is required for cellulose synthesis in extracts of *Acetobacter xylinum*, has been shown to be present in *A. tumefaciens*

[8,9]. However, the possibility exists that our extracts already contained a saturating amount of cyclic diguanylic acid. No incorporation of UDP-glucose into material which is insoluble in base and ethanol was obtained with extracts from *cel chvB* double mutants.

CHARACTERIZATION OF *CEL* GENES

We have identified a clone from a library of *A. tumefaciens* DNA constructed in the vector pCP13 by S. Farrand which complemented all of our *cel* mutants. The region of DNA involved appeared to be about 15kb. We have used mutagenesis with the promoter probe transposon Tn3HoHo1 to identify the location and direction of open reading frames. This transposon carries a promoterless b-galactosidase gene at one end and since *A. tumefaciens* does not make b-galactosidase we can examine the control of the expression of the *cel* genes using Tn3HoHo1. Beta-galactosidase expression from Tn3HoHo1 insertions in some locations in the *cel* genes appeared to be increased by the addition of plant extracts to the bacterial culture medium. The active substance in the extracts was able to pass through a dialysis membrane with a 1,000 d retention size. We are currently in the process of identifying this inducing molecule and of sequencing the *cel* genes.

ACKNOWLEDGEMENT

This work was funded by a grant from NSF, number DCB-8916586.

REFERENCES

1. See, *e.g.* Ream, W. *Ann. Rev. Phytopathol.* **1989**, *27*, 583.
2. Matthysse, A. G.; Holmes, K. V.; Gurlitz, R. H. G. *J. Bacteriol.* **1981**, *145*, 583.
3. Matthysse, A. G. *CRC Crit. Rev. Microbiol.* **1986**, *13*, 281.
4. Matthysse, A. G. *J. Bacteriol.* **1987**, *169*, 313.
5. Matthysse, A. G. *J. Bacteriol.* **1983**, *154*, 906.
6. Minnemeyer, S. L.; Lightfoot, R.; Matthysse, A. G. *J. Bacteriol.* **1991**, *173*, 7723.
7. Robertson, J. L.; Holiday, T.; Matthysse, A. G. *J. Bacteriol.* **1988**, *170*, 1408.
8. Amikam, D.; Benziman, M. *J. Bacteriol.* **1989**, *171*, 6649.
9. Ross, P.; Aloni, Y.; Weinhouse, C.; Michaeli, D.; Weinberger-Ohana, P.; Mayer, R.; Benzimaan, M. *FEBS Lett.* **1985**, *186*, 191.

2

Fungal cellulose biosynthesis

M. Fèvre, V. Bulone and V. Girard - Laboratoire de Biologie Cellulaire
Fongique, Centre de Génétique Moléculaire et Cellulaire, UMR 106-CNRS,
Université Claude Bernard Lyon 1, 43, Boulevard du 11 Novembre 1918, 69622
Villeurbanne Cedex, France.

ABSTRACT

The structural components of hyphal cell walls from *Saprolegnia* are 1,3-ß-D-glucans
and cellulose which is poorly crystalline in spite of its microfibrillar structure. Particulate
or solubilized enzymes synthesize *in vitro* 1,4-ß-D- or 1,3-ß-D-glucans from UDP-D-
glucose. 1,4-ß-D-Glucan synthase activity but not 1,3-ß-D-glucan synthase, is
stimulated by nucleotides (ATP, GTP) and c-di-GMP, the activator of *Acetobacter
xylinum* cellulose synthase. Purified 1,4-ß-D- and 1,3-ß-D-glucan synthases are
respectively characterized by polypeptides of 52, 58, 60 kDa and 34, 48, 50 kDa. Due
to their differences in polypeptide composition and regulation, fungal 1,4-ß-D- and
1,3-ß-D-glucan synthases appear to be distinct enzymes but fungal and bacterial
cellulose synthases seem to share a similar organization or structure.

INTRODUCTION

Fungal cell walls have a very simple organization and composition compared to those
of higher plant cell walls. They have been divided into two groups according to their
composition : chitinous fungi which contain chitin and 1,3-ß-D-glucans in their cell
walls and cellulosic fungi (i.e. Oomycetes) which have cellulose instead of chitin. Cell
walls of Oomycetes have cellulose in common with higher plants and certain bacteria
and callose type glucans in common with other fungi and higher plants.
The gram negative bacteria *Acetobacter xylinum* produced cellulose as an
extracellular product and the bacterial cellulose synthases have been well
characterized at the biochemical and molecular level (1-2-3). The demonstation of *in*

vitro 1,4-ß-D-glucan synthesis by enzymatic preparation from higher plants has been difficult. In contrast, 1,3-ß-D-glucans synthases exhibit a high level of activity *in vitro* while in intact cells these enzymes are fully latent.

As cellulose and 1,3-ß-D-glucans are the constitutive cell wall components of *Saprolegnia monoica*, this Oomycete fungus provides the opportunity to study both enzyme activities and to try to understand the differences and similarities in cellulose synthesis that may exist between procaryotic and eucaryotic cells.

MATERIALS AND METHODS

Isolation of Solubilized Enzymes

Membrane-bound enzymes were isolated by differential centrifugation as described previously [4]. Enzymes were solubilized by treating intracellular membranes with an equal volume of Chaps (10 mg.ml^{-1}). After 30 min at 4°C, the suspension was centrifuged at 48 000 g for 1 h. The supernatant was used as the solubilized enzymes. Chaps-solubilized enzymes were entrapped following incubation in assay mixture for 1,3-ß-D-glucan synthesis, then collected by low speed centrifugation [4,5].

When enzymes were assayed for 1,4-ß-D-glucan synthesis, the complete assay mixture contained 0,3 nmol UDP-[^{14}C]-D-glucose, 0,1 µmol DTT, 2 µmol of cellobiose, 5.5 µmol of MgCl$_2$, 25 mM Pipes-Tris buffer (pH 6.0). When enzymes were assayed for 1,3-ß-D-glucan synthesis, MgCl$_2$ was replaced by 400 nmol UDP-[^{12}C]-D-glucose [4].

Glycerol density gradient centifugation

Linear gradients (30 ml) were made from 1,025 and 1,113 g.cm^{-3} glycerol is 0.01 M Tris HCl buffer pH 7.4 ; 2 ml enzyme samples were layered on the gradient and centrifuged at 120 000 g for 30 min to 4 h in a Beckman SW 27 rotor [5]. 2 ml fractions were collected and assayed for enzyme activities.

RESULTS

Polypeptide composition of 1,3-ß-D- and 1,4-ß-D-glucan synthases.

Membrane-bound glucan synthases have been difficult to purifiy because these enzymes are unstable, however using Chaps as a detergent, enzymes were efficiently solubilized retaining a high activity [5]. Separation of the enzyme activities was achieved when Chaps-solubilized enzymes were centrifuged on glycerol gradients. 1,3-ß-D- and 1,4-ß-D-Glucan synthases sedimented as sharp overlapping peaks, ahead of the bulk of solubilized proteins, 1,4-ß-D-glucan synthase exhibiting a higher apparent mol wt.

Improvement of the enzyme purification was achieved using an entrapment procedure and gradient centrifugation. Microfibrillar glucans were synthesized *in vitro* using solubilized enzymes, collected by centrifugation then resuspended in extraction buffer. Most of the 1,3-ß-D-glucan synthases were recovered in the supernatant. On the contrary 1,4-ß-D-glucan synthase activity were still associated with the aggregated glucans and collected in the pellet. Glycerol gradient centrifugation achieved a better purification of each enzymatic fraction (Figure 1). SDS-PAGE analysis of the most purified fractions were characterized by three main polypeptides of different mol wt : 34, 48 and 50 kDa for 1,3-ß-D-glucan synthases and 52, 58 and 60 kDa for 1,4-ß-D-glucan synthases [5,6]. Thus, in our system, the enzymes may have an oligomeric structure composed of different subunits.

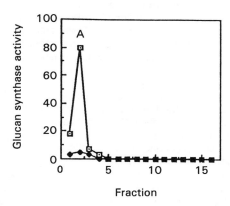

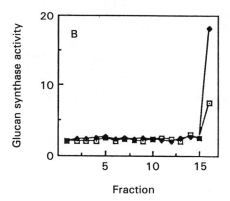

Figure 1. Glycerol gradient centrifugation of enzymes released (A) from aggregated glucans and of enzymes associated (B) to aggregated glucans. Chaps-solubilized enzymes were incubated in 1.2 mM UDP-Glc, then entrapped enzymes were collected by centrifugation. The pellet was resuspended in extraction buffer then centrifuged at 13 000 g for 15 min giving two fractions : enzymes released in the supernatant (A), and enzymes still associated with the glucans (B). Each fraction was layered onto 1.025 to 1.110 g.cm^{-3} glycerol gradient. Fraction 1, top of the gradient. Fraction 16, bottom of the gradient ; ▫1,3-ß-D-glucan synthase activity ; ◆ 1,4-ß-D-glucan synthase activity.

The enzyme complexes were also shown to be different. Polyclonal antibodies of the polypeptides characterizing 1,3-ß-D-glucan synthase did not recognize or immunoprecipitate the 1,4-ß-D-glucan synthase polypeptides [6]. This difference between the enzymatic complexes was also confirmed by Western blot analysis using antibacterial cellulose synthase antibodies as a probe. These antibodies recognized polypeptides of about 75 and 60 kDa but not the bands characterizing the 1,3-ß-D-glucan synthase activity. [Benziman, Fèvre and Bulone, unpublished].

Regulation of 1,4-ß-D-glucan synthase activity

As the fungal cellulose synthase exhibited structural homology with the bacterial enzymes, it was interesting to know if the fungal enzyme activities were regulated by the bacterial cellulose synthase activator c-di-GMP. This activator had no effect on 1,3-ß-D-glucan synthase activity. In contrast, 1,4-ß-D-glucan synthase activity was increased in the presence of increasing concentration of the activator (figure 2). The effects of c-di-GMP on fungal enzymes were similar to those reported for bacterial cellulose synthase as the V max was increased without modifying the apparent Km for UDP-glucose [7]. This effect was exerted directly on the enzymes as the activity of the 1,4-ß-D-glucan synthases purified by glycerol gradient centrifugation were stimulated by c-di-GMP [7].

ATP and GTP stimulated membrane-bound 1,4-ß-D-glucan synthase but not membrane-bound 1,3-ß-D-glucan synthase (figure 3). Phosphorylation by the γ phosphate is probably involved in this process as the non-phosphorylating analogues AMP-PCP or GMP-PCP did not enhance the enzymatic activity [8]. The possible involvement of a phosphorylated compounds in the regulation of 1,4-ß-D-glucan synthase activity was confirmed by treating particulate enzymes with phosphatases prior to the assays. Following alkaline phosphatase treatment, glucan synthesis was slightly reduced but stimulation by nucleotide was greatly reduced. Acid phosphatase treatment produced a strong inhibition of glucan synthesis even in the presence of ATP [9].

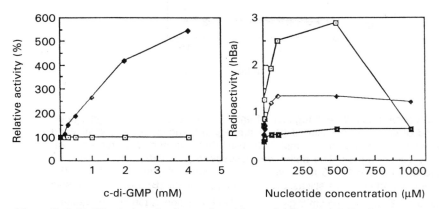

Figure 2 (left). Effect of cyclic diguanylic acid on solubilized 1,4-ß-D -glucan synthase (◆) and 1,3-ß-D- glucan synthase (▫) activities. The relative activity is expressed as the percentage of the activity observed without c-di-GMP.

Figure 3 (right). Effects of nucleotides and nucleotide analogues on membrane bound 1,4-ß-D--glucan synthase activities. ▫ ATP ; ◆ ATP-S ; ■ AMP-PP.

DISCUSSION

In the fungus *Saprolegnia*, 1,4-ß-D- and 1,3-ß-D-glucan synthases must have an oligomeric structure composed of different subunits as several polypeptides characterized each of the purified enzymes. As antibodies to the purified *Acetobacter* cellulose synthase cross-react with proteins of the 1,4-ß-D-glucan synthase preparation, this indicates that fungal and bacterial enzyme may share a similar organization or structure. This homology between fungal and bacterial enzymes was also found at the level of the regulatory mechanism of their activity. The nucleotide c-di-GMP activator which is an allosteric effector of purified bacterial enzymes stimulated the activity of partially purified enzymes from *Saprolegnia*. Another mechanism involving ATP regulates the activity of 1,4-ß-D-glucan synthase and is specific of membrane bound enzymes as solubilized enzymes are not stimulated. It is possible that a direct phosphorylation of these enzymes or of an unknown compound which activates the enzymes could be involved in the activity of the glucan synthases. These enzymes have a transmembrane orientation. We can speculate that *in vitro* stimulation produced by ATP simulates the regulation which occurs in intact cells. ATPase or nucleoside triphosphate binding proteins may be associated with the glycosyl transferases in the membranes providing energy or maintaining a transmembrane proton gradient to preserve the enzymatic activity. Dissociation of these complexes during enzyme extraction could explain the low activity generally observed in cell free extracts. We can hypothesize that in addition to the involvement of a direct activator, such as c-di-GMP, to regulate fungal cellulose synthase, the environmental proteic equipment of the synthases in the membranes plays an important role in the glycosyl transferase activity.

REFERENCES

1. Ross, P.; Mayer, R.; Benziman, M. *Microbiol Rev.* **1991**, *55*, 35.
2. Wong, H.C.; Fear, A.L.; Calhoon, R.O.; Eichinger, G.H.; Mayer, R.; Amikam, D.; Benziman, M.; Gelfand, D.; Meade, J.H.; Emerick, A.W.; Bruner, R.; Ben-Bassat, A.; Tal, R. *Proc. Natl. Acad. Sci.* USA **1990**, *87*, 8130.
3. Saxena, I.M.; Lin, F.C.; Brown Jr., R.M. *Plant Mol. Biol.* **1990**, *15*, 673.
4. Fèvre, M.; Rougier, M. *Planta* **1981**, *157*, 232.
5. Bulone, V.; Girard, V.; Fèvre, M. *Plant Physiol.* **1990**, *94*, 1748.
6. Bulone, V.; Girard, V.; Fèvre, M. **1992**, This volume.
7. Girard, V.; Fèvre, M.; Mayer, R.; Benziman, M. *FEMS Microbiol Lett* **1991**, *82*, 293.
8. Fèvre, M. *J. Gen. Microbiol.* **1984**, *130*, 3279.
9. Girard, V.; Fèvre, M. *FEMS Microbiol. Lett.* **1991**, *79*, 285.

3

Gene manipulation for cellulose-producing bacterium *Acetobacter xylinum*

M. Fujiwara, K. Maruyama, M. Takai, and J. Hayashi - Department of Applied Chemistry, Faculty of Engineering, Hokkaido University, Japan.

ABSTRACT

A shuttle vector pUF106, which had double hosts: *Acetobacter xylinum* and *Escherichia coli*, was constructed. Under optimum conditions, pUF106 was introduced into *A. xylinum* at a practical level. β-Isopropylmalate dehydrogenase gene of an *Acetobacter* gene were cloned in *E. coli* on a plasmid vector pULH1.

INTRODUCTION

Acetobacter xylinum is a gram-negative aerobe, which has the ability to produce pure cellulose extracellularly. This cellulose is called 'Bacterial Cellulose'. The mechanical or physical properties of Bacterial Cellulose will be superior to those of other forms of cellulose. Bacterial Cellulose has also high moldability with network-structure of very fine microfibrils. However, the price of Bacterial Cellulose is very high at the present time. Consequently, Bacterial Cellulose may have quite limited application to high value-added products. Considerable cost reduction will be possible by the improvement of Bacterial Cellulose microorganism productivity. Recombinant DNA technologies seems to be the most efficient methods for improvement of strains. For elucidation of the genetic background of this organism, the development of host-vector systems for cloning genes in *A. xylinum* cells was required.

We have already developed a genetic transformation method for *Acetobacter xylinum* host. We have also constructed several vector DNAs, which are composed of both the cryptic plasmids of *A. xylinum* and drug resistance plasmids from *Escherichia coli*.

On the basis of these studies, we tried to clone the β–isopropylmalate dehydrogenase (β–IPMDH) gene of *A. xylinum* in *E. coli*.

MATERIALS AND METHODS

Acetobacter xylinum (*A. xylinum*) ATCC10245 was used as a recipient strain for transformation.
A *leuB* mutant of *E. coli* C600 lacking β–IPMDH was used as an *E. coli* host. Plasmid pUC18 were used for cloning. For cultivation of *A. xylinum*, YPG medium (5g yeast extract, 3g polypepton, 30g glucose per liter, pH 6.5)[1] was employed and incubation was carried out at 30°C for 24h with shaking (120 strokes/min). M9 minimal medium [2] with added 1% thiamine and 30mg/ml of Ampicillin (Ap) was used for selection of *E. coli* transformants. Southern hybridization were carried out by the method of Sambrook, *et al.*[3].

RESULTS

Host-vector system for cloning genes of *Acetobacter xylinum*.

We have constructed a shuttle vector pUF106 (Fig.1)[4]. Since this vector is composed of both *Acetobacter* plasmid pFF6 and *E. coli* plasmid pUC18, it can be maintained in both hosts. It also carries an ampicillin resistance gene of pUC18 as a genetic marker. It has various possible cloning sites by representative restriction endonucleases. From these facts, we thought this vector suitable for cloning genes of *A. xylinum*.

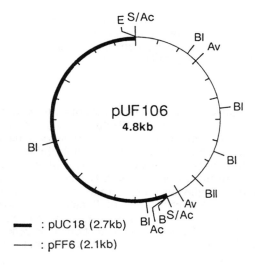

Fig. 1   Physical map of pUF106.
Ac: *Acc*I, Av: *Ava*I, B: *Bam*HI, Bl: *Bgl*I, Bll: *Bgl*II, E: *Eco*RI, S: *Sma*I
pUC18: a plasmid of *E. coli*, pFF6: a plasmid of *A. xylinum* IFO3288

We tried to transform *A. xylinum* with the plasmid vector pUF106. *A. xylinum* ATCC10245 was treated PCPII buffer (KCl 250mM, MgCl$_2$ 5mM, Tris-HCl 5mM; pH 7.5) to be competent cells. They were mixed with vector pUF106. During incubation time (0°C, 90min), 30 to 50 percent of polyethylene glycol 4,000 was added. Under these conditions, we can obtain more than 10 to the fourth transformants per microgram DNA. This transformation efficiency is a practical level for cloning a gene of microorganisms.

```
Cloning of the Acetobacter xylinum β-IPMDH gene in E.
coli.
```

As a model experiment, we tried to clone one of the genes in *A. xylinum*. The gene we have tried to clone is the β-isopropylmalate dehydrogenase gene (β-IPMDH). The gene coding for β-IPMDH is involved in the pathway of synthesis of leucine in microorganisms.

Fig. 2 shows the experimental scheme of shotgun cloning of β-IPMDH gene of *A. xylinum* ATCC10245. Total DNA of *A. xylinum* ATCC10245 was digested with *Bam*HI, *Hin*dIII or *Sal*I. Plasmid vector pUF106 or pUC18 was also digested with the same restriction enzymes. Then the total DNA fragment was ligated to the linear plasmid vector. The ligation mixture was introduced by

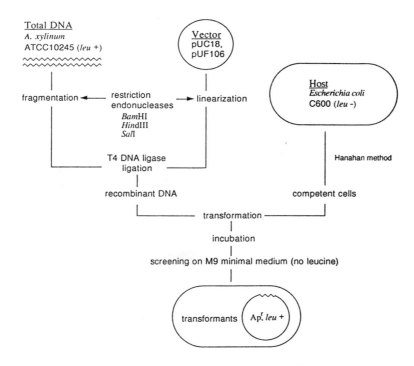

Fig. 2 Scheme of shotgun cloning of β-isopropylmalate dehydrogenase gene from *A. xylinum*.

transformation into *E. coli* C600 cells. This strain has a mutation of *leuB* gene, so it cannot synthesize leucine intracellularly. The only transformants complementing *leuB* gene mutation can appear on M9 minimal medium plates (no leucine contained) as colonies. We can obtain the β-IPMDH gene of *A. xylinum* in these transformants. The appearance frequency of Ap^r leu+ transformant was about one-three thousandths of the number of Ap^r transformants. All the Ap^r leu+ transformants were found to contain the same size of plasmid DNAs, which was 9.2kb. When the plasmid was digested with *Bam*HI, it produced pUC18 and an insert of 6.5kb.

Restriction mappings of the plasmid DNA designated as pUL1 are shown in Fig. 3 (a). When we reintroduced pUL1 into *E. coli* C600, it gave a lot of Ap^r leu+ transformants on M9 minimal medium. This indicates that the β-IPMDH gene of *A. xylinum* ATCC 10245 was cloned and expressed in *E. coli* C600.

Localization of the β-IPMDH gene in pUL1

To determine the coding region for the β-IPMDH gene in the cloned insert in pUL1, we constructed various plasmids carrying deleted inserts, and we examined their ability to complement the *leuB* mutation of *E. coli* C600. The results are

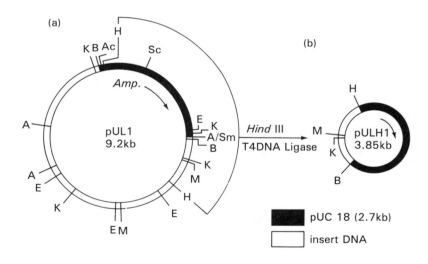

Fig. 3   Physical maps of pUL1 and pULH1.
Ac: *Acc*I, Av: *Ava*I, B: *Bam*HI, E: *Eco*RI, H: *Hind*III, K: *Kpn*I,
M: *Mlu*I, Sc: *Sac*I, S: *Sma*I

summarized in Fig. 4. pULA1 were deleted from this region with *Ava*I and had the ability to complement the *leuB* mutation of *E. coli* C600. pULH1 was deleted with *Hind*III and also had the same ability. However, pULM1 deleted with *Mul*I did not complement the *leuB* mutation. From these results, about 1.15kb of pULH1 was sufficient to complement the *leuB* mutation. This result indicates that the β-IPMDH gene is located in the fragment of pULH1 (Fig. 3 (b)), which is between the *Bam*HI and *Hind*III sites in the fragment of pULl.

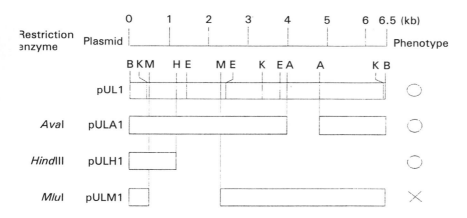

Fig. 4 Ability to complement the *leuB* mutation of *E. coli* by
a set of deleted plasmids.
A: *Ava*I, B: *Bam*HI, E: *Eco*RI, H: *Hind*III, K: *Kpn*I, M: *Mlu*I

Identification of the cloned β-IPMDH gene was done by the Southern hybridization method using a biotin-labeled probe.

The cloned β-IPMDH gene intensely hybridized with the original gene of *Acetobacter xylinum* ATCC 10245.

This is one example of the cloning and expression in *E. coli* of an *Acetobacter* gene. The goal of our study is obtaining high efficiency strains by direct genetic manipulation of the genes involved in cellulose synthesis. It will facilitate commercializing Bacterial Cellulose production.

REFERENCES

1. Ohmori, S., Uozumi, T., Beppu, T. *Agric. Biol. Chem.* **1982**, *46*, 381.
2. Sambrook, J., Fritsch, E.F., Maniatis, T. in Molecular Cloning: a laboratory manual (Cold Spring Harbor Laboratory Press, New York, 1989), A.3.

3. Sambrook, J., Fritsch, E.F. & Maniatis, T. in Molecular Cloning: a laboratory
    manual (Cold Spring Harbor Laboratory Press, New York, 1989), 9.31-
    9.57.
4. Fujiwara, M., Fukushi, K., Takai, M., Hayashi, J. in CELLULOSE
    structural and functional aspects, eds. Kennedy, J.F., Phillips, G.O.,
    Williams, P.A. (Ellis Horwood Limited, Chichester, 1989) pp153-158.

4

The first in vitro synthesis of cellulose via enzymatic polymerization and characterization of the "synthetic cellulose"

S. Kobayashi*, S. Shoda and K. Kashiwa - Department of Molecular Chemistry and Engineering, Faculty of Engineering, Tohoku University, Aoba, Sendai 980, Japan.

ABSTRACT

The first <u>in vitro</u> synthesis of cellulose via a non-biosynthetic path has been achieved by condensation reaction of β-cellobiosyl fluoride as substrate for cellulase, a hydrolysis enzyme of cellulose, in a mixed solvent of acetonitrile/acetate buffer (pH 5) (5:1). The water-insoluble part of the product is "synthetic cellulose", structure of which was confirmed by comparison with an authentic natural cellulose sample by using solid ^{13}C NMR and IR spectroscopy as well as by a hydrolysis experiment. The present synthetic cellulose was converted to the corresponding triacetate whose molecular weight was at least 6.3×10^3 (degree of polymerization, DP>22). X-ray as well as ^{13}C analysis showed that its crystal structure is of type II with high crystallinity. The mechanism of the present enzymatic polymerization is discussed in comparison with that of the biosynthetic reaction.

INTRODUCTION

There are three important classes of naturally occurring biopolymers ; nucleic acids (DNA and RNA), polypeptides (protein), and polysaccharides. The synthesis of the former two biopolymers has been developed extensively and even automatic equipment for their synthesis has become commercially available. The synthesis of polysaccharides is far less developed due to several difficulties in constructing the glycosidic bonds between sugar moieties. The control of stereochemistry of anomeric carbons, control of regioselectivity of various hydroxy groups having a similar reactivity, and selective blocking and deblocking of the hydroxy groups have become serious problems, and, therefore, the development of an efficient, general method of polysaccharide synthesis is strongly required.

Cellulose, one of the most typical polysaccharides, is the most abundant organic substance occurring in nature. Some 10^{15} Kg of cellulose are photo-synthesized and degraded per year [1]. For over a century, many researchers have been attracted by cellulose, and very many studies have been performed in view of both fundamental sciences and practical applications [2]. In vitro synthesis of cellulose, therefore, has long been one of the most difficult, yet important challenging topics from the early stage of macromolecular science. Many efforts have been devoted to regio- and stereo-selective preparations of cellulose, i.e. construction of a stereoregular polysaccharide having (1→4)-β-D-glycosidic linkage. The chemical approaches so far attempted, however, have failed to solve the problem in spite of the remarkable development of modern synthetic methods. The condensation of glucose 2,3,6-tricarbanilate with phosphorus pentoxide·in a mixture of chloroform-dimethyl sulfoxide gave branched products, and the molecular weight of the resulting polysaccharide after removing the protecting group was low [3]. 1,4-Anhydro-2,3,6-tri-0-benzyl-α-D-glucose has been polymerized using a Lewis acid as catalyst giving rise to polymers having mixed structures of (1→4)-β-D- (cellulose type) and (1→4)-α-D- (amylose type) linkages [4]. Uryu and co-workers investigated the possibility of synthesizing polysaccharides having (1→4)-β-D- linkages (cellulose) by the cationic ring-opening polymerization of 1,4-anhydro-glucose derivatives. However, stereoregular polysaccharides having the desired structure were not obtained due to the lack of regio-selective ring-opening [5]. Concerning a stepwise-synthesis of cellooligosaccharide derivatives, several oligomers up to octamer have been synthesized starting from allyl 2,3,6-tri-0-benzyl-4-0-(4-methoxybenzyl) β-D-glucoside, however, elimination of the protecting group to an oligomer, e.g., cello-octaose, has not been achieved yet [6].

This paper is concerned with a completely novel synthesis of cellulose by condensation polymerization of β-cellobiosyl fluoride 1 (substrate monomer) catalyzed by an enzyme: cellulase (enzymatic polymerization)(eq.1). Cellulase is an extracellular hydrolysis enzyme of cellulose, and therefore, in reaction (1) the enzyme promotes the reverse reaction of the hydrolytic decomposition in an aqueous organic solvent medium. The polymerization via a non-biosynthetic path has solved the above problems encountered with the polysaccharide synthesis.

ENZYMATIC POLYMERIZATION

As substrate, β-cellobiosyl fluoride (334 mg, 0.97 mmol) was dissolved in a mixture of acetonitrile (33 mL) and 0.05 M acetate buffer (pH 5, 5.6 mL). To this solution was added a 0.05 M acetate buffer (1.0 mL) solution of cellulase (17 mg, 26 units, 5.1 wt% for the substrate) from <u>Trichoderma viride</u>, and the mixture was shaken at 30 °C for 12 h. As the reaction proceeded, the initially homogeneous solution gradually became heterogeneous with a white precipitation of the polysaccharide product. The resulting suspension was then heated at 100 °C for 10 min to inactivate the enzyme. This procedure is essential for the reaction to prevent the resulting polysaccharides from further hydrolysis by the enzyme. After evaporating the acetonitrile, the residue was poured into an excess amount of methanol-water (5:1) in order to remove the catalyst cellulase, lower molecular weight products like D-glucose and D-cellobiose, and a trace amount of inorganic compounds derived from the buffer solution. The precipitate was collected by filtration giving rise to 212 mg of methanol-water (5:1)-insoluble part (64%) after drying in vacuo. The insoluble part was then suspended in distilled water and the mixture was filtered. The white powder was dried in vacuo affording 180 mg of the water-insoluble part (54%) which is "synthetic cellulose" (<u>vide infra</u>). On the other hand, the filtrate was concentrated in vacuo to give 32 mg of the white powdery water-soluble part (10%), which is a mixture of cellooligosaccharides. In solvent systems examined, a mixture of acetonitrile and buffer (5:1) gave the best results in terms of producing synthetic cellulose (Table I).

Table I Solvent effects on the polymerization of $\underline{1}$ with cellulase[a]

No.	solv.	yield[c] (%)
1	CH_3CN/buffer[b](5:1)	64(54)
2	C_2H_5CN/buffer(5:1)	10(3)
3	CH_3NO_2/buffer(5:1)	15(1)
4	CH_3OH/buffer(5:1)	17(6)
5	C_2H_5OH/buffer(5:1)	20(7)
6	$(CH_3)_2CO$/buffer(5:1)	13(7)
7	1,4-dioxane/buffer(5:1)	4(~0)
8	DMF/buffer(5:1)	~0(0)
9	DMSO/buffer(5:1)	~0(0)

a)Polymerized at 30°C for 12 h:[$\underline{1}$]=2.5×10^{-2} mol/L
cellulase;5.0 wt% for $\underline{1}$
b)Acetate buffer(0.05M,pH 5).
c)Methanol:water (5:1) insoluble part.
   In parentheses the yield of the water insoluble part
   is given.

The CP/MAS solid ^{13}C NMR spectrum of the water-insoluble part (Figure 1 (a)) is very similar to that of natural cellulose (Figure 1 (b)), and hence, each signal can be assigned to the corresponding carbon atom of glucose unit (C1-C6)[7]. The IR spectrum of the water-insoluble part exhibited also a very similar pattern to that of natural cellulose [8]. Treatment of the water-insoluble part with cellulase in 0.05 M acetate buffer afforded a hydrolysis product of D-glucose quantitatively. These results clearly indicate that the glycosidic

bond formation occurred in a regio- and stereo-selective manner between cellobiose units to a stereoregular polysaccharide having (1→4) linkages.

In order to determine the molecular weight of the water-insoluble part, it was
converted to the corresponding acetylated derivative using acetic
anhydride/perchloric acid. The gel permeation chromatographic (GPC) analysis of
the product revealed that the number average molecular weight is 6.3×10^3 which
corresponds to the degree of polymerization, DP=22 [9].

The X-ray diffractogram of the synthetic cellulose (the water-insoluble part)
showed three peaks at $2\theta = 22.0^O$, 19.9^O, and 12.2^O characteristic to the type II
cellulose (Figure 2 (a)), which clearly differentiates the product from type I
natural cellulose (Figure 2 (c))[11]. In addition, the sharp peaks of the
present cellulose compared with that of mercerized cellulose of type II (Figure
2 (b)) indicate that the synthetic cellulose has higher crystallinity than the
latter. These results are consistent with the clear splitting of the signal due
to the C-1 carbon atom in the CP/MAS ^{13}C NMR spectrum (Figure 1 (a)).

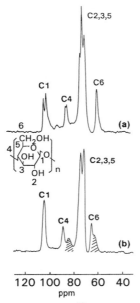

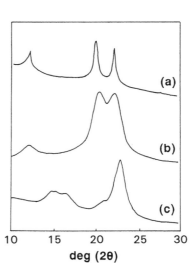

Figure 1. CP/MAS ^{13}C NMR spectra of
(a) the water-insoluble part of the
product (synthetic cellulose) and
(b) natural cellulose.

Figure 2. X-ray diffractograms of
(a) synthetic cellulose (type II),
(b) mercerized natural cellulose
(type II), and (c) natural cellulose (type I).

MECHANISM

At present, there are two conceivable reaction mechanisms concerning the chain
propagation for the present polymerization. The first one involves the formation of an active intermediate of disaccharide unit (GG-Enz) by reaction of <u>1</u>

(GG-F) and the cellulase (Enz$^-$) followed by an attack of the terminal 4'-hydroxy group of the propagating polymer ((GG)$_n$-F) giving rise to the product (GG)$_{n+1}$-F (activated monomer mechanism:(i) and (ii)). In the second mechanism, an active intermediate (GG)$_n$-Enz is formed on the chain-end and the propagating process is realized by the attack of 4'-hydroxy group of disaccharide unit (GG-F) to this intermediate (active chain-end mechanism:(iii) and (iv)).

$$\text{GG-F} \;+\; \text{Enz}^- \;\rightleftharpoons\; \text{GG-Enz} \;+\; \text{F}^- \qquad \text{(i)}$$

$$\text{GG-Enz} \;+\; \text{(GG)}_n\text{-F} \;\rightleftharpoons\; \text{(GG)}_{n+1}\text{-F} \;+\; \text{Enz}^- \qquad \text{(ii)}$$

$$\text{(GG)}_n\text{-F} \;+\; \text{Enz}^- \;\rightleftharpoons\; \text{(GG)}_n\text{-Enz} \;+\; \text{F}^- \qquad \text{(iii)}$$

$$\text{(GG)}_n\text{-Enz} \;+\; \text{GG-F} \;\rightleftharpoons\; \text{(GG)}_{n+1}\text{-F} \;+\; \text{Enz}^- \qquad \text{(iv)}$$

GG-F ≡

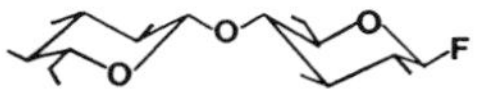

The formation of the stereoregular polysaccharides may be explained by assuming the following two processes. The first step involves the formation of a glycosyl-enzyme intermediate (Figure 3(a)) or a glycosyl oxocarbenium ion (Figure 3(b)) at an active site of cellulase with the elimination of fluoride anion [12]. This reactive intermediate is then attacked by 4'-hydroxy group of another polysaccharide which locates in a subsite of the enzyme leading to the stereoselective formation of the (1→4)-β-D- linkage. Consequently, the stereochemistry of the product is retention of configuration via "double inversion" concerning the anomeric carbon atom of the β-cellobiosyl fluoride 1.

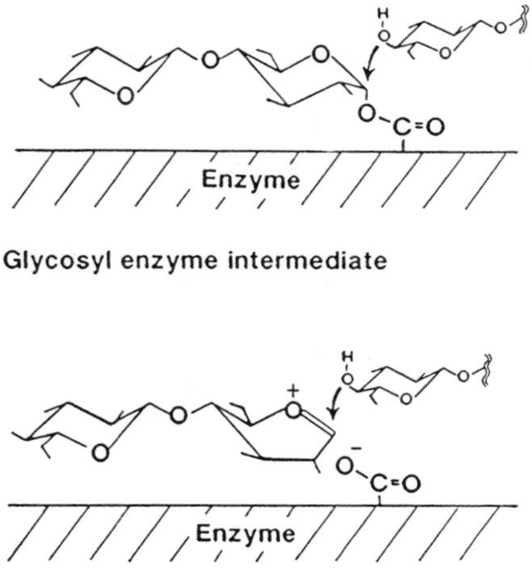

(a) Glycosyl enzyme intermediate

(b) Glycosyl oxocarbenium ion intermediate

Figure 3 . Proposed models in the stereoselective formation of the (1→4)-β-D-glycosidic linkage via a substrate-enzyme complex involving double inversion of configuration at C1 carbon.

The present reaction mechanism is to be interestingly compared with a biosynthetic pathway of cellulose which involves the "inversion" of the configuration concerning the C1 carbon atom of the substrate of uridine diphosphate glucose (UDP-glucose) (eq. 2).

UDPglucose

REFERENCES

1.   See, for example, L. Stryer, Biochemistry, 3rd. Eds., W. H. Freeman and Company, NY, 1988 p342.
2.   H. Mark, Cellulose Chem. Technol., 1980, 14, 569.
3.   E. Husemann and G. J. M. Müller, Makromol. Chem., 1966, 91, 212.
4.   F. Micheel and O-E. Brodde, Liebis Ann. Chem., 1974, 702.
5.   T. Uryu, C. Yamaguchi, K. Morikawa, K. Terui, T. Kanai, and K. Matsuzaki, Macromolecules, 1985, 18, 599.
6.   F. Nakatsubo, T. Takano, T. Kawada, and K. Murakami, CELLULOSE, Structural and Functional Aspects, J. F. Kennedy, G. O. Phillips, and P. A. Williams, Eds., Ellis Horwood, NY, 1989, p201.
7.   R. H. Atalla, J. C. Gast, D. W. Sindorf, V. J. Bartuska, and G. E. Maciel, J. Am. Chem. Soc., 1980, 102, 3249.
8.   F. H. Forziati and J. W. Rowen, J. Res. Natl. Bur. Stand., 1951, 14, 38.
9.   Generally in the carbohydrate chemistry, a "polysaccharide" designates a substance having more than ten monosaccharide units.[10] According to this terminology, it is allowable to refer to the water-insoluble part as "cellulose".
10.   J. F. Kennedy and C. A. White, Comprehensive Organic Chemistry, D. Barton, W. D. Ollis, and E. Haslam, Eds., Pergamon Press, Oxford, Vol. 5, 1979, p755.
11.   J. Hayashi, T. Yamada and K. Kimura, Appl. Polym. Symp., 1976, 28, 713.
12.   See, for example, K. Nishizawa and Y. Hashimoto, The Carbohydrates, Chemistry and Biochemistry, Vol. 2A, W. Pigman and D. Horton Eds., Academic Press, NY, p241.

5

Immunological characterization of 1,3-β-D-glucan synthase from *Saprolegnia monoica*

V. Bulone, V. Girard and M. Fevre - Laboratoire de Biologie Cellulaire Fongique, Centre de Génétique Moléculaire et Cellulaire, UMR 106-CNRS, Université Claude Bernard Lyon 1, 43, Boulevard du 11 Novembre 1918, 69622 Villeurbanne Cédex, France.

ABSTRACT

Polyclonal antibodies were raised against each polypeptide (34, 48 and 50 kDa) characterizing the 1,3-ß-D-glucan synthase complex from the fungus *Saprolegnia*. The serum directed against the 34 kDa polypeptide recognizes specifically this protein, immunoprecipitates the 1,3-ß-D-glucan synthase activity and labels the entrapped enzyme associated to its reaction product. Our results indicate that the 34 kDa polypeptide is associated with the 1,3-ß-D-glucan synthase activity, but not with the 1,4-ß-D-glucan synthase activity and that fungal 1,3-ß-D-glucan synthases are not related to 1,4-ß-D-glucan synthases.

INTRODUCTION

The properties of higher plants and fungal cell walls, i.e. their mechanical strength, morphological features and biological activities, reside in their particular chemical composition. So, polysaccharides, which represent the major cell wall components, fulfill undoubtedly many of the specific cell wall functions.

In green plants, cell wall polysaccharides are cellulose and hemicellulose. Callose, i.e. 1,3-ß-D-glucan chains, occurs in the wall of some specialized cells and cell parts, i.e. pollen mother cells, sieve tube members, and plasmodesmatal openings. Its synthesis occurs in defense mechanisms against pathogen attack or mechanical stress. In fungi, 1,3-ß-D-glucans are normal constitutive cell wall polymers representing up to 80% of the hyphal wall.

Cell-free preparations from higher plants and fungi have been shown to incorporate glucose from UDP-D-glucose into glucans with 1,3-ß-D- or 1,4-ß-D- linkages. However, little progress has been made in the purification and characterization of the enzymes which participate to the synthesis of these cell wall polysaccharides. Consequently, molecular mechanisms for the synthesis of these important biological products and the molecular biology of glucan synthases are not well known compared to those of other cellular components. Nevertheless, several reports describing the use of various techniques (gradient density centrifugation, column chromatography, electrophoresis) have shown that glucan synthases are large protein complexes (>450 kDa), and several protein subunits ranging from 18 to 83 kDa might be involved in glucan synthesis [1, 2, 3].

In the fungus *Saprolegnia*, cellulose and 1,3-ß-D-glucans are integral parts of the cell wall. 1,3-ß-D- and 1,4-ß-D-glucans can be produced *in vitro* by isolated membrane fractions, according to the assay conditions [4]. Our recent findings provide a few promising polypeptide candidates for 1,3-ß-D- and 1,4-ß-D-glucan synthases [5, 6]. In the present paper, we extend the separation and partial purification of 1,3-ß-D- and 1,4-ß-D-glucan synthases from *Saprolegnia* to immunological approaches in order to identify the function of the selected polypeptide candidates. Polyclonal antibodies were raised against polypeptides of 34, 48 and 50 kDa previously characterized as components of the 1,3-ß-D-glucan synthase complex. Their ability to recognize 1,3-ß-D- and 1,4-ß-D-glucan synthases was tested by immunoprecipitation and immunosensitivity.

MATERIALS AND METHODS

Isolation of Solubilized Enzymes and ß-D-Glucan Synthase Assays

Membrane-bound enzymes were isolated by differential centrifugation as described previously [4]. Chaps prepared in extraction buffer (0.01 M Tris-HCl, 0.5 M sorbitol, pH 7.2) was added to an equal volume of particulate enzymes at the final detergent concentration of 5 mg/ml. After vigorous vortex mixing and solubilization for 30 min at 4°C, the suspension was centrifuged at 48,000g for 1 h. The supernatant was used as the solubilized enzymes.

ß-D-Glucan synthases were assayed as previously described [5]. 1,4-ß-D-Glucan synthase assay mixtures contained 0.3 nmol of UDP-$[^{14}C]$-glucose, 0.1 µmol of DTT, 2 µmol of cellobiose, 5.5 µmol of $MgCl_2$, 25 mM Pipes-Tris buffer (pH 6.0), and 100 µl of freshly prepared enzymes in a final volume of 370 µl. When enzymes were assayed for 1,3-ß-D-glucan synthesis, $MgCl_2$ was omitted and UDP-$[^{12}C]$-D-glucose levels increased to 400 nmol.

Immunological Characterization of the 1,3-ß-D-Glucan Synthase Polypeptides

Polyclonal antibodies were raised against the 34, 48 and 50 kDa polypeptides from the 1,3-ß-D-glucan synthase fraction prepared according to Bulone *et al.* [5]. Polypeptides were electroeluted from SDS-PAGE gel in 200 mM Tris-acetate (pH7.4), 1 % SDS and 100 mM of DTT per 0.1g of wet polyacrylamide gel. The running buffer was 50 mM Tris-acetate (pH 7.4), 0.1% SDS and 0.5 mM sodium thioglycolate. After 3 h at 100 V, SDS and Tris were removed from the protein solution by electrodialysis for 3 h at 100 V in 0.01 M sodium bicarbonate as running buffer. The proteins recovered were freeze-dried, resuspended in PBS then injected into different rabbits (3 injections made at 2 weeks interval).

For Western blots, proteins from glycerol gradient fractions transferred from PAGE gels to nitrocellulose sheets were incubated for 2 h at room temperature with each polyclonal serum in a 1/1000 dilution in TBS-milk (5%). As a second antibody, an anti-rabbit immunoglobulin peroxidase conjugate, was used with diaminobenzidine and H_2O_2 as peroxidase substrates.

Immunoprecipitations were carried out by incubating 200 µl of protein samples with an equal volume of each dilution of polyclonal antibodies. Antisera were diluted in 0.01 M Tris-HCl, pH 7.2, in order to preserve the enzymatic activities. After 2 h at 4°C the antibody-protein complexes were adsorbed during 30 min at 4°C to a 2.5% suspension of *Staphylococcus* cells. Adsorbed complexes were pelleted (30 s at 13,000g), washed 3 times with 0.01 M Tris-HCl, pH 7.2 then assayed for glucan synthase activities.

The effects of antibody binding on glucan synthase activities were tested by incubating for 2 h at 4°C 200 µl of the enzymatic sample with an equal volume of antisera diluted in 0.01 M Tris-HCl, pH 7.2. Glucan synthase activities were then assayed as previously described.

Immunolabelling of entrapped 1,3-ß-D-glucan synthase was conducted on Chaps solubilized enzymes. The enzymes were first incubated at 24°C for 15 min in assay mixtures for 1,3-ß-D-glucan synthesis. Mixtures were then centrifuged at 13,000 g for 15 min and pelleted glucans with entrapped enzymes were resuspended in 100 µl 0.01 MTris-HCl buffer, pH 7.2. One hundred µl of a 1/1000 dilution of the anti-34 kDa antibodies were incubated with the suspension of entrapped enzymes for 1 h at 24°C. Immune complexes were washed 3 times in PBS and incubated with a colloïdal gold conjugate anti-rabbit IgG fraction. Suspensions were washed 3 times, deposited on colloïdon coated grids and shadowed with Au/Pd. Observations and photographs were made by using a Hitachi HU12A electron microscope.

RESULTS

Polyclonal antibodies to each polypeptide characterizing the 1,3-ß-D-glucan synthase complex were raised in rabbits. The specificity of each serum was tested by Western blots realized on the enzymatic proteins separated by linear glycerol gradient centrifugation. The antiserum directed against the 34 kDa polypeptide was very specific of this band as the corresponding Western blot revealed a single polypeptide of 34 kDa (Figure 1). The antisera directed against the 48 and 50 kDa polypeptides are not specific of these bands as numerous proteins present in the gradient fractions were revealed after Western blotting (Figure 1). However, the three polypeptides of 34, 48 and 50 kDa exhibited the strongest signals with these antibodies.

The ability of the three sera to immunoprecipitate the 1,3-ß-D-glucan synthase activity was tested on an enriched 1,3-ß-D-glucan synthase fraction isolated following glycerol gradient centrifugation. The serum directed against the 48 kDa polypeptide was not able to precipitate the 1,3-ß-D-glucan synthase activity of the fraction. Antisera directed against the 34 and the 50 kDa polypeptides were able to immunoprecipitate respectively 17% and 12% of the 1,3-ß-D-glucan synthase activity of the fraction. Moreover, the antibodies directed against the 34 kDa polypeptide immunoprecipitated up to 20% of the 1,3-ß-D-glucan synthase activity distributed in a glycerol gradient.

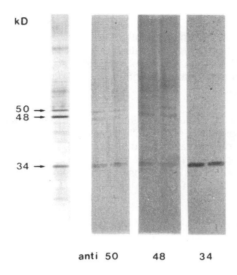

Figure 1. Western blots of the solubilized ß-D-glucan synthase complex isolated by glycerol gradient centrifugation. Two µg proteins were probed with the different antibodies.

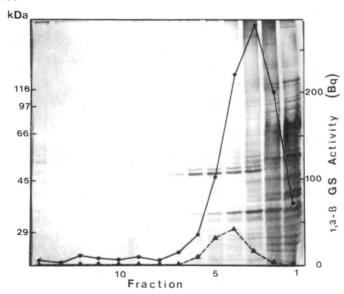

Figure 2. Immunoprecipitation by the anti-34 kDa antibodies of the 1,3-ß-D-glucan synthase recovered in fractions following ultracentrifugation of solubilized enzymes on a linear glycerol gradient. Fraction 1, top of the gradient; Fraction 16, bottom of the gradient. ▲ , Immunoprecipitated 1,3-ß-D-glucan synthase activity with the anti-34 kDa antibodies. ●, 1,3-ß-D-glucan synthase activity in the different fractions of the gradient. SDS-PAGE (10% gel) of the gradient fractions was silver stained.

This was correlated with the distribution of the 34 kDa polypeptide in the gradient (Figure 2). These results tend to confirm that the 34 kDa band is involved in the 1,3-ß-D-glucan synthase complex.
The three sera were unable to immunoprecipitate the 1,4-ß-D-glucan synthase activity.

Among the three sera tested, those directed against the polypeptides of 34 and 50 kDa did not affect the 1,3-ß-D-glucan synthase activity. On the contrary, the anti-48 kDa polypeptide antibodies stimulated 1,3-ß-D-glucan synthase activity : the enzyme activity was twice fold higher in the presence of a 1/128 dilution of the sera. This may be due to the preservation of the antigen stability, i.e., enzyme activity.
The three sera obtained had no effect on the 1,4-ß-D-glucan synthase activity.

Solubilized enzymes synthesize *in vitro* microfibrillar glucans. Following incubation in UDP-glucose and low speed centrifugation, EM examination showed that globular structures of 10 nm diameter were associated with single microfibrils and could represent the glucan synthase complexes [4]. Gold immunolabelling of the enzymatic complexes was performed on the entrapped proteins. Figure 3 shows that the antibodies directed against the 34 kDa polypeptide are able to recognize the globular structure associated with the 1,3-ß-D-glucan synthase reaction product.

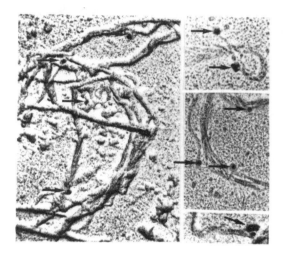

Figure 3. Electron micrograph of the immunolabelled and entrapped 1,3-ß-D-glucan synthase. After entrapment, the enzyme associated to the reaction product was labelled (arrows) with the antibodies directed against the 34 kDa polypeptide (see materials and methods).

DISCUSSION

Polyclonal antibodies were produced against each polypeptide of 34, 48 and 50 kDa characterizing the 1,3-ß-D-glucan synthase complex. Western blots showed that the serum directed against the 34 kDa polypeptide was very specific of this polypeptide.

The anti-34 kDa antibodies were able to immunoprecipitate 1,3-ß-D-glucan synthase activity in correlation with the distribution of the 34 kDa band in a linear glycerol density gradient. Moreover, the antibodies recognized the globular structure associated with the reaction product and corresponding to the entrapped enzyme. These results indicate that the polypeptide of 34 kDa is implicated in the proteic complex responsible for the 1,3-ß-D-glucan synthase activity. However, the incubation of these antibodies with a fraction enriched in 1,3-ß-D-glucan synthase had no effect on the enzyme activity, perhaps because the antisera were raised against SDS-denatured polypeptides.

Antibodies directed against polypeptides of 48 and 50 kDa are less specific : they recognize numerous polypeptides on Western blots. Nevertheless, the anti-50 kDa polyclonal antibodies are able to immunoprecipitate a part of the 1,3-ß-D-glucan synthase activity present in an enriched fraction. So, this polypeptide may also be involved in the synthesis of the 1,3-ß-D-glucans in *Saprolegnia*. The anti-48 kDa polypeptide antibodies did not immunoprecipitate the 1,3-ß-D-glucan synthase activity but they were able to increase twice fold this activity after incubation with the enzyme. This serum contains probably antibodies able to stabilize the 1,3-ß-D-glucan synthase or to activate an effector of the glucan synthase activity. These results confirm the oligomeric structure of the 1,3-ß-D-glucan synthase. The antibodies obtained did not react in immunoprecipitation and immunosensitivity with 1,4-ß-D-glucan synthases. This absence of cross reaction between 1,3-ß-D- and 1,4-ß-D-glucan synthases indicates that   these enzymes are not related in fungi.

ACKNOWLEDGEMENT

The authors thank R. Pépin and I. Jouffret for their help with EM experiments.

REFERENCES

1. Eiberger, L.L.; Wasserman, P.B. *Plant Physiol.* **1 9 8 7**, *83*, 982.
2. Lin, F.C.; Brown Jr., R.M.; Drake, R.D.; Haley, B.E. *J. Biol. Chem.* **1 9 9 0**, *265*, 4782.
3. Read, S.M.; Delmer, D.P. *Plant Physiol.* **1 9 8 7**, *85*, 1008.
4. Fèvre, M.; Rougier, M. *Planta* **1 9 8 1**, *157*, 232.
5. Bulone, V.; Girard, V.; Fèvre, M. *Plant Physiol.* **1 9 9 0**, *94*, 1748.
6. Fèvre, M.; Bulone, V.; Girard, V. **1 9 9 2**, This volume.

6

Incorporation of GlcNAc residue into bacterial cellulose by *A. xylinum* – characterization of a novel BC

R. Ogawa, M. Sato, Y. Miura, M. Fujiwara*, M. Takai* and S. Tokura - Department of Polymer Science, Faculty of Science. *Department of Applied Chemistry, Faculty of Engineering, Hokkaido University, Sapporo 060, Japan.

ABSTRACT

Bacterial cellulose containing N-acetylglucosamine residue (N-AcGBC) was produced at the surface of liquid medium by *A. xylinum* strains adapted to N-acetylglucosamine (GlcNAc) and showed susceptibility for chitinolytic enzymes such as lysozyme and chitinase. The higher plane orientation and Young's modulus than those of bacterial cellulose (BC) were observed. The FT-IR analyses suggest that intra- or inter-molecular hydrogen bonds would contribute to form a strong N-AcGBC membrane through induced acetamido groups of GlcNAc residues.

INTRODUCTION

In order to produce a novel bacterial cellulose (BC) with new functions (especially with chitinous properties) and also to investigate the specificity of BC synthesis on glucose analogues, we have concentrated on efforts to introduce N-acetylglucosamine (GlcNAc) residues into BC through the biosynthesis of *A. xylinum* strains, and found that *A. xylinum* strains adapted to GlcNAc produced BC containing GlcNAc (N-AcGBC) at the surface of the liquid medium [1]. The GlcNAc contents of N-AcGBC estimated by amino acid analysis of acid hydrolysates were 0.5-4.0 mol%, and susceptibility for lysozyme[EC 3. 2. 1. 17] of N-AcGBC was investigated by turbidimetry [1]. The incorporation of GlcNAc residues into BC main chain through 1,4-ß-D-glycoside linkages was confirmed by the facts that (i) the susceptibilities for lysozyme of N-AcGBC were proportional to GlcNAc contents and (ii) the rate of turbidity decrease due to lysozymic hydrolysis depended on lysozyme concentrations [1, 2]. N-AcGBC also showed slight susceptibility to chitinase [EC 3. 2. 1. 14]; accordingly, it was found that N-AcGBC was a novel BC which had susceptibility to chitinolytic enzyme, in addition to cellulolytic enzyme[1.2].

This paper examines the effects of incorporation of a few mol% of GlcNAc on the properties of BC by X-ray, FT-IR analyses and the measurement of tensile strength. As a result, N-AcGBC membrane showed higher plane orientation and Young's modulus than those of BC, and the crystalline structure of N-AcGBC seemed to vary slightly from that of BC.

MATERIALS AND METHODS

Preparation of BC and N-AcGBC membrane
BC and N-AcGBC pellicles were prepared by the methods described previously [1]. The washed pellicles were dried at 105 °C in an air oven on glass plates and then were peeled off and subjected to X-ray, IR analyses and the measurement of tensile strength.

Analytical methods
X-ray diffractograms of membranes were obtained by reflection methods with a Rigaku automatic diffractometer operated in the ω - 2θ scanning mode between 5° and 30° (2θ). Slits of 1° and 0.15 nm for the tube and detector, respectively, were used. Tensile strengths of membranes (2.5 × 0.5 cm) were measured. FT-IR spectra were obtained by the use of a Nicolet 5D × B FT-IR spectrophotometer at a resolution of 4 cm^{-1}.

RESULTS AND DISCUSSION

X-ray diffractograms
According to Sisson [3], the preferred orientations can be classified into five types: random orientation, uniplaner orientation, selective uniplaner orientation, uniaxial orientation, and selective uniaxial orientation. On the other hand, Takai *et al*. showed that membranes of BC dried with a flat surface parallel to a glass plate provide a typical selective uniplanar orientation of the ($1\bar{1}0$) crystallographic plane [4].

The X-ray diffractograms of BC and N-AcGBC obtained by the reflection method are shown in Fig. 1, and demonstrate that (i) in BC, the intensity of an interference indexed in the ($1\bar{1}0$) plane is equivalent with that of the (020) plane, whereas, in N-AcGBC, the peak of ($1\bar{1}0$) is more intensive than that of (020), (ii) the diffraction of (110) disappears in N-AcGBC, (iii) the peaks of ($1\bar{1}0$) and (020) of N-AcGBC shift slightly to higher angle positions compared with those of BC, and (iv) no significant difference in half width of the ($1\bar{1}0$) or (020) interference between BC and N-AcGBC is observed.

Since the crystallite size of cellulose could be obtained from the half width of the ($1\bar{1}0$) or (020) interferences [4], the crystallite size of N-AcGBC seems to be similar to that of BC. However, the crystalline structure of N-AcGBC might differ slightly from that of BC because of the fact (iii) described above.

It has been known that the peak intensity ratio ($1\bar{1}0$) to (020) in X-ray diffractograms has suggested a pseudoquantitative estimation of selective uniplanar orientation and that the diffraction of (110) has appeared sharply as a result of disorientation of (1$\bar{1}$0) [4]. Therefore, higher orientation intensity of the N-AcGBC membrane than that of the BC membrane is suggested by the observation of (i) and (ii) described above. The peak intensity ratios (1$\bar{1}$0) to (020) of N-AcGBC were higher compared to BC as shown in Table 1. The higher orientation of N-AcGBC than that of BC seems to be attributed to the strong hydrogen bonds in (1$\bar{1}$0) plane formed by acetamido groups of GlcNAc residues.

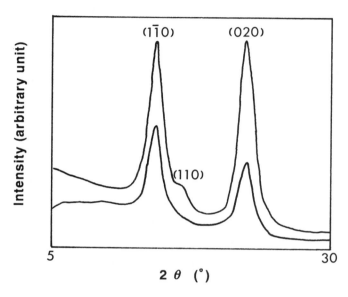

Figure 1. X-ray diffractograms of BC (a) and N-AcGBC (b) membranes.

Table 1. Peak intensity ratios (1$\bar{1}$0) to (020) in X-ray diffractograms of BC and N-AcGBC membranes.

membrane		GlcNAc content (mol%)	(1$\bar{1}$0 / 020)[a]
BC	No. 1	0.0	1.07
	No. 2	0.0	0.95
	No. 3	0.0	1.08
N-AcGBC	No. 1	0.8	1.22
	No. 2	2.0	1.34
	N0. 3	2.8	1.45

(a) : the ratio for peak intensity

The tensile strength
The preliminary results of tensile strength measurements (shown in Fig. 2) showed that the N-AcGBC membrane was not appreciably stretched and was severed sooner than the BC membrane. Since this property was exhibited in a chitin membrane, N-AcGBC seemed to be endowed with a chitin property by the introduction of GlcNAc residues. Young's modulus of N-AcGBC and BC membranes estimated from Fig. 2 were 8.45 GPa and 6.03 GPa, respectively. The higher Young's modulus of N-AcGBC might be due to the strong intra- and inter-molecule hydrogen bonds formed by GlcNAc residue. These results might show that modification of BC to have both bacterial cellulosic and chitinous properties is achieved by introduction of a few mol% of GlcNAc into the BC main chain.

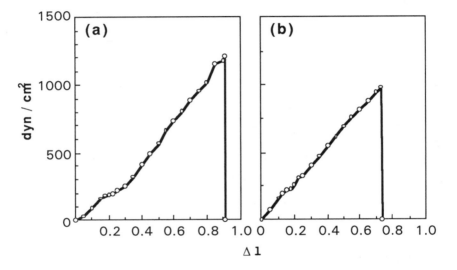

Figure 2.   Tensile strength of BC (a) and N-AcGBC (b) membranes.

FT-IR spectra
The infrared spectra of BC and N-AcGBC membranes are shown in Fig. 3; it is hard to find the difference between the spectra of BC (a) and N-AcGBC (b). The absorption at 1650 cm^{-1} (amide I) and 1560 cm^{-1} (amide II), attributed to amide group of GlcNAc, are faintly observed on the spectrum of N-AcGBC membrane, but the quantification of acetyl groups was not achieved due to the overlapping of adsorptions of amide I and adsorbed H2O (the absorption at 1650 cm^{-1} is also shown in the spectrum of BC).

Marrinan & Mann [5] and Liang & Marchessault [6] obtained two types of band patterns in OH and CH stretching regions (3μ region; 3600-2800 cm^{-1}) for native cellulose. Marchessault & Liang [7] extended the study to the region 1700-640 cm^{-1}. The polarized ($\perp$) IR spectra were measured (Fig. 4), which also showed no major differences in OH and CH stretching regions between the

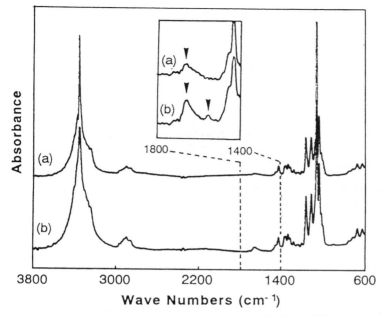

Figure 3. The FT-IR spectra of BC (a) and N-AcGBC (b) membranes.

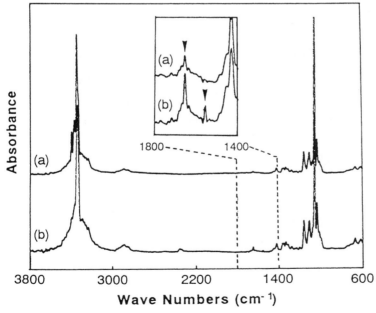

Figure 4. The polarized (⊥) FT-IR spectra of BC (a) and N-AcGBC (b)
membranes.

spectra of BC (a) and N-AcGBC (b) membrane. However, the absorptions at 1650 and 1560 cm^{-1} of N-AcGBC shown in Fig. 4 became sharper and stronger than those in Fig. 3. This result suggests that acetamido groups of GlcNAc residues are involved in intra- or inter-molecular hydrogen bonding.

CONCLUSIONS

Our results suggest that the introduction of a few mol% of GlcNAc residue into the BC main chain influenced the crystalline structure of BC. A study on hydrogen-deuterium exchange applying NMR or IR would confirm this suggestion. On the other hand, we found a method of production of bacterial cellulose by stirred culture. The BC had higher orientation of microfibrils than that by static culture. Attempts to produce N-AcGBC with heterogeneous distributions of GlcNAc residues are currently in progress by applying this method.

ACKNOWLEDGEMENTS

We are very much indebted Mr. Kawamura and Mr. Fukaya of central research laboratory, Nakano Vinegar Co. Ltd. for the supplying bacteria. This work supported by the Grant-in-Aid of the Ministry of Education (No. 02555183).

REFERENCES

1. Ogawa, R.; Tokura, S., *Carbohydr. Polym.*, in press.
2. Ogawa, R.; Miura, Y; Koriyama, T; Tokura, S., *Int. J. Biol. Macromol.*, submitted.
3. Sisson, W. A.; Susich, G. V., *Z. Phys. Chem. Abt.*, 1929, (**B**) 4, 431
4. Takai, M.; Tsuta, Y.; Hayashi, J.; Watanabe, S., *Polym. J.*, 1975, 7, 157
5. Marrinan, H. J.; Mann, J., *J. Polym. Sci.*, 1956, 21, 301
6. Liang, C. H.; Marchessault, R. H., *J. Polym. Sci.*, 1959, 37, 385
7. Marchessault, R. H.; Liang, C. H., *J. Polym. Sci.*, 1959, 39, 269

7

Molecular dynamics of cellulose models

B.J. Hardy* and A. Sarko⁺ - Department of Chemistry and the Cellulose Research Institute, College of Environmental Science and Forestry, State University of New York, Syracuse, NY 13210, USA.

ABSTRACT

Molecular dynamics simulations offer the newest and, perhaps, the best tool for modeling cellulose and cellulosic systems. Such a study is now in progress, encompassing cellulose models ranging in size from glucose to cellooctaose and attempting to determine the effects of various force-fields, the size of the model molecule, presence of water as solvent, and the duration of simulation on the results. Comparisons with static modeling are drawn in all cases. The results obtained to date show that all of the variables under study exert an effect on the outcome of the dynamics simulations, to greater or lesser extents.

INTRODUCTION

The structure of cellulose in its many forms remains even today one of the most studied areas of investigation in polymer science. Crystalline polymorphs of cellulose, transformations between them through processes such as alkali mercerization, cellulose in solution and in complexes with other materials, and the many cellulose derivatives are included in these studies. Information on various structural levels is being developed, for example, on what is the shape of the molecule in

---

*Current address: Biophysics Lab, Center for Biologics Evaluation & Research, National Institutes of Health, 8800 Rockville Pike, Bethesda, MD 20892

⁺To whom correspondence should be addressed.

its various chemical and physical states, on how these molecules interact or crystallize, and on how cellulose is synthesized by various living systems. A variety of structural tools are brought to bear in these investigations—from x-ray diffraction for the crystalline systems to NMR for the solution studies to the techniques of biochemistry for the living systems. Relationships between structural characteristics on the molecular scale and the bulk physical properties remain of very high interest in these investigations.

Molecular modeling, as it has developed and progressed in sophistication during the past two decades, has become of increasingly significant help in many of the studies. A prime example can be found in the crystallography of cellulosics: without the integrated diffraction and exhaustive molecular model analysis that we currently employ we would not have been able to develop the extensive structural knowledge base that we have on these molecules. The key in the success of this type of analysis has been the ability to validate modeling predictions with experimental data, and diffraction methods can produce experimental data of very high quality and reliability. The successes in this area are now naturally leading to attempts to characterize the more difficult states, such as solutions, in a similar fashion. Various scattering and spectroscopic techniques, including NMR, yield information only on *average* molecular characteristics and are often unable to draw distinctions between variant interpretations of the data. Molecular modeling can be successfully employed in aid of these studies in a number of ways. The simplest is the usual *static* modeling in which low-energy molecular structures are predicted for comparison with the average structures suggested by an experiment. Although in some instances such comparisons may be sufficient, in many other cases static modeling cannot produce enough information. The principal drawback of static modeling lies in its inability to predict *all* of the conformations that contribute to the average structure and the residence time of the molecule in each one.

With the increasing availability of very fast computers, static modeling is beginning to give way to *dynamic* modeling. With the latter, the motions of individual atoms of the molecule can be simulated leading to the description of the structural (conformational) changes in the molecule with time. From such trajectories, desired average characteristics of the molecule can be calculated for a better comparison with experimental results.

Having these objectives in mind, we began some time ago a systematic investigation of molecular dynamics simulations of models of cellulose. The immediate objectives were to assess the differences between the results obtained by static and dynamic modeling, and the effects of different variables such as force-fields, chain length, the length of the simulation, and modeling in vacuum and in solvent (water). Following preliminary results reported earlier,[1] additional, more quantitative data resulting from these studies are presented herein. Full, detailed reports of the results will appear elsewhere in due course.[2,3]

EXPERIMENTAL

The underlying principle of molecular dynamics (MD) simulation is a simple one—the evolution of the position vector r_j (of the atom j) in time as described by Newton's classical equation of motion:

$$F_j = m_j a_j = \nabla_j U = m_j \frac{d^2 r_j}{dt^2} \qquad (1)$$

(In this equation, F_j is the total force on atom j, m_j is its mass, a_j the acceleration, U is the total potential energy, ∇_j is the gradient, and t is time.) A typical expression of U may have the following form:

$$U = \Sigma A(b-b_o)^2 + \Sigma B(\theta-\theta_o)^2 + \Sigma C(1 + s \cdot \cos n\phi) + \Sigma D[(r^*/r)^{12} - 2(r^*/r)^6]$$
$$+ \Sigma q_i q_j / r\epsilon \qquad (2)$$

where the first three terms describe bond length, bond angle, and torsion angle energies, respectively, the fourth term is for nonbonded contact energies, and the final term is an electrostatic energy term. Many variants of this expression are in use.

The position vector r_j and the velocity of the atom, v_j, are calculated by suitably integrating eq. 1. The leap-frog integration algorithm[4] was used in all our simulations, with a time-step of 1 femtosecond. The total potential energy U was calculated using various force-field descriptions: MM2 (1985 version),[5] CHARMm,[6] and AMBER.[7] (The details of the simulation methodology can be found in ref. 2.) All simulations were carried out on an Alliant FX/80 computer using a suitably parallelized version of the molecular modeling program, DISCOVER.[8] The models examined in these simulations were glucose, cellobiose, cellotetraose, and cellooctaose.

RESULTS

The trajectories accumulated in most simulations spanned a time from several hundred picoseconds (ps) to 1 nanosecond (ns). As expected, the largest motions observed were the rotations of the O(6) hydroxyl groups and all of the hydroxyl hydrogens, as well as the ϕ,ψ-rotations in all models larger than glucose. Typical O(6) rotations are shown in Fig. 1 from where it is evident that the observed rotational positions are the staggered ones commonly referred to as gt, gg, and tg.[9] Similarly, for cellobiose and larger oligomer models, the ϕ,ψ-rotations generally occur as shown in Fig. 2, with sharp and frequent transitions taking place between two (or more) potential energy minima, with considerable synchrony in the two transitions. A different view of these transitions is shown in Fig. 3 for cellobiose, exhibiting two of the principal potential energy minima and the considerable range of motion within these minima.

In assessing the effects of method variations on the results, significant differences were observed. Likewise, differences existed between the results obtained by time-averaging over dynamic simulations and

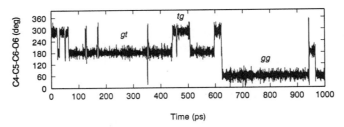

Fig. 1 Typical O(6) torsion transitions observed in glucose during a 1 ns dynamics simulation at T = 300 K.

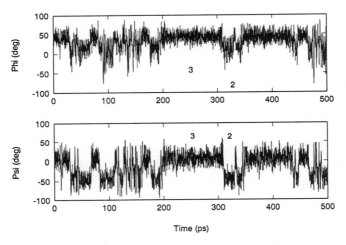

Fig. 2 Typical φ,ψ-rotation transitions observed in cellobiose during a 500 ps dynamics simulation at T = 300 K.

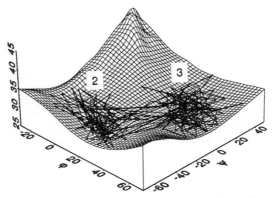

Fig. 3 Transitions shown in Fig. 2 superimposed on the total potential energy surface obtained from constrained minimization (minima 2 and 3 are indicated; energy in kcal/mole).

those obtained in static energy minimizations. These results are briefly
summarized below.

Static vs. dynamic modeling. — Typical of the differences observed in
these two types of modeling are those in the occupancies of the three
O(6) rotational positions of β-D-glucose, as shown in Table 1. The
static modeling results were obtained by separately minimizing the
energies of each of the three O(6) rotamers, then calculating the
fractions of each from a Boltzmann equation. The occupancy fractions
for the dynamic simulations were obtained simply by averaging over
the entire time span of the simulation. As is evident from these results,
considerable differences can occur with any given method of
calculation. However, it should also be noted that due to the relatively
short simulation lengths (1 ns) and the high torsional energy barrier
for O(6) rotation, the statistical uncertainty in the dynamics results is
quite large. Longer trajectories, of the order of 10 ns, are required
before these results can be validated.

Table 1. Fractions of O(6) rotamers in β-D-glucose predicted from
dynamic and static modeling using variations of MM2(85)
force-field.

| Method[a] | Probability of O(6) Conformation | | | | | |
| | Dynamics | | | Static Minimization | | |
	gg	*gt*	*tg*	*gg*	*gt*	*tg*
No LP, no charges	0.52	0.36	0.12	0.30	0.59	0.11
With LP, no charges	0.91	0.08	0.01	0.34	0.56	0.10
No LP, charge set CA	0.54	0.38	0.08	0.48	0.48	0.04
LP + CA	0.35	0.50	0.15	0.56	0.40	0.04
No LP, charge set CB	0.52	0.31	0.17	0.23	0.54	0.23

[a]MM2(85); LP = lone pairs

Effects of force-field. — As is also evident from the data in Table 1, the
type of force-field used can exert considerable influence on the
outcome of the calculation. The overall force-field used to obtain the
data of Table 1 was very similar to that of MM2(85). The differences
shown resulted from the incorporation or elimination of lone pairs of
electrons on oxygen atoms (LP) or the presence or absence of an
electrostatic energy term (the last term of eq. 2) in the force-field
expression. The effects of two different charge sets (CA,[10,11] CB[12]) are
shown.

Differences due to method are further illustrated in Tables 2 and 3. In
Table 2 are shown the differences in total energies predicted for the α-
and β-anomers of D-glucose ($\Delta E = E_\alpha - E_\beta$) and the effects of LP and
charge terms on them. From ΔE, the fractions of α- and β-anomers in
an equilibrium mixture of the two can be calculated, as shown in the
third column of the table. These values are seen to fluctuate widely
with the method of calculation. They can be compared with the

experimentally determined anomer ratio in water,[13] $\alpha:\beta$ = 39:61. For comparison, values of ΔE and %β calculated from static energy minimizations are also shown in Table 2 (columns 4 and 5). Reasonable agreement between the results from dynamic and static modeling exists in this case, primarily because the anomeric energy differences were not very sensitive to O6 conformation.

Table 2. Anomeric energy differences and the resultant fractions of β-anomer from dynamic and static modeling using variations of MM2(85) force-field.

Method	Dynamics		Static Minimization	
	ΔE (kcal/mole)[a]	%β	ΔE (kcal/mole)[a]	%β
No LP, no charges	-0.25	36-44	-0.5	30
With LP, no charges	-0.74	18-28	-0.7	24
No LP, charge set CA	0.5	58-79	0.3	62
LP + CA	-0.3	32-43	-0.3	38
No LP, charge set CB	0.06	40-65	0.03	51

[a]$\Delta E = E_\alpha - E_\beta$

Additional comparisons between results calculated using different force-fields in dynamic simulations are shown in Table 3. In this case, dynamics-averaged values of ϕ,ψ-rotation angles of cellobiose were converted into optical rotation[14] (Λ) and NMR coupling constants[15] (Jϕ, Jψ) for a comparison with corresponding experimental values.[14,16] Again, method-induced differences can be seen to be considerable, with the MM2 force-field being generally better than those of CHARMm or Amber.

Table 3. Values of optical rotation (Λ) and coupling constants (J) predicted for cellobiose from dynamic and static modeling using different force-fields. (Static minimization values given in parentheses.)

Method[a]	Λ (deg)	Jϕ (Hz)	Jψ (Hz)
MM2	101 (84)	4.6 (4.3)	4.1 (3.9)
MM2+LP	71 (37)	4.2 (3.8)	4.3 (4.7)
MM2+LP+CA (3)	51 (38)	3.8 (3.7)	4.6 (4.9)
MM2+LP+CA (1)	19 (21)	3.2 (3.3)	5.2 (5.3)
Amber+LP+CA (1)	33 (40)	2.8 (2.9)	5.1 (5.3)
CHARMm+CB (1)	33 (106)	2.9 (5.2)	5.1 (5.0)
Experiment	92	4.2	4.3

[a]Values in parentheses after charge set designations are for dielectric constant.

Effects of model size. — Sizable differences were seen in a number of conformational features of the models, which, as noted earlier, ranged from glucose to cellooctaose. In an example shown in Table 4, the distributions of O(6) rotamers are apparently greatly affected by the length of the oligomer, reversing the *gt* preponderance in cellobiose to that of *gg* in cellooctaose. (As in the case of glucose, the statistical uncertainty of these distributions is still large and much longer dynamics simulations are needed to verify the observed results.) Another, more complex effect of model size can be seen in Fig. 4 which depicts the changes in ϕ,ψ-rotations of cellotetraose and the consequent transitions of conformation of the molecule between three energy minima, one of which (min. 4) is separated by a relatively high barrier from the other two. This behavior of cellotetraose should be compared with that of cellobiose (*cf.* Fig. 2) which is seen to move only between two of the lowest-energy minima (min. 2 and 3). These differences in conformational transitions stem partly from the more cooperative motions seen in longer oligomers as well as more prominent effects exerted on a given transition by those occurring in neighboring glucose residues. An example of the latter is illustrated in Fig. 5 in which an inner glucose residue of the tetramer is shown in transition from the low-energy chair conformation through a number of high-energy boat and twist forms during a time span of 21 ps. This behavior was not observed to take place in glucose and cellobiose models.

Table 4. Fractions of O(6) rotamers in different oligomers predicted from dynamics.

Model	Probability of O(6) Conformation		
	gg	*gt*	*tg*
Cellobiose	0.19	0.65	0.16
Cellotetraose	0.31	0.55	0.14
Cellooctaose	0.53	0.39	0.08

Another example of the effects of cooperative motions is shown in Fig. 6 for cellooctaose. Over a time span of *ca.* 350 ps, the normally more extended molecule assumes a nearly completely folded conformation and remains in it for a considerable length of time before reverting to the more normal shape. This behavior was seen only in the octamer and not the tetramer.

Effects of water as solvent. — Because a molecule in solution is generally surrounded by solvent molecules, the latter can be expected to play an important part in modulating the dynamic behavior of the solute molecule. The solute-solvent collisions may affect all torsional librations, making them more diffusive in solution. Also, of particular importance in carbohydrates, the strength of intramolecular hydrogen bonds can be expected to decrease due to hydrogen bonding of sugar hydroxyls with water. Both effects were, in fact, seen in simulations of cellobiose in water—the only model studied in solution thus far. In other respects, the average ϕ,ψ torsional angles of cellobiose in solution were little affected in comparison with vacuum, as illustrated in Table 5.

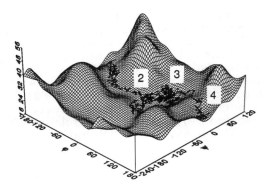

Fig. 4 The φ,ψ-rotations in cellotetraose superimposed on the total energy surface.

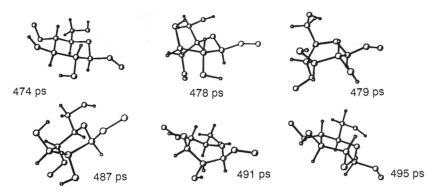

Fig. 5 Conformations of an inner glucose residue of cellotetraose observed during a dynamics simulation run.

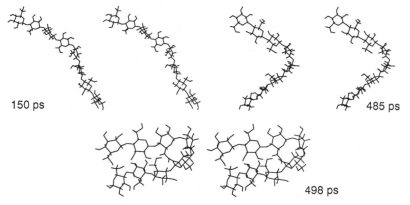

Fig. 6 Conformations of cellooctamer observed during a dynamics simulation run (stereo views).

Table 5. Comparison of ϕ,ψ torsions of cellobiose in three energy minima, predicted from dynamics simulations in solution and vacuum systems.

Cellobiose Torsions	Solution Ave (rms)	Vacuum Ave (rms)
Min. 1:		
ϕ	44.2° (10.5)	26.2° (24.8)
ψ	-177.4° (9.5)	-175.2° (10.8)
Min. 2:		
ϕ	30.2° (13.4)	30.3° (20.0)
ψ	-47.2° (13.6)	-42.8° (15.2)
Min. 3:		
ϕ	54.4° (12.0)	51.3° (12.9)
ψ	2.9° (13.2)	1.1° (11.0)

Effects of simulation length. — Whenever dynamics-averaged structural features of a model are needed, the simulations must be run for long enough times to yield a sufficiently large number of transitions, thus assuring a view of the equilibrium system. For vacuum simulations, based on 1 ns trajectories, it appears that longer runs of 10 ns or more, may be necessary. In solution models, however, only a few hundred ps simulations are available at this time and it is not yet clear to what length they must be extended.

ACKNOWLEDGMENTS

These studies were supported by the National Science Foundation, under grants DMB8320548 and DMB8703725. All computations were conducted at the Northeast Parallel Architectures Center (NPAC) at Syracuse University, which is funded by and operates under contract to DARPA and the Air Force Systems Command, Rome Air Development Center, Griffiss AFB, NY.

REFERENCES

1. Sarko, A.; Chen, C.-H.; Hardy, B.J.; Tanaka, F. *ACS Symp. Ser.* **1990**, *430*, 345.
2. Hardy, B.J. Ph.D. Dissertation, Syracuse University, Syracuse, NY, 1990.
3. Hardy, B.J.; Sarko, A. In preparation.
4. Berendsen, H.J.C.; van Gunsteren, W.F. Barnes, A.J.; Orville-Thomas, W.J.; Yarwood, J. (eds.) in *Molecular Liquids: Dynamics and Interactions*, Klewer Academic Publishers, Hingham, MA, 1983, p. 475.
5. MM2 (85), *Quantum Chemistry Program Exchange*, Indiana University, Bloomington, IN 47405. Also: Allinger, N.L. *J. Amer. Chem. Soc.* **1977**, *99*, 8127.
6. Brooks, B.R.; Bruccoleri, R.E.; Olafson, B.D.; States, D.J.; Swaminathan, S.; Karplus, M. *J. Comput. Chem.* **1983**, *4*, 17.

7. Weiner, S.J.; Kollman, P.A.; Nguyen, D.T.; Case, D.A. *J. Comput. Chem.* **1986**, *7*, 230.
8. Biosym Technologies, Inc., 10065 Barnes Canyon Road, San Diego, CA 92121.
9. Sarko, A.; Marchessault, R.H. *J. Polym. Sci., Part C* **1969**, *28*, 317.
10. Melberg, S.; Rasmussen, R.; Scondamaglia, R.; Tosi, C. *Carbohdyr. Res.* **1979**, *76*, 23.
11. Melberg, S.; Rasmussen, K. *J. Mol. Struct.* **1979**, *57*, 215.
12. Ha, S.N.; Giammona, A.; Field, M.; Brady, J.W. *Carbohydr. Res.* **1988**, *180*, 207.
13. Dunfield, L.G.; Whittington, S.G. *J. Chem. Soc., Perkin Trans. 2* **1977**, 654.
14. Rees, D.A. *J. Chem. Soc., Sect. B* **1970**, 877.
15. Hoch, J.C.; Dobson, C.M., Karplus, M. *Biochemistry* **1985**, *24*, 3831.
16. Hamer, G.K.; Balza, F.; Cyr, N.; Perlin, A.S. *Can. J. Chem.* **1978**, *56*, 3109.

8

Conformational analysis of cellobiose with MM3

Alfred D. French and Michael K. Dowd* - Southern Regional Research Center, US Department of Agriculture, P.O. Box 19687, New Orleans, LA 70179. *Chemical Engineering Department, Iowa State University, Ames, IA 50011, USA.

ABSTRACT

The present work follows earlier studies with the computer program MMP2(85) on the conformations of cellobiose, and by implication, cellulose. Improvements over the previous work include the MM3 force field itself, the use of 16 different starting geometries, and investigation at dielectric constants of 1.5 (vacuum), 4 (crystal field) and 80 (aqueous solution). As before, the MM3 maps had 4 minima within 5 kcal/mol of the global minimum, but the two minima that lead to hollow helical conformations or folds are now predicted to be less likely. Also, a 2-fold screw structure is somewhat more likely.

INTRODUCTION

Many workers have examined the conformation of cellobiose with calculations of potential energy [1-7]. Such calculations supplement available experimental work, such as the crystal structures of cellobiose [8] and of the methyl cellobioside - methanol complex [9]. Questions regarding the shape of cellulose in amorphous zones or on the likelihood of folded or hollow helical structures are readily addressed by such computational methods, even though experimental treatment of such subjects is often difficult.

The classic conformational analysis is based on rotation of the two glucose residues of cellobiose about their bonds to the mutually held glycosidic oxygen atom. Resulting cellobiose energy surfaces have four main minima (A - D). The two central minima have roughly the same conformations as the crystal structures of cellobiose [8] and its methylated derivative [9], respectively, and are about equal in energy. The other two minima are usually higher in energy; perhaps so high that one or the other may not be included in the contours presented from studies in which the glucose residues are held rigid.

When we used MMP2(85) some four years ago for relaxed-residue conformational analyses [10,11], we were concerned with the effects of such analyses and the procedures for carrying out such analyses. At present our computer resources have increased considerably, and more extensive studies can be carried out. Also, the MM3 program [12,13] has been released, and we wanted to see how relaxed-residue conformational surfaces would appear when calculated by this improved force field.

ϕ = H1-C1-O4'-C4'
Ψ = H4'-C4'-O4'-C1

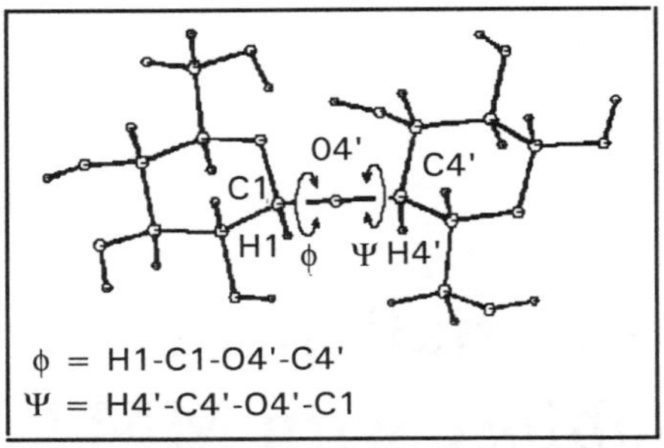

Figure 1. The linkage torsion angles of cellobiose. OH groups are r (left residue) and c (right residue). O6 is gt on both.

Although there are numerous changes in MM3, one of the most important is that lone pairs of electrons are not treated explicitly. With fewer "atoms", the calculations on carbohydrates are twice as fast as with MMP2. This expands further the possible scope of problems to be examined.

A supplemental hydrogen bonding function that depends on dielectric constant is used with MM3. Since there are so many hydroxyl groups in sugars, the results of such MM3 modeling studies are quite sensitive to changes in dielectric constant. The work described herein examines the effect of varied dielectric constant on cellobiose maps, and is an update on the work presented at the Tenth Syracuse Cellulose Symposium in 1988 [11]. A parent study by Dowd, French and Reilly [14] covers all the equatorial-equatorial linked disaccharides of glucose. That paper includes comparison of cellobiose models at a dielectric constant of 4 with experimental results.

Although the form of the sugar ring is usually assumed, we have extensively analyzed the ring shape and plan to publish the results elsewhere. Stated briefly, those MM3 analyses showed that $^4C_1$ was the only likely ring form. The best alternative had an energy nearly 6 kcal/mol higher than the usual chair, even when incorporated into a disaccharide. Conversion of this energy difference into probability at 300° suggests that at least 99.99% of the glucose rings will have the usual shape. The minimum alternative energy, 6.36 kcal/mol for a ^2S_O form, converts by the Boltzmann equation to a population of 0.002% at 300°. At 400°, its population is increased by a factor of 15. Other force fields can give smaller energy differences among the forms. If those force fields are used, the predicted probability of non-$^4C_1$ forms will increase, especially at higher temperatures.

METHODS

The methods used have been described thoroughly in a recent contribution on conformational analyses of the trehaloses [15]. In the present paper, both DecStation computers at Iowa State University and VAX computers at the Southern Regional Research Center were used. The January 1990 release of MM3, with the April 1991 corrections, was unmodified, and we used the energy minimization routine rather than geometry optimization. Only the block diagonal minimization method was used for the work discussed in the present paper. For each conformation examined, a separate input file was created instead of using the dihedral driver supplied with MM3. The rationale for using

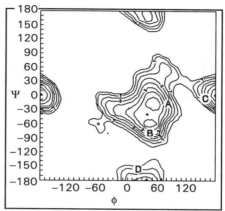

Figure 2. MM3 energy for cellobiose at dielectric constant of 1.5. The minima A–D are shown, and contours are at 1 kcal/mol.

either an alternate driver or separate input files is discussed extensively in a chapter by French, Tran and Perez [16].

Starting models were sketched with either CHEM-X or PCMODEL. Primary alcohol groups were arranged in either gt or gg orientations, with the secondary hydroxyl groups arranged either c or r [17]. If each residue can have four combinations of rotating groups (ggr, ggc, gtr, gtc), the disaccharide has 16 possible combinations. For each combination, the torsion angles ϕ (H1-C1-O4'-C4') and ψ (C1-O4'-C4'-H4') were varied in 20° increments, giving 361 points on the ϕ - ψ grid. One of the starting models is shown in Fig. 1. These searches were carried out at dielectric constants of 1.5 (vacuum), 4 (crystal field) and 80 (aqueous solution). As described below, a final map was needed for the low-energy, central area when the dielectric constant was 1.5. For that map, 5° increments covered the range of ϕ = 10 to 70, ψ = -10 to 70. Only seven combinations of side groups were

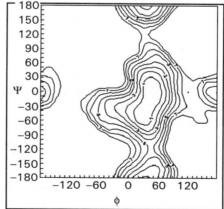

Figure 3. As in Fig. 2 but with dielectric of 80.

used for that final map. In all, 18,693 structures were minimized (compared to 497 used in the first study with relaxed models).

Energy surfaces shown in this work were made with SURFER package for IBM-PC compatibles. The values used at each grid point were the lowest energy of the 16 calculated at the point. The use of 16 starting models (7 in the final map) always yielded at least one useful energy value at each grid point. Therefore, unlike the work by French, Tran and Perez [16], no compensation for missing energy values was needed.

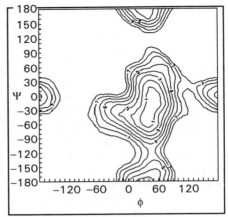

Figure 4. As in Fig. 2, but with dielectric constant of 4.

RESULTS AND DISCUSSION

The energy maps for cellobiose in dielectric fields of 1.5, 4 and 80 are shown in Figs. 2, 3 and 4, and the values for the normalized energies at the minima and barriers are given in the Table. The areas within the energy contour lines increase as the dielectric constant increases. The larger area inside the 1 kcal/mol contour on the dielectric = 80 map suggests that cellobiose will be more flexible in water than in solvents of low dielectric constant.

Fig. 2 shows a small peak near the point $\phi = 40$, $\psi = -40$. Visual comparisons of the model cellobiose structures at this point and structures at neighboring points showed that the O6 hydroxyl hydrogen atoms had rotated in the neighbors, permitting a lower energy. At the $\phi = 40$, $\psi = -40$ point, however, the

hydrogen had not rotated, leaving a structure with higher energy. This is an example of the behavior that led us to characterize energy surfaces constructed with MM3 as somewhat treacherous, especialy when using a dielectric constant of 1.5 or less.

To solve this problem, a 17th starting model was created with the geometry shown in Fig. 5, and a map with 5° increments of ϕ and ψ was constructed with the new model and the six starting models that had given low energies in the central area. Fig. 6

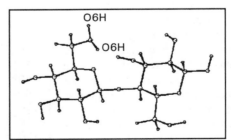

Figure 5. The 17th starting model with O6H in the new (upper) position as well as the problematic old (lower) position.

shows the map, from which the value of the barrier shown in the Table was taken. On Fig. 6, a diagonal line indicates the structures that would correspond to cellulose in a 2_1 screw axis conformation. The maximum energy barrier between the two low energy minima in the central area is on a parallel line (not shown) that intersects the ψ axis about 15° above the 2_1 axis line. Part of the increased energy of 2_1 structures is thought to arise from repulsion between H1 and H4'. With MM3, this distance is 2.27 Å, an acceptable if not completely satisfactory value.

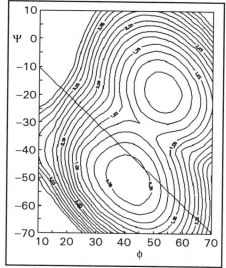

Figure 6. Cellobiose with 17th starting geometry, dielectric of 1.5 Contours at 0.25 kcal/mol.

These maps all show a lower barrier that separates the two central minima than was found with MMP2(85), which was about 2 kcal/mol [10,11]. Using a better (lower-energy) starting model, Sarko et al. found the barrier to be 1.3 kcal/mol [7], also with MMP2(85). They also tested the PS-79 [18] and LALS [19] programs. Those programs, based on relatively simple potential functions, showed that the 2_1 structures correspond to a minimum in the energy. The presence of a 2_1 axis is controversial, being indicated by diffraction studies and being contradicted by nmr studies. The present work with MM3 shows that 2_1 structures could easily exist under the influences of a crystal field.

As shown by Sarko [20], the modeling potential function can strongly affect the prediction of favored conformers. Here we have varied only the dielectric constant, and as shown in the above maps and the Table, there is a substantial effect. In all cases, however, the structures near the centers of the maps had the lowest energies.

RELATIVE MM3 ENERGIES (KCAL/MOL) OF THE MAIN MINIMA AND SCREW-AXIS BARRIER OF CELLOBIOSE

Dielectric Const.	A 50°, -10°	B 35°,-50°	C 180°,0°	D 0°,180°	Barrier Energy
1.5	0.32	0.00	0.98	2.27	0.90
4	0.03	0.00	3.59	2.08	0.20
80	0.00	0.21	4.08	1.92	0.43

REFERENCES

1.   Melberg, S.; Rasmussen, K. *Carbohydr. Res.* **1979**, *71*, 25.
2.   Tvaroska, I. *Biopolymers* **1984**, *23*, 1951.
3.   Lipkind, G.M.; Verovsky, V.E.; Kochetkov, N.K. *Carbohydr. Res.* **1984**, *133*, 1.
4.   Pizzi, A.; Eaton, N. *J. Macromol. Sci.* **1984**, *Chem. A21(11&12)*, 1443.
5.   Henrissat, B.; Perez, S.; Tvaroska, I.; Winter, W.T. in Atalla, R.H. (Ed.) *The Structures of Cellulose - Characterization of the Solid State*, ACS Symposium Series *#340*, ACS Books, Washington, DC **1987**, 38.
6.   Simon, I.; Scheraga, H.A.; St. John Manley, R. *Macromolecules* **1988**, *21*, 983.
7.   Sarko, A.; Chen, C.-H.; Hardy, B.J.; Tanaka, F. in French, A.D.; Brady, J.W. (Eds.) *Computer Modeling of Carbohydrate Molecules*, ACS Symposium Series *#430*, ACS Books, Washington, DC **1990**, 345.
8.   Chu, S.S.G.; Jeffrey, G.A. *Acta Crystallogr.* **1968**, *B24*, 830.
9.   Ham, J.T.; Williams, D.G. *Acta Crystallogr.* **1970**, *B26*, 1373.
10. French, A.D. *Biopolymers* **1988**, *27*, 1519.
11. French, A.D. in Schuerch, C. *Cellulose and Wood - Chemistry and Technology*, Wiley Interscience, New York **1989**, 103.
12. Allinger, N.L.; Yuh, Y.H.; Lii, J.-H. *J. Amer. Chem. Soc.* **1989**, *111*, 8551.
13. Allinger, N.L.; Rahman, M.; Lii, J.-H. *J. Amer. Chem. Soc.* **1990**, *112*, 8293.
14. Dowd, M.K.; French, A.D.; Reilly, P.J., *Carbohydr. Res.* in press.
15. Dowd, M.K.; French, A.D.; Reilly, P.J., *J. Comp. Chem.* **1992**, *13*, 102.
16. French, A.D.; Tran, V.; Perez, S. in French, A.D.; Brady, J.W. (Eds.) *Computer Modeling of Carbohydrate Molecules*, ACS Symposium Series *#430*, ACS Books, Washington, DC **1990**, 191.
17. Ha, S.N.; Madsen, L.J.; Brady, J.W. *Biopolymers* **1988**, *27*, 1927.
18. Zugenmaier, P.; Sarko, A. in French, A.D.; Gardner, K.H. (Eds.) *Fiber Diffraction Methods*, ACS Symposium Series *#141*, ACS Books, Washington, DC **1980**, 225,
19. Smith, P.J.C.; Arnott, S. *Acta Crystallogr.* **1978**, *A34*, 3.
20. Sarko, A.; Hardy, B.J. This book.

9

Antiparallel molecular models of crystalline cellulose

R.-H. Mikelsaar and Alvo Aabloo* - Tartu University, Institute of General and Molecular Pathology, 34 Veski Street, 202400 Tartu, Estonia. *Tartu University, Department of Experimental Physics, 4 Tähe Street, 202400 Tartu, Estonia.

KEYWORD: molecular modelling, structure of cellulose I.

ABSTRACT

Antiparallel molecular structure models are proposed for cellulose I having the same geometric parameters which were established by Sugiyama et al. [1] for cellulose α and ß phases.

INTRODUCTION

Although the presence of two crystalline phases in native cellulose (I) was demonstrated by NMR spectroscopy in 1984 [2,3], the exact unit cell geometries of the α and ß phases were established only recently [1]. Sugiyama et al. [1] interpret their experimental data according to a parallel molecular chain hypothesis.

DISCUSSION

The aim of the present paper is to show that antiparallel cellulose molecular structures may exist having the same unit cell dimensions. Molecular modelling was carried out by Tartu plastic space-filling atomic models [4]. The results of the modelling experiments allowed us to present following schemes of cellulose structure (Fig. 1-5).

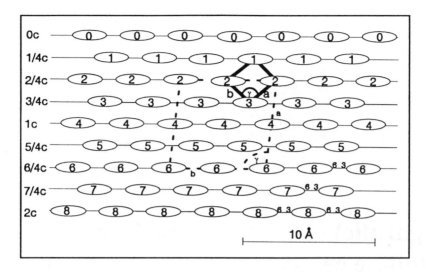

Figure 1. A structure of cellulose Iα phase according to Sugiyama et al [1]. Symbols: white ovals - parallel chains; a,b,c,γ - unit cell dimensions; numbers in ovals - the amount of c/4 shifts in the direction of c- axis of this unit cell; 6-3 - bonds O6H..O3.

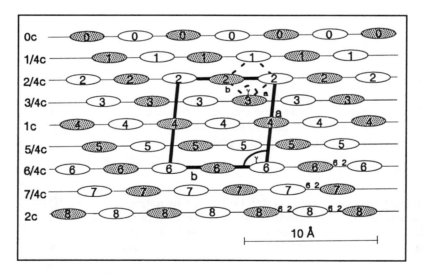

Figure 2. Antiparallel cellulose structure model "Iα-V". Symbols: white ovals - parallel chains; grey ovals - antiparallel chains; a,b,c,γ - unit cell dimensions; number in ovals - the amount of c/4 shifts in direction of c - axis of this unit cell; 6-2 - bonds O6-H..O2. Each chain(oval) contains two intra molecular bonds (O3-H..O5' and O2'-H..O6) and oxymethyl groups in tg or gg conformation.

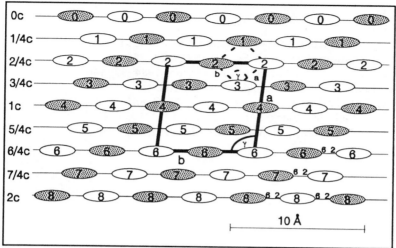

Figure 3. Antiparallel cellulose structure model "Iα-P". Symbols see Fig.2.

The analysis of schemes presented in current paper allow us to suppose that antiparallel molecular structures of cellulose may have unit cell parameters corresponding to those given in the work of Sugiyama et al.[1]. Antiparallel structure models should be considered in the interpretation both of cellulose I and cellulose II investigations [5-7].

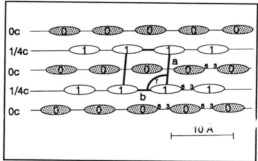

Figure 4. Antiparallel model of structure of cellulose Iß phase. Symbols see Fig.1 and 2.

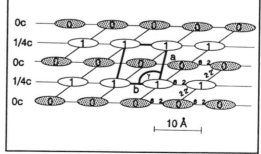

Figure 5. Antiparallel model of structure of cellulose II. Symbols see Fig.1. and 2.

REFERENCES

1. Sugiyama,J., Vuong,R., Chanzy, H. *Macromolecules* **1991**, 24, 4168.
2. Atalla, R. H., VanderHart, D. G. *Science* **1984**, 223, 283.
3. VanderHart, D. G., Atalla, R. H. *Macromolecules* **1984**, 17, 1465.
4. Mikelsaar, R. *Trends in Biotechnology* **1986**, 6, 162.
5. Stipanovic, A. J., Sarko, A. *Macromolecules* **1976**, 9, 851.
6. Kolpak, J. F., Weih, M., Blackwell, J. *Polymer* **1978**, 19, 123.
7. Takahashi, Y., Matsunaga, H. *Macromolecules* **1991**, 24, 3968.

10

Calculations of potential energy of the cellulose crystal structure

Alvo Aabloo, Aleksandr J. Pertsin* and R.-H. Mikelsaar** - Tartu University, Department of Experimental Physics, Tähe 4 Street, 202400 Tartu, Estonia. *Institute of Element-Organic Compounds, Vavilova 28 Street, GSPI, V-334, 117813 Moscow. **Tartu University, Institute of General and Molecular Pathology, Veski 34 Street, 202400 Tartu, Estonia.

ABSTRACT

Many aspects of cellulose crystal structure remain without satisfactory explanation. We have calculated objective functions based on diffraction intensities and potential energies of the cellulose II crystal using "rigid model" calculations. We used different geometries of the glucose rings and found that this does not shift the results of potential energy calculations very much. The second part of this work contains preliminary results of potential energy calculations for the cellulose Iα crystal structure. Models with the lowest energy are in the tg conformation. However, the best gg model has very complicated hydrogen bonding that includes two hydrogen bonds between sheets.

INTRODUCTION

In recent years there have been only a few attempts to determine the structure of cellulose by combined potential energy and X-ray diffraction calculations [1]. A typical X-ray diagram of cellulose contains only few dozen reflections, a number that is clearly insufficient to refine all atomic parameters by standard crystallographic methods. To increase the ratio of observations to refineable parameters we have used various stereochemical and packing constraints and non-bonded contacts calculated by the atom-atom potential method. For calculating the most probable models of a crystal of cellulose we have minimized the objective function

$$F=U+W^*R''$$

where U is the potential energy of the system, R'' is the crystallographic discrepancy factor based on the cellulose II intensity data from Kolpak et al [2]. The potential energy consists of

$$U=U_{mon}+U_{junct}+U_{inter'}$$

in which U_{mon} is the conformational energy of the monomer residues, U_{junct} is the conformational energy between two successive residues along both of the two crystallographically distinct chains, U_{inter} is the intermolecular energy which includes non-bonded and H-bond energy between the atoms of different chains. Calculations have been executed in internal coordinates. There are two different extents of atomic adjustments that can be used to reduce F. The first possibility is to leave all bond lengths etc. flexible. The second way is to fix some structural components that do not vary much during minimization. This method allows us to decrease the number of variables. In earlier work [3] thirteen most probable models for the cellulose II crystal structure were calculated.

CALCULATIONS OF THE POTENTIAL ENERGY OF THE CRYSTAL STRUCTURE OF CELLULOSE II

To learn the importance of variable residue geometry, we have calculated all thirteen most probable models of cellulose II crystal structure [3] by the use of the "rigid model" method described in [3,4], applying five different initial glucose rings. We used β-glucose rings constructed by chiral inversion of residues found in cyclohexaamylose-KOAc/CHA2/, planteose/PLA/ and glucose-urea/GUR/ [5]. We also used reducing ß-D-cellobiose/CBIO/ and ß-D-glucose/ßGLU/ [6]. We presumed symmetry $P2_1$ for the chain of cellulose II. During minimization we fixed the bond lengths, bond angles and glucose ring geometry. We varied the torsional angles of the hydroxymethyl and three hydroxyl groups and the torsional and bond angles describing the junction between two successive monomer residues. We also varied rotations and shift of chains in the unit cell w cellulose II [2]. As for the most probable models, their variable parameters do not shift remarkably when the ^Xxfferent geometries are used. The only models that shifted were ones that ranked last in the initial preference list [3] (using "average Arnott-Scott" ring). When the model A11 was based on PLA and ßGLU rings, it did not find the same energy minimum. Instead the A11 shifted to the A1 model. In Figs. 1 and 2 we present the value of the objective function and the potential energies of the thirteen most probable models of crystal structure II. It can been seen that the models of lowest values of energy do not change remarkably. We deduce that "rigid" model calculations are quite correct for the initial calculations of cellulose structure. For more exact calculations it is reasonable to use "flexible" methods (as MM3) in which "rigid model" results can be used as initial variable sets.

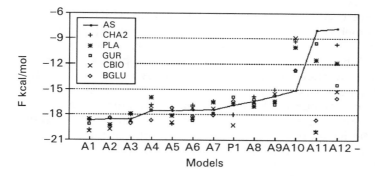

Figure 1. Objective functions of most probable models of cellulose II crystal. Glucose rings: AS - average Arnott-Scott; CHA2- cyclohexaamylose; PLA - planteose; GUR - glucose-urea; CBIO - cellobiose; βGLU - β-D-glucose. A1..A12,P1 - most probable models of the unit cell of cellulose II [3].

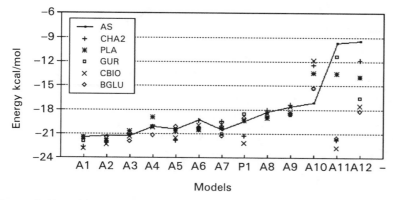

Figure 2. Potential energies of most probable models of the cellulose II crystal. For abbreviations see Fig.1.

PRELIMINARY CALCULATIONS OF THE POTENTIAL ENERGY OF
THE CELLULOSE Iα CRYSTAL STRUCTURE

For a long time there were many questions about the structure of
crystalline cellulose I. Several authors [7,8] tried to solve the problem, but
it has up to now been unsolved. Later, using NMR data [9,10] it was
found that the crystal of cellulose I consists of two phases: cellulose Iα
which must have triclinic symmetry, and cellulose Iβ which must have
monoclinic symmetry. This year Sugiyama and collaborators [11] reported
a new structure of the crystal of cellulose I. They used a new precision
electron diffraction apparatus which allows them to apply a very narrow
initial electron beam. They found that cellulose Iα has a one-chain triclinic
unit cell with parameters: α=117°, β=113°, γ=81°, a=6.74Å, b=5.93Å,
c=10.36Å. The aim of our work is to examine that unit cell from the
energetic viewpoint. Unfortunately, X-ray diffraction data of phase Iα do
not exist yet. We have calculated the potential energy of the crystal of
cellulose Iα using "Arnott-Scott" glucose ring geometry for "rigid model"
calculations. This time we presumed symmetry P1 for the chain of
cellulose Iα. We varied torsional angles of the two hydroxymethyl and six
hydroxyl groups and the two torsional and bond angle between two
successive units along the chain (Figure 3.). Table 1 shows the 19 models
having lowest energy for phase Iα. The best "up" model (U1) has an
energy of -19 kcal/mol in tg conformation. It is remarkable that at
energies -14 to -16 kcal/mol there exist 6 "up" models with different
hydrogen bonds. The best "down" model (D1) has an energy of -17
kcal/mol. The best gg model has an energy of -14 kcal/mol. It has very
complicated hydrogen bonding. There are two hydrogen bonds between
sheets along the a-axis.

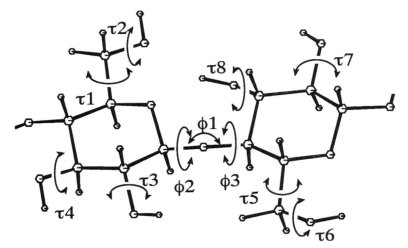

Figure 3. Variable torsional and bond angles in the chain of cellulose Iα.

Table 1. The most probable models of the cellulose Iα triclinic unit cell.

U_ - "up" models; D_ - "down" models; τ_,φ_ - variable bond and torsional angles (see Fig. 3.); ε - angle describing chain rotation; [1] - H-bonds in a chain; [2] - H-bonds in a sheet <110>; [3] - H bonds between sheets <100>.

Energy kcal/mol	τ1	τ2	τ3	τ4	φ1	φ2	φ3	τ5	τ6	τ7	τ8	ε	Conf.	H-bonds
U1 -19.5	71	166	-52	175	118	93	146	73	166	-52	175	-77	tg	O5..HO3'^{1} O2H..O6'^{1} O6H..O3^2
D1 -16.8	59	166	-53	168	117	87	143	74	163	-53	170	-1	tg	O5..HO3'^{1} O2H..O6'^{1} O6H..O3^2
U2 -16.3	69	-61	164	168	119	92	148	72	-63	170	171	-72	tg	O5..HO3'^{1} O1;O2..HO6'^{1}
U3 -16.2	68	102	-56	161	115	94	147	69	102	-56	161	-78	tg	O5..HO3'^{1} O2H..O6'^{1}
U4 -15.7	64	60	-79	177	116	95	146	66	57	-79	177	-79	tg	O5..HO3'^{1} O2H..O6'^{1}
U5 -15.0	69	34	-54	72	116	94	146	71	35	-54	72	-78	tg	O2H..O6'^{1} O3H..O6^2
U6 -14.3	101	162	-52	161	115	94	147	101	162	-52	161	-79	tg	O5..HO3'^{1} O6H..O2;O3^2
D2 -14.3	50	108	-55	168	119	83	143	70	109	-56	171	-6	tg	O5..HO3'^{1} O2H..O6'^{1} O6H..O3^2
D3 -14.2	53	-53	-150	175	119	83	138	72	-50	-150	177	-1	tg	O5..HO3'1 O1;O2..HO6'^{1}
U7 -14.1	-35	-62	-45	185	115	100	138	-17	-113	24	186	-97	gg	O5..HO3'^{1} O2H..O6'^{1} O6H..O5^{2}O6H..O2^3
U8 -13.6	76	172	183	167	117	96	145	76	172	183	167	-78	tg	O5..HO3'^{1} O6H..O3^2 O2H..O6^2
U9 -13.1	73	-49	-156	174	117	95	141	74	-49	-157	174	-79	tg	O5..HO3'^{1} O1;O2..HO6'^{1}
D4 -12.9	46	61	-67	175	118	84	140	74	67	-69	177	-3	tg	O5..HO3'^{1} O2H..O6'^{1} O2H..O6^3
D5 -12.6	-29	81	184	174	115	95	141	-102	61	184	175	18	gg	O5..HO3'^{1} O1..HO6'^{1}
D6 -12.4	47	35	-54	66	119	81	141	76	39	-55	65	-6	tg	O2H..O6'^{1} O3H..O6^2
U10 -12.3	63	-51	-148	177	116	95	139	69	-50	-147	177	-86	tg	O5..HO3'^{1} O1;O2..HO6'^{1}
U11 -12.3	-156	-61	-37	169	117	94	147	-159	-61	-36	170	-78	gt	O5..HO3'^{1} O3H..O6^2 O6H..O2^2
D7 -12.2	187	175	-112	181	117	97	128	192	171	-114	181	25	gt	O5..HO3'^{1} O6H..O2;O3^2
D8 -12.1	100	177	174	175	115	98	142	94	174	171	173	3	tg	O5..HO3'^{1} O6H..O3^2 O2H..O6^2

Comparing these crystal energies with energies of a crystal of cellulose II which are calculated by the same method, it can seen that cellulose Iα has slightly higher energy minima than cellulose II. It can explained by the conversion of cellulose I into cellulose II. This conversion must be energetically profitable.

REFERENCES

1. Millane, R. P.; Narasaiah,T. V.; *Polymer* **1989**, 60, 1763.
2. Kolpak,F. J.; Weih, M.; Blackwell, J.; *Polymer* **1978**, 19, 123.
3. Pertsin, A. J.; Nugmanov, O. K.; Marchenko, G. N.; Kitaigorodsky, A. I.; *Polymer* **1984**, 25, 107.
4. Pertsin, A. J.; Kitaigorodsky A. I.; *The Atom - Atom Potential Method Applications to Organic Molecular Solids*, Springer Ser. Chem. Phys., V43, Springer-Verlag, Berlin Heidelberg New-York London Paris Tokyo, 1987.
5. French, A. D.; Murphy, V. G.; *Carbohydrate Research* **1973**, 27, 391.
6. Chu,S. S. C.; Jeffrey, G. A.; *Acta Cryst. B* **1970**, 26, 1373.
7. French, A. D.; *Carbohydrate Research* **1978**, 61, 67.
8. Gardner, K. H.; Blackwell, J.; *Biopolymers* **1974**, 13, 1975.
9. Horii, F.; Hirai, A.; Kitamaru, R.; *Macromolecules* **1987**, 20, 2117.
10. VanderHart, D. L.; Atalla,R. H.; *Macromolecules* **1984**, 17, 1465.
11. Sugiyama,J.; Vuong, R.; Chanzy,H.; *Macromolecules* **1991**, 24, 4108.

11

Supermolecular structure of cellulose suggested from the behaviour of chemical reactions

Y. Shimizu, K. Kimura, S. Masuda and J. Hayashi - Department of Applied Chemistry, Faculty of Engineering, Hokkaido University, Sapporo, 060 Japan.

ABSTRACT

In DMF, chloral and pyridine solvent systems ramie easily dissolved but rayon and mercerized ramie did not though they had lower crystallinity and DP. The reactivity of native cellulose decreased by mercerization. It looks as if the crystallite form of cellulose II gives rise to these results, but it was shown that the cause was based on their supermolecular structure which prevented penetration of nonaqueous chemical reagents. By pretreatment their solubility and reactivity increased with the increasing ability to relax hydrogen bonds in amorphous regions of the solution used. The amorphous regions were between microfibrils and had different levels of intermolecular hydrogen bonding. The distribution of the level in cellulose fibers were measured. For ramie it had a maximum at the lower level of hydrogen bonds and ramie allowed penetration of the reagents. For rayon it was broad and had a maximum at the higher level, and the reactivity increased successively with increasing ability of the solution used for pretreatments to relax them. Mercerization reconstructed the hydrogen bonds in amorphous regions more strongly.

INTRODUCTION

Rayon and mercerized native cellulose did not dissolve in some organic solvents which easily dissolved native cellulose(1, 2). On nitration and acetylation of native cellulose mercerized with NaOH aq. solution of various concentration, the reactivity

decreased at the concentration that cellulose I changed into II
for each kind of cellulose. The cause of the phenomena was
investigated, and the supermolecular structure of them was
studied based on their behavior to the reactions.

EXPERIMENTAL

Samples: Ramie, linter, pulp (NBKP) and those mercerized with
NaOH aq. solution of 3 to 20% at 20 °C for 24 h.
Solubility Test: DMF (5 moles), chloral (2 moles) and pyridine
(2.4 moles) per mole of glucose residues were added to cellulose
samples which were then shaken at 20 °C for 12 h. Three kinds
of pretreatment were used. 1) Soaking in water at 20 °C for 3
h, pressing, replacing with DMF and pressing. 2) Soaking in
water, replacing with acetone and air drying. 3) Soaking in DMF
at 20 °C for 3 h and pressing. The amount of DMF in the samples
were corrected in the test.
Esterification: Nitration was carried out with a mixed acid of
HNO_3, Ac_2O and AcOH in equivalent moles at 0 °C for 24 h.
Acetylation was carried out with a mixed acid of Ac_2O of 49.5%,
AcOH of 49.5 % and H_2SO_4 of 1.0% at 25 °C for 5 h.
Degree of Penetration of Chemicals: The weighed cellulose
samples were soaked in DMF, AcOH or water at 20 °C for 12 h,
pressed and put in a centrifuge tube equipped with a wire net
of 200 mesh in the middle of it. After centrifugation at 3,500
rpm for 10 min the weight of the chemicals remaining in the
samples was measured. The increasing ratio of the weight shows
the degree of penetration of the chemicals into the inside of
the fibers.

RESULTS AND DISCUSSION

The results of the solubility test are shown in Table 1. Ramie
and pulp swelled strongly in DMF, chloral and pyridine systems
and dissolved without pretreatment, but rayon, mercerized ramie
and pulp changed into cellulose II did not swell and did not
dissolve even after leaving for one year. The mercerization
brought a change into the fine structure preventing swelling.
Also ramie mercerized with 3 to 9% NaOH remained as cellulose
I swelled much but did not dissolve. Pulp mercerized with 6 to 9%
NaOH dissolved because of the dissolution of hemicellulose. By
pretreatment(1) with water-DMF all samples easily dissolved
after strong swelling. The results showed that the dissolution
did not depend on the crystallite forms but on the fine
structure allowing penetration of the solvents. Pretreatment(2)
with water-acetone did not bring dissolution of the mercerized
ramie and pulp but pretreatment(3) with DMF did. In the former
the drying counteracted the effect of the pretreatment. On the
other hand rayon did not dissolve and only became swollen by
both pretreatments. The degree of penetration of DMF is shown
in Table 1 as degree of swelling. The small value of 1.076 in
rayon indicated the fine structure prevented penetration of non
aqueous solutions. The decrease in the degree of swelling by

mercerization provided evidence for the change of supermolecular structure as above mentioned.

Table 1 Solubility of ramie, pulp, mercerized ramie and pulp and rayon in DMF, chloral and pyridine systems.

Cellulose Sample	Crystal Form	Degree of Swelling	Solubility Test Pretreatment			
			Without	H$_2$O-DMF	H$_2$O-Ace.	DMF
Ramie	I	1.334	○	◎	○	◎
Mercerized Ramie						
3% NaOH	I		△	◎		
6% NaOH	I		△	◎		
9% NaOH	I+(II)		△	◎		
11% NaOH	I + II		×	◎		
13% NaOH	II		×	◎		
16% NaOH	II		×	◎		
20% NaOH	II	1.253	×	◎	△	○
Pulp(NBKP)	I	1.601	○	◎	○	
Mercerized Pulp						
3% NaOH	I		△	◎		
6% NaOH	I		○	◎		
9% NaOH	I + II		○	◎		
11% NaOH	(I)+II		×	◎		
13% NaOH	II		×	◎		
16% NaOH	II		×	◎		
20% NaOH	II	1.520	×	◎	△	○
Rayon	II	1.076	×	○	×	×

◎:Easily soluble, ○:Soluble, △:Strong swelling, ×:Insoluble

Fig. 1 and Fig. 2 show the results of nitration and acetylation of the cellulose fibers mercerized with NaOH solution of various concentration, respectively. Linter, ramie and pulp changed into cellulose II by mercerization with NaOH solution over 14.0, 12.0 and 11.5%, respectively. The reactivity in the reactions of the fibers decreased at these concentrations of NaOH solution for each fiber, and they showed constant reactivity in the concentration region where they were pure cellulose I and II. In the intermediate region the reactivity showed proportional values between the two values according to the ratio of cellulose I and II. This suggests the reactivity depends on the crystallite forms, but this is not the cause since it was increased remarkably by the pretreatments though the crystallite forms did not change. Fig. 3 shows the results of the penetration test with AcOH and water on linter and ramie mercerized with NaOH solution of various concentration. The curves showing the relation between the degree of penetration of AcOH and the concentration of NaOH were analogous to those

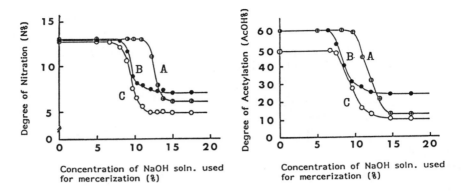

Fig.1 Nitration of mercerized linter(A), pulp(B) and ramie(C).

Fig.2 Acetylation of mercerized linter(A), pulp(B) and ramie(C).

between the reactivity and the concentration. The degree of penetration decreased at the same concentration that the reactivity decreased for each fiber. The results showed that the decrease in the reactivity was brought from the decreasing penetration of the chemical reagents. Intermolecular hydrogen bonds in amorphous regions between microfibrils were relaxed with NaOH soln. above a certain concentration and reformed more strongly than before during drying after washing with water thus preventing the penetration of the reagents.

On the other hand the curves relating the degree of penetration of water and the concentration of NaOH were just the reverse to those of AcOH. The degree of penetration of water increased at the critical concentration of NaOH for each fiber. When the pretreatment of soaking in water and replacing with AcOH was used the reactivity increased at the critical concentration. Water, having high ability to relax the hydrogen bonds, could penetrate into the amorphous regions reformed more strongly and the effect of increasing the amorphous regions during the mercerization brought an increase in the degree of penetration.

Fig. 4 shows the effect of pretreatment on the acetylation of ramie and rayon. The pretreatment was carried out as follows: the fibers were soaked in 40% AcOK aq. solution, pressed, dried at 80 °C and washed with AcOH. AcOK formed a complex with cellulose and the effect of pretreatment was not counteracted by the drying. Without a pretreatment ramie showed moderate low crystallinity. By the pretreatment the reactivity of rayon increased remarkably but ramie reactivity did not increase so much and was lower than the pretreated rayon.

Fig. 5 shows the effects of pretreatment on the acetylation of ramie mercerized with 18% NaOH. The pretreatments were carried out by soaking in 0 to 100% AcOH aq. solution and 40% AcOK aq. solution, pressing and replacing with AcOH. In the case of AcOK

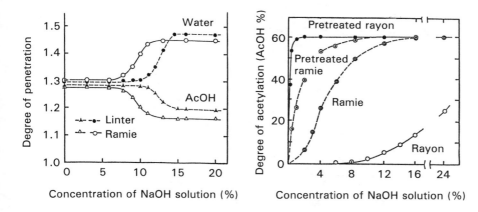

Fig.3 Degree of penetration of water and   Fig.4 Effect of pretreatment with 40% AcOK on
AcOH on mercerized cellulose.   the acetylation of ramie and rayon.

the fibers were replaced with AcOH after drying at 80 °C. The
reactivity of the mercerized ramie was low without pretreatment
but increased with increasing degree of penetration of the
solution used for the pretreatments.

By the pretreatments the crystallinity of the fibers did not
change. The solutions relaxed the hydrogen bonds in the amorphous
regions between microfibrils and bundles and allowed penetration
of the reaction reagents into the inside of the fibers to
increase the reactivity. Namely, the penetration was the rate-
determining step. The amorphous regions contained different
levels of hydrogen bonding and they were relaxed only by the
soaking solution having higher ability to relax them. The
distribution of the levels in the amorphous regions was measured
by successively increasing the reactivity by pretreatment with
AcOH aq. solutions of 0 to 100%. The results are shown in Fig.
6. The distribution of ramie had a maximum in the lower degree.
In ramie the chemical reagents could penetrate inside without a
pretreatment and the reactivity was rather high. The increasing
effect of the reactivity by the pretreatments was not large
because there were few hydrogen bonds to relax. The distribution
of rayon was broad and had a maximum in the higher degree. The
chemical reagents could not penetrate inside without a pretreatment
and the reaction started from the surface of the fiber or large
structural units. The reaction rate not only decreased for the
small reaction surface area but stopped at some stage because
the width of the reaction unit was too long to supply chemical
reagents to the inside by a concentration gradient. The reactivity
of rayon increased successively with increasing ability of the
pretreatment used. The distribution of mercerized ramie was
intermediate to both. Mercerization changed the supermolecular

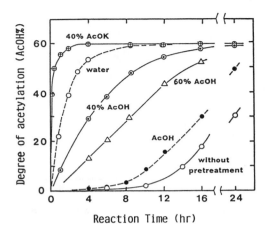

Fig.5 Effect of various pretreatments on the acetylation of ramie
       mercerized with 18% NaOH.

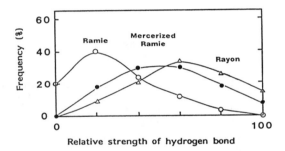

Fig.6 Distribution of amorphous regions having different strength
       of intermolecular hydrogen bond. 0 and 100 in relative
       strength are ones relaxed with 100% AcOH and only with only
       water, respectively.

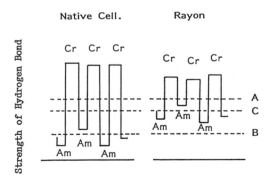

Fig.7 Schematic models of supermolecular structure of native
       cellulose and rayon. Am: Amorphous region, Cr: Microfibril.

structure of native cellulose into that of rayon by the relaxation and reconstruction of hydrogen bonds during airdrying. When an effective pretreatment was used the reactivity depended on the crystallinity and crystallite size of the fibers, and rayon showed higher reactivity (Fig. 4).

We have reported that microfibrils of cellulose fibers, even of rayon, were in a single crystallite state(3). They had defects in the whole, and had no so-called amorphous regions. Also microfibrils had small periodical areas with concentrated defects and high reactivity of the area to hydrolysis caused Level off DP.(4). Amorphous regions were between microfibrils or their bundles. The characteristics of supermolecular structure of native cellulose and rayon are illustrated in Fig. 7. Line A shows the border line of the crystallite regions. The microfibril structure advances in native cellulose, but it is underdeveloped in rayon. Line B and C show the levels of the ability to relax the hydrogen bonds of the chemical reagents without and with pretreatment, respectively. The models show that the levels of hydrogen bonds in the most amorphous regions in native cellulose are under line B and the reagents can penetrate into the regions without a pretreatment. However, in rayon the levels in the most amorphous regions are over line B and the reagents cannot penetrate. By the pretreatment the reagents can penetrate into the regions having the level under the line C. The reactivity increases remarkably in rayon but not so much in native cellulose.

REFERENCES

(1) K. Kamide, K. Okajima, T. Matsui and S. Manabe, Polymer J.,
    12, 521 (1980)
(2) S. Yamazaki and O. Nakao, Sen-i Gakkaishi, 30, T-234 (1974)
(3) J. Hayashi, Proceedings of The International Symposium on
    Fiber Science and Technology 1985 Japan, 84 (1985)
(4) T. Yachi, J. Hayashi and M. Takai, J. Appl. Polym. Sci.,
    37, 325 (1983)

12

Alkali-swollen structures of native cellulose fibers

N.-H. Kim*, J. Sugiyama** and T. Okano** - *Department of Wood Science and Technology, Faculty of Forestry, Kangweon National University, Chuncheon 200-701, Korea. **Department of Forest Products, Faculty of Agriculture, The University of Tokyo, Yayoi, Bunkyo-ku, Tokyo 113, Japan.

ABSTRACT

Alkali-swollen structures of native cellulose fibers were investigated by X-ray and electron diffraction analyses. The ramie fibers treated in a 3.5 N NaOH solution for 384 seconds gave a typical Na-cellulose I diagram, whereas less treated samples showed residual reflections of cellulose I. By washing and drying, however, a large amount of cellulose I was regenerated. The higher the temperature was, the more cellulose I was regenerated. In a fiber wall the regenerated cellulose I was localized heterogeneously along the direction of wall thickness and was rich in the inner parts of the wall. This implied that the swelling was less in the inner parts than in outer parts because it proceeded centripetally and there were not enough space for full swelling. In another words, the fully swollen and completed Na-cellulose I could neither be readily formed in the early stage of mercerization nor be localized evenly in a fiber wall. In addition, such metastable Na-cellulose I was transformed to Na-cellulose IV by low temperature wash (hydration) and converted back to cellulose I by high temperature wash (dehydration). Consequently, once fully swollen Na-cellulose I or fully hydrated Na-cellulose IV were formed, they did not transform back to cellulose I.

INTRODUCTION

A fibrous transformation from cellulose I to cellulose II occurs during mercerization, and this process has been known to be irreversible.

On the phase transformation mechanism of cellulose I to cellulose II, two hypotheses have been proposed : namely, chain-polarity

transformation (parallel to antiparallel)[1-3] and chain-conformation transformation (bent to bent-twisted)[4,5].
The question on reversibility of Na-cellulose I is also important. Hayashi and others[4,5], and Fink and others[6,7] reported that Na-cellulose I(I$_T$) can be reconverted to cellulose I. Okano and Sarko[2], however, proposed that Na-cellulose I cannot be reconverted to cellulose I. However, it is still not known when chain packing or chain conformation changes to the other form, and the studies on this question mainly used X-ray diffraction.
Therefore, to provide further information on this unsolved problem, attention is directed towards formation of the Na-cellulose I in the early stage of alkaline swelling, reconversion of Na-cellulose I to cellulose I after washing and drying of Na-cellulose I, and the effect of washing temperature on the structure of Na-cellulose I by combined analysis by X-ray and electron diffraction methods.

EXPERIMENTS

All the procedures of sample preparation, mercerization, X-ray diffraction analysis and electron microscopy were reported previously[8,9].

RESULTS

Formation of Na-cellulose I

Freezing the sample rapidly in liquid nitrogen, together with the use of the cooling sample-holder, allowed us to stop the process of Na-cellulose I formation at any desired stage and to obtain a fiber diagram of it. Typical mixed Na-cellulose I + cellulose I and Na-cellulose I diffraction diagrams were shown in Fig. 1. The amount of Na-cellulose I was found to increase with the alkali-swelling duration, and after 384 seconds only Na-cellulose I appeared on the diagrams.

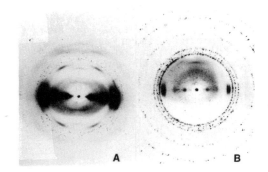

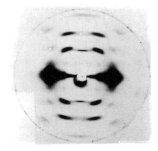

Fig. 1. X-ray diffraction diagrams of Na-cellulose I + cellulose I (A) and Na-cellulose I (B) with a low-temperature camera. Fiber axis is vertical.
Notes: (A) alkali swelling for 12 sec.
       (B) alkali swelling for 384 sec.

Fig. 2. X-ray diagram of regenerated cellulose obtained from Na-cellulose I (Fig. 1B) after washing with water.

The effects of washing temperature on the structure of Na-
cellulose I

A typical diffraction diagram of a regenerated cellulose sample is
shown in Fig. 2. Whereas the diffraction diagram of the 384 sec
treatment showed only Na-cellulose I (Fig. 1B), when it was
washed and dried, it clearly showed that cellulose I and
cellulose II coexisted in the sample. The variations of the
cellulose I content during the alkaline swelling and after the
washing and drying are shown in Fig. 3. During alkaline swelling,
cellulose I decreased with increasing swelling duration and
completely disappeared after 384 sec of treatment. A cellulose I
diagram of about 27 % of the initial intensity, however, re-
appeared after washing and drying.

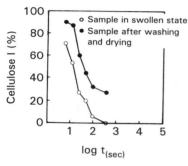

Fig. 3. Changes of cellulose I content
        as a function of alkali-swelling
        duration.

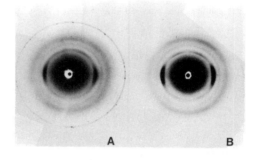

Fig. 4. X-ray diffraction diagrams of Na-cellulose
        I_I (A) and Na-cellulose I_{II} (B).
Note : The fiber axis is vertical.

When the ramie cellulose I was mercerized with 3.5 M sodium
hydroxide solution for 2 hours, the X-ray diffraction diagrams did
not show any reflections assigned to that of cellulose I, and
showed only Na-cellulose I_I (Fig. 4-A). Similarly cellulose II
was integrally converted to Na-cellulose I_{II} (Fig. 4-B) by the
same treatment.
The X-ray diffraction diagrams in Fig. 5 were recorded from Na-
cellulose I_I which was washed at 0 °C(A), 20 °C (B) and at 100 °C
(C). The diagrams were characterized clearly as Na-cellulose
IV(A), a mixture of cellulose I, cellulose II and Na-cellulose
IV(B) and a mixture of cellulose I and cellulose II(C),
respectively. The content of cellulose I increased with raising
washing temperature. On the contrary, washing at 0 °C followed by
drying resulted in complete formation of cellulose II and
disappearance of cellulose I. In addition, the Na-cellulose IV
obtained after washing at 0 °C could not be reconverted into
cellulose I even after long washing at 100 °C.
The recovery percentage of cellulose I due to increasing washing
temperature are shown in Fig. 6. Cellulose I was not observed on
the X-ray diffraction diagram obtained from washing at 0 °C, but
the amount of cellulose I clearly increased with increasing
washing temperature. On the other hand, the recovery of cellulose
I from aged fibers decreased. The X-ray diffraction diagram is
shown in Fig. 7.
Fig. 8 was recorded from Na-cellulose I_{II} which was washed at
0 °C(A), 20 °C (B) and at 100 °C (C).

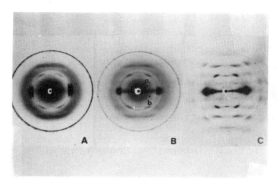

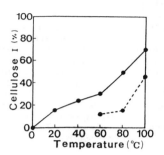

Fig. 5.

X-ray diffraction diagrams obtained after washing

Na-cellulose I₁ with water at 0°C (A), 20°C (B), and
100°C (C).

a and b of diagram (b) indicate (1Ī0) reflections of
Na-cellulose IV and cellulose II, respectively.

Fig. 6.

Variation of the intensity of

cellulose I at various washing

temperatures.

——— : Samples treated for 2 hrs,
······ : Samples treated for 2 days.

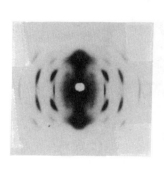

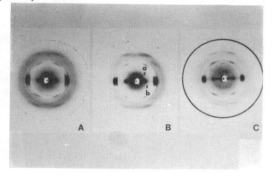

Fig. 7.

An X-ray diffraction diagram

obtained from the sample treated

for 2 days and washed at 100°C

for 1 hour.

Fig. 8.

X-ray diffraction diagrams obtained after washing

Na-cellulose I₁₁ with water at 0°C (A), 20°C (B), and
100°C (C).

a and b of diagram (B) indicate (1Ī0) reflections of
Na-cellulose IV and cellulose II, respectively.

The diagrams were characterized clearly as Na-cellulose IV(A), a
mixture of cellulose II and Na-cellulose IV(B) and cellulose
II(C). The content of cellulose II increased with raising washing
temperature. These results indicated that Na-cellulose IV once
completely formed cannot be reconverted into cellulose I.
Certainly, it was clear that when the Na-cellulose I₁ and I₁₁ are
washed at 100 °C, they are directly transformed into a mixture of
cellulose I and cellulose II, and cellulose II only, respectively,
without the formation of Na-cellulose IV.

Localization of reversible and irreversible Na-cellulose I in a fiber wall

Electron diffraction diagrams recorded of the external (A) and internal (B) layers of regenerated ramie fibers after a 384 sec treatment were presented in Fig. 9. The crystal structure was distinguishable clearly between the external and internal layers: namely, the diffraction diagram (A) from outer layers, exhibited a mostly cellulose II pattern, whereas the diffraction diagram (B) from inner layers showed a mixed pattern of cellulose I and cellulose II. Electron diffraction diagrams (Fig. 10) were obtained from the external (a) and internal (b) layers of the fiber which was washed with water at 100 °C. The diagrams showed the presence of regenerated cellulose I in both inner and outer walls. The amount of regenerated cellulose I was rich in the inner wall.

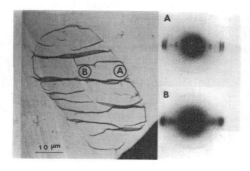

Fig. 9. Electron diffraction diagrams of the external (A) and internal (B) layers of a regenerated ramie fiber after a 384 sec. treatment.

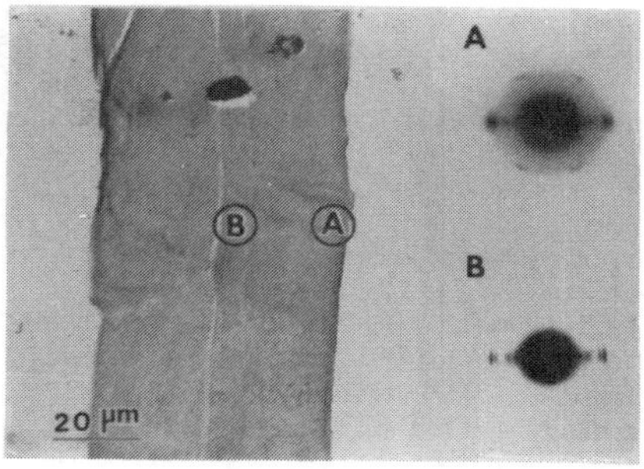

Fig.10. Electron diffraction diagrams of external (A) and internal (B) layers after washing Na-cellulose I₁ with water at 100°C.

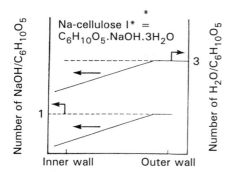

Fig.11. A possible model of the reversible Na-cellulose I in terms of the sodium hydroxide and water concentration gradient due to insufficient swelling of a fiber.

Legend : ←, depend on NaOH treatment time.
          *, the numbers of NaOH and H₂O molecules in a unit cell were taken from results of Sakurada and Okamura and of Sobue and others.

Na-cellulose I* = $C_6H_{10}O_5.NaOH.3H_2O$

DISCUSSION

The results of the X-ray diffraction analysis suggest that the conversion of Na-cellulose I to cellulose I, in part, is reversible. The amount of regenerated cellulose I increased with raising washing temperature.

The heterogeneous localization of reversible and irreversible parts in the cell wall, shown by this study, however, gave a new aspect. That is, when a fiber wall is swollen, compression or tension stresses occur in the fiber wall and restrict sodium-cellulose formation in the inner layer. Then a part of the sodium-cellulose in that region can reconvert to cellulose I. Accordingly, it is proposed that Na-cellulose I within a fiber wall should be characterized not only as two distinct crystalline polymorphs but also as a transitional material existing in between perfect and imperfect states. This concept is outlined schematically in Fig. 11. Because the alkali solution intrudes from the outer part of a wall into its inner part, and a wall swells rather centripetally, a gradient of the swelling is likely to take place. Focusing on the inner wall, although cellulose I swells to cause complete disappearance of a cellulose I diagram, our results proposed the existence of metastable sodium-cellulose which has incomplete composition ((less than $C_6H_{10}O_5:NaOH:H_2O=1:1:3$ reported by Sakurada and Okamura[10], and Sobue and others[11])) of NaOH and H_2O for complete Na-cellulose I caused by restricted swelling. Thus, it is considered that the materials in such a state are imperfect Na-cellulose I and that they partly return to cellulose I upon washing. In the case of a constant condition of the washing temperature, both lines in Fig. 11 slide to the left side in proportion to the NaOH duration, and the recovery of cellulose I becomes less. Therefore, by introducing the concept of metastable Na-cellulose I, the heterogeneous localization of regenerated cellulose I within a fiber wall can be explained.

As a result, metastable sodium-cellulose could be converted to cellulose I with washing at high temperature (dehydration) and Na-cellulose IV with washing at low temperature (hydration). By increasing the alkaline swelling duration however, the reversible region in a fiber wall was decreased; namely, metastable Na-cellulose I (parallel packing or bent conformation) slowly converted to stable Na-cellulose I (antiparallel packing or bent and twisted conformation) by aging it in alkali solution. The reason of the prolonged time for the formation of stable Na-cellulose I might be due to insufficient swelling caused by a gradient of swelling within fiber walls.

REFERENCES

1)Sarko, A.: "Wood and Cellulosics", J. F. Kennedy, G. O. Phillips, P. A. Williams, Eds., Ellis Horwood Ltd., 1987, P. 55-69.
2)Okano, T.; Sarko, A.: J. Appl. Polm. Sci., 30, 325-332 (1985).
3)Nishimura, H.; Sarko, A.: J. Appl. Polym. Sci., 33, 867-874 (1987).
4)Hayashi, J.; Yamada, T.; Watanabe, S.: Sen-i Gakkaishi, 30, T 190-198 (1974).
5)Hayashi, J.; Yaginuma, Y.: "Wood and Cellulosics", J. F. Kennedy, G. O. Phillips, P. A. Williams, Eds., Ellis Horwood Ltd., Chichester, 1987, P. 47-53.
6)Fink, H.-P.; Fanter, D.; Loth, F.: Acta Polym., 33, 241-245 (1982).
7)Fink, H.-P.; Fanter, D.; Philipp, B.: Acta Polym., 36, 1-8 (1985).
8)Kim, N. H.; Sugiyama, J.; Okano, T.: Mokuzai Gakkaishi, 36, 120-125 (1990).
9)Kim, N. H.; Sugiyama, J.; Okano, T.: Mokuzai Gakkaishi, 37, 637-643 (1991).
10)Sakurada, I.; Okamura, S.: Kolloid-Z., 81, 199-208 (1937).
11)Sobue, H.; Kiessig, H.; Hess, K.: Z. Phys. Chem., B43, 309-328 (1939).

13

FTIR spectra of celluloses – use of the second derivative mode

Anthony J. Michell - CSIRO Division of Forest Products, Private Bag 10, Clayton, 3168, Australia.

ABSTRACT

The value of IR spectroscopy as a technique for obtaining information about cellulose and its structures has been limited by its inability to resolve overlapping bands. The second-derivative technique, one means of improving this resolution, is used here to follow changes in the structure of a purified wood cellulose resulting from room temperature treatments with solutions of sodium hydroxide ranging from 0 to 20 percent and from decrystallisation by ball milling. The better resolution enabled more bands to be identified as being associated with the mercerisation of cellulose. The derivative spectrum of ball milled cotton cellulose was found to resemble that of cellulose II more closely than that of native cellulose except in the OH stretching region suggesting that the decrystallised cellulose retains largely the same type of hydrogen bonding as the parent cellulose.

INTRODUCTION

Infrared spectroscopy, one of the earliest techniques used to examine the crystalline nature of the celluloses [1], does not determine structures of substances with the same precision as X-ray, neutron and electron diffraction methods, where single crystal samples are available but gives useful information about aspects of the structures of substances of lower crystallinities where results from diffraction methods are less conclusive. The early work was summarised by Blackwell [2]. The usefulness of infrared spectroscopy for studying cellulose structure has been limited by difficulties in resolving overlapping bands arising from large numbers of

OH and CH groups with similar absorption frequencies. New techniques such as deconvolution [3,4] and differentiation to give the second derivative [5,6] have been shown to assist in sharpening and in resolving the bands in the spectra of celluloses.

Important recent advances in understanding the structures of the major cellulose polymorphs were the discovery by VanderHart and Atalla [7-9] and Sugiyama et al. [10] that cellulose I was a composite of two crystalline allomorphs Iα and Iβ. and a scheme for classifying all native celluloses into two families [11]. Very recently Sugiyama et al. [12] used FTIR spectroscopy to show that the Iα phase was unstable and that an annealing treatment could be used to convert it to the stable Iβ phase.

The technological celluloses such as wood and cotton tend to be less crystalline than the algal and bacterial celluloses and less amenable to study by diffraction methods. Two transformations studied here are the conversion of native cellulose into mercerised cellulose and into an amorphous form. The second derivative technique has been used to follow changes in the FTIR spectra of cellulose following (i) treatments of cellulose with alkali up to twenty percent concentration and (ii) decrystallisation induced by ball milling .

EXPERIMENTAL
The wood cellulose was a highly purified eucalypt pulp (pentosan = 1.0 per cent) supplied by APM Ltd. Details of the purification of the cotton and of the alkali treatments have been given [13]. The decrystallised cellulose was purified cotton which had been ground in a ball mill for 112 h. The sample of cellulose II was prepared by regeneration from a solution of Avicel in phosphoric acid after standing for 6 weeks [14]. Details of sample preparation and spectroscopic conditions have been published [5].

RESULTS AND DISCUSSION
The even derivatives of profiles of functions describing the shapes of bands in infrared spectra have the same abscissal value as that of the parent peak but are considerably sharper. However, the effect of noise on the profile of higher derivatives increases sharply. The use of the second-derivative is a good compromise giving improved resolution with acceptable noise.

Spectra of Alkali-treated Wood Cellulose
The 3800-3000 cm^{-1} region of the normal and second derivative spectra of wood cellulose treated with various concentrations of sodium hydroxide from 0 to 20 percent is shown in Figure 1. In the second derivative spectra "peaks" are oriented downwards and the peak heights of the bands in the derivative spectra depend on the widths as well as on the intensities of the bands.

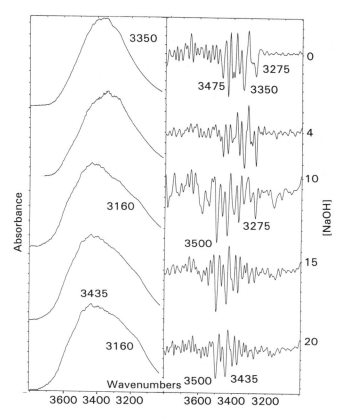

Figure 1　Spectra (3800-3000 cm^{-1}) of alkali-treated wood cellulose
LHS Normal.　RHS Second derivative mode (peaks inverted).

The derivative spectrum of wood cellulose shows a band near 3275 cm^{-1} but not a band near 3245 cm^{-1} [6] and the cellulose is therefore in the Iß phase.　The peak intensity shifts from near 3350 cm^{-1} in the normal spectrum of the control sample to 3435 cm^{-1} in the spectrum of the sample treated with 20　percent alkali and a weak　peak also appears　near 3160 cm^{-1}. Strong bands near 3350 and 3275 cm^{-1} in the derivative spectra of the control sample of cellulose and of the sample treated with 4 percent alkali become weaker in the spectra of samples treated with higher concentrations of alkali. A band at 3500 cm^{-1} becomes prominent in the spectrum of the sample treated with 10 percent alkali and continues to be intense in the spectra of samples treated with higher concentrations of alkali. Detailed assignment of the bands is difficult because, as discussed previously [5], the number of bands exceeds the number of distinguishable OH groups which shows that interchain coupling of the OH stretching modes occurs. Nevertheless, in the spectrum of the OD stretching region of mildly deuterated cellotetraose, a model compound for cellulose II, analogues of the 3500 and 3435 cm^{-1} bands appear suggesting that these bands arise from uncoupled vibrations. Marchessault and Liang [15] assigned these bands to intramoleculary hydrogen bonded OH groups.   Bands in the region 3000-2700 cm^{-1} often arise from strongly coupled vibrations and this led to them being thought of as being of little interest in relation to structure. More recently, however, it has been shown [19,20] that CH stretching modes isolated by isotopic substitution are sensitive to conformation. Bands in this region affected by alkali treatment of the cellulose include those near 2985, 2967, 2956, 2920, 2894 and 2855 cm^{-1}. Assignments of these bands have been proposed [15,21] but greater certainty would result from studying the second derivative modes of decoupled vibrations of C-deuterated samples of the type prepared by Dechant [22].

TABLE 1.   Frequencies (cm^{-1}) of Bands in the Second Derivative Spectra of Cellulose Sensitive to Mercerisation

Intensity Decrease	Intensity Increase	Assignment	Ref
	3500	OH stretching(Intra H bond)	15
3475		OH stretching	
3350		OH stretching	
3275		OH stretching(Iß phase)	6
	2985	CH stretching	
2967		CH stretching	
	2956	CH stretching	
	2935	CH$_2$ antisym. stretching	15
	2894	CH stretching	
	1464		
1428		CH$_2$ bending	15
1371		CH bending	
	1362		
	1264		
	1223		
1108		antisym.in-phase ring str.	16
1058		CO stretching	16
1032		CO stretching	16
	1024	CO stretching	15
	998	CO stretching	15
	965	CO stretching	15
	899	ß-link indicator	17,18
710		CH$_2$ rocking (Iß phase)	6,16
560			23

Changes in bands in the remaining regions of the spectra and assignments from the literature are summarised in Table I. The number of affected bands is far greater than the four bands at 1428, 1111, 990 and 893 cm^{-1} identified earlier by McKenzie and Higgins [13].

Spectra of Crystalline and Decrystallised Celluloses

The infrared spectra of native cotton cellulose, decrystallised cellulose and cellulose II are shown with the region 3800 - 3100 cm^{-1} being depicted in Figure 2.

The normal spectra are rather featureless but the derivative spectra show the spectrum of the decrystallised cellulose being more akin to the spectrum of the native cellulose having bands common to both but having only a weak band near 3500 cm^{-1} where a stronger band is characteristic of the cellulose II spectrum. This result would suggest that the hydrogen bond structure of the native cellulose is retained to some degree in the decrystallised cellulose. It is consistent with the finding of Fengel and Ludwig [3] that differences in this region remained between the deconvoluted spectra of native and regenerated fibres after ball milling. In the remaining regions both the derivative and normal spectra of the decrystallized cellulose resembled the spectrum of cellulose II more closely than they do the spectrum of native cellulose.

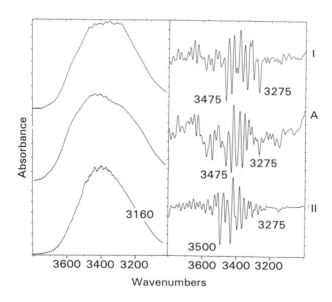

Figure 2   Spectra (3800-3000 cm^{-1}) of cotton cellulose (I), ball-milled cotton cellulose (A) and cellulose II(II).

Deconvolution and differentiation are both techniques which can be used in combination with other methods. Better resolution still might be obtained by deconvoluting or differentiating spectra observed with polarised radiation. These techniques together with selective C-deuteration give promise of giving new useful information about the conformations of CH groups in the celluloses.

CONCLUSIONS

Use of the second derivative mode resulted in better resolution of bands in the spectra of eucalypt wood cellulose subjected to mercerisation by solutions of alkali of varying strength enabling known changes in band intensities to be confirmed and the discovery of new more subtle changes.

Similarly, the spectrum of decrystallised cotton cellulose was found to resemble that of cellulose II more strongly except in the OH stretching region where it resembled that of the native cellulose.

ACKNOWLEDGEMENTS

The author thanks Mr A.W.McKenzie and Dr A.F.A. Wallis for providing him with samples of alkali-treated eucalypt wood cellulose and cellulose II.

REFERENCES

1.   Marrinan, H.J. and Mann, J. *J. Polym Sci*, **1956, 21**, 301.
2.   Blackwell, J. Infrared and Raman Spectra of Cellulose in Arthur, J.D. (Ed.) *Cellulose Chemistry and Technology;* ACS Symposium Series 48; American Chemical Society, Washington, DC, **1977**, p206.
3.   Fengel, D. and Ludwig, M. *Das Papier* **1991, 45**, 45.
4.   Fengel, D. *Das Papier* **1991, 45**, 97.
5.   Michell, A.J. *Carbohyd.Res.* **1988, 173**, 185.
6.   Michell, A.J. *Carbohyd.Res.* **1990, 197**, 53.
7.   Atalla, R.H. and VanderHart, D.L. *Science* **1984, 223**, 283.
8.   VanderHart, D.L. and Atalla, R.H. *Macromolecules* **1984, 17**, 1465.
9.   VanderHart, D.L. and Atalla, R.H. in Atalla, R.H. (Ed.) *The Structure of Cellulose*; ACS Symposium Series 340; American Chemical Society, Washington, DC, **1987**, p88.
10.   Sugiyama, J., Okano, T., Yamamoto, H. and Horii, F. *Macromolecules* **1990, 23**, 3196.
11.   Horii, F., Hirai, A. and Kitamura, R. *Macromolecules* **1987, 20**, 2117.
12.   Sugiyama, J., Persson, J. and Chanzy, H. *Macromolecules* **1991, 24**, 2461.
13.   McKenzie, A.W. and Higgins, H.G. *Svensk Papperstid.* **1958, 61**, 893.
14.   Evans, R., Wearne, R.H. and Wallis, A.F.A. *J.Appl.Polym. Sci.* **1989, 37**, 3291.
15.   Marchessault, R.H. and Liang, C.Y. *J.Polym. Sci.* **1960, 43**, 71.
16.   Marchessault, R.H. and Liang ,C.Y. *J.Polym Sci.* **1959, 39** ,269.
17.   Barker, S.A., Bourne, E.J., Stacey, M. and Whiffen, D.H. *J.Chem.Soc.* **1954**, 171.
18.   Stacey, M., Moore, R.H., Barker, S.A., Weigl, H., Bourne, E.J. and Whiffen, D.H.   2nd United Nations Internat. Confce. on the Peaceful Uses of Atomic Energy, Geneva, **1958**, p251.
19.   McKean, D.C. *Chem.Soc. Rev.* **1978, 7**, 399.
20.   Abbate, S., Conti, G. and Naggi, A. *Carbohyd.Res.* **1991, 210**, 1.
21.   Liang, C.Y. and Marchessault, R.H. *J.Polym Sci.* **1959, 37**, 385.
22.   Dechant, J. *Faserforsch.Textiltechnik* **1968, 19**, 491.
23.   Schneider, B. and Vodñansky, J. *Coll. Czech. Chem. Comm.* **1963, 28**, 2080.

14

Molecular structures of some branched cellulosics

R.P. Millane - Whistler Center for Carbohydrates Research, Purdue University, West Lafayette, Indiana 47907-1160, USA.

ABSTRACT

X-ray fiber diffraction has been used to study the molecular structures of some polysaccharides that have cellulose backbones substituted with various oligosaccharide sidechains. The molecules studied are xanthan, variants of xanthan with a variety of other sidechains, and a xyloglucan from tamarind seeds. Whereas the backbones of xanthan and related polymers adopt a five-fold helical structure, the backbone of the xyloglucan adopts a two-fold ribbon structure as in cellulose. These results enable one to examine the effects of branching on the conformation of cellulose, and on the properties of the molecules in solution.

INTRODUCTION

Xanthan is an anionic extracellular polysaccharide produced by the microorganism *Xanthomonas campestris*. Its primary structure is a cellulosic mainchain with a trisaccharide sidechain attached to the 3-position of every second glucose unit (Fig. 1a), the mannoses being specifically and variably acetylated and pyruvylated [1]. In aqueous solution, xanthan undergoes a thermally reversible, cooperative order-disorder transition [2], but the details of the ordered structure are unknown [3]. Xanthan has a wide variety of industrial applications because of its pseudoplasticity and the stability of its rheological properties over wide ranges of temperature and pH.

Variant xanthan polysaccharides with truncated sidechains have been produced using mutant bacteria [4]. These have rheological properties similar to, but different from, xanthan. We have studied variant polysaccharides in which the sidechain terminal mannose is absent (which we call the "polytetramer", Fig. 1b), the terminal mannose and glucorante residues are absent ("polytrimer", Fig. 1c), and the O-acetyl group is absent ("acetate-free xanthan"). We have also obtained diffraction patters from "acetan", a polysaccharide from *Acetobacter xylinum*, that has a similar chemical structure to xanthan with a pentasaccharide sidechain (Fig. 1d) [5]. Structural analysis of these polymers provides an opportunity to study the sidechain functionality of xanthan.

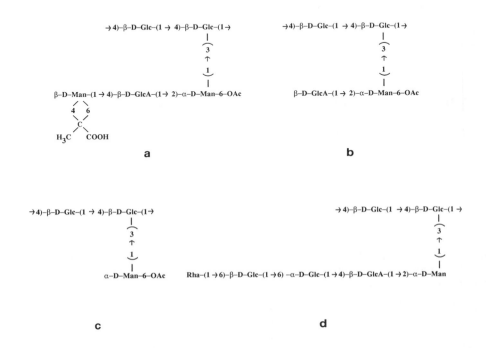

Figure 1. Primary structures of (a) xanthan, (b)polytetramer, (c) polytrimer, and (d) acetan.

Tamarind seed polysaccharide (TSP) is a xyloglucan from the endosperm of seeds of the tamarind tree, *Tamarindus indica*, that is common in southeast Asia [6]. The chemical structure of TSP is shown in Fig. 2 [7]. TSP forms viscous solutions that are relatively pH and temperature stable. Addition of sucrose or glucose to TSP solutions increases the viscosity markedly, and a gel forms at sugar concentrations exceeding 40%. Partial removal of the galactose residues from TSP increases the viscosity of solutions, and gel-formation occurs if more than about 50% are removed [8].

METHODS

Samples of Na$^+$ xanthan, Na$^+$ acetate-free xanthan, Li$^+$ polytetramer, polytrimer, and Li$^+$ acetan, were dry spun using standard methods and X-ray diffraction patterns obtained at various relative humidities (r.h.) using a flat film pinhole camera and Ni filtered CuKα radiation. The optical densities on the diffraction patterns were digitized using an Optronics Photoscan P-1000 rotating drum microdensitometer. The center and orientation of the diffraction patterns, fiber tilt to the beam and c-repeat were determined using standard methods [9].

Tamarind kernel powder obtained from TIC Gums Inc. (Belcamp, MD) was gently boiled for 20 minutes, centrifuged, the supernatant passed through a column containing IR-120 resin, and the compound within the pH range 5.1-6.4 collected the lyophilized. Fibers were prepared and X-ray diffraction patterns obtained as described above.

Molecular models of an isolated chain of designated pitch and chirality were generated and refined using the linked-atom-least-squares (LALS) technique [10] and standard pyranose ring geometry.

RESULTS

Spacings measured from X-ray diffraction patterns from xanthan and the related polymers are listed in Table 1. In all cases, systematic absences on the meridians of the diffraction patterns show that all of the molecules have five-fold helix symmetry [11]. The diffraction pattern from xanthan indicates that the molecules are well oriented but are not organized laterally. The layer line spacing corresponds to a molecular repeat distance (c-repeat) of 47.2 Å. A sharp reflection at low resolution on the equator indicates some limited lateral ordering with a spacing of 18.0 Å.

The diffraction pattern from acetate-free xanthan is very similar to that from xanthan. The molecules are also well oriented, and have the same repeat distance and helix symmetry as xanthan. The spacing of the first reflection on the equator is 1.5 Å less than that of xanthan.

Table 1. Spacings for xanthan and related polysaccharides.

	xanthan	acetate-free xanthan	polytetramer		polytrimer	acetan
Salt	Na^+	Na^+	Li^+	Li^+	--	Li^+
r.h. (%)	92	66	75	15	66	75
c-repeat (Å)	47.2	47.2	47.5	47.5	47.7	47.9
helix symmetry	5-fold	5-fold	5-fold	5-fold	5-fold	5-fold
lateral spacing (Å)	18.0	16.5	14.4	13.4	12.0	19.0

A diffraction pattern from the polytetramer at 75% r.h. shows that the molecules are not as well oriented as in xanthan. The c-repeat is almost identical to that of xanthan, but the first diffraction maximum on the equator indicates some lateral ordering with a smaller spacing of 14.4 Å (Table 1). A diffraction pattern from a polytetramer fiber at 15% r.h. shows a further 1.0 Å reduction in the lateral spacing, but the c-repeat is unchanged.

A diffraction pattern from the polytrimer indicates little orientation of the molecules, but the c-repeat is very similar to that of xanthan. The lateral spacing is further reduced to 12.0 Å.

Diffraction patterns from acetan show that the molecules are well oriented, but with limited lateral organization. The molecular repeat distance is 0.7 Å longer, and the lateral spacing 11.0 Å larger, than that of xanthan. The diffraction pattern is, overall, rather similar to that of xanthan.

A diffraction pattern from tamarind seed polysaccharide at 15% r.h. shows limited orientation and no lateral crystallinity [12]. The pattern indicates a molecular repeat distance of 20.6 Å, and a lack of systematic absences on the meridian indicates one-fold helix symmetry [12]. The pattern is dominated by diffraction on the even layer lines, indicating an approximate repeat distance of 10.3 Å, showing that the backbone is probably in the cellulose conformation. Molecular models of an isolated chain were constructed and sterically refined to explore the possible molecular conformations and intramolecular interactions. The backbone glycosidic conformation angles were fixed, and the five conformation angles (χ, ϕ_1, ψ_1) and (ϕ_2, ψ_2) at the $(1 \rightarrow 6)$ and $(1 \rightarrow 2)$ linkages respectively, and the conformation of the primary hydroxyl group, were variable parameters. Molecular models could be generated free of steric compression for χ in each of three staggered domains $(\chi \simeq -60°, 60°, 180°)$. The conformations of these three models are shown in Figure 3.

DISCUSSION

Conservation of the c-repeat and symmetry of xanthan and the related polysaccharides indicates that the conformations of the molecules are all rather similar. This indicates that the $(1 \rightarrow 3)$ branch point, rather than the detailed primary structure of the sidechain, is the important determinant of their ordered structures. Molecular modeling and refinement against the X-ray data shows that the most likely molecular structure of xanthan is a right-handed 5-fold single helix, or a double helix made up of two such molecules [13]. Although the O-acetyl group is located close to the branch point, its removal does not appear to affect the ordered structure, suggesting no critical involvement. The smaller lateral spacing for deacetylated xanthan (Tale 1) is probably a result of the lower humidity. The lateral spacing for xanthan is greater than for the variants. This is probably due, at least in part, to a reduction in the molecular diameter on removal of the sidechain sugar residues. The reduction in the lateral spacing of the polytetramer on drying is probably due to removal of water between the molecules. Conversely, the longer sidechain for acetan appears to increase the molecular diameter. Overall, these results allow us to conclude that branching at alternate 3-positions in the xanthan backbone is the critical factor controlling the xanthan conformation (i.e., formation of a 5-fold, rather than a cellulosic 2-fold, helix), and that any interactions involving the terminal two sugar units do not play an essential role in stabilizing the structure.

Tamarind seed polysaccharide forms semi-ordered structures in the solid state, in which the sidechains do not modify the conformation of the cellulosic backbone. TSP is similar to galactomannans in this respect. Three

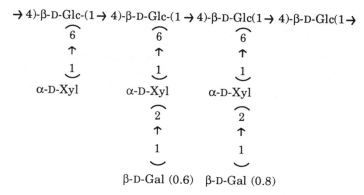

→ 4)-β-D-Glc-(1 → 4)-β-D-Glc-(1 → 4)-β-D-Glc(1 → 4)-β-D-Glc(1 →

6	6	6
↑	↑	↑
1	1	1
α-D-Xyl	α-D-Xyl	α-D-Xyl
	2	2
	↑	↑
	1	1

β-D-Gal (0.6)   β-D-Gal (0.8)

Figure 2.   Average chemical repeat of tamarind seed polysaccharide.   The numbers in parentheses denote the average DS of each galactose residue.

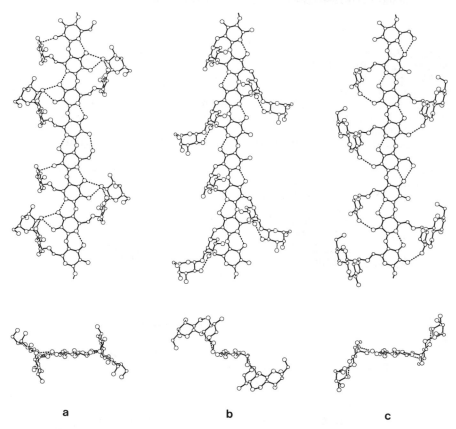

a   b   c

Figure 3.   Mutually perpendicular view of molecular models of tamarind seed polysaccharide for χ ≃ (a) -60°, (b) 60°, and (c) 180°.   The broken lines denote hydrogen bonds.

conformations of the sidechain, corresponding to the three staggered domains at the (1→6) linkage, are sterically feasible. The three models are somewhat different in the juxtaposition of the sidechains relative to the cellulose ribbon (Fig. 3). The sidechains solubilize the material by preventing close-packing of the cellulose chains. The (1→6) and (1→2) linkages are stabilized by hydrogen bonds in all three models. In all cases there are mainchain-sidechain interactions that would increase the characteristic ratio over that for similar polysaccharides such as CMC, as has been observed [14]. Similar results have been obtained for the primary cell wall xyloglucan extracted from pea stems [15].

ACKNOWLEDGEMENTS

I am grateful to the Synergen/Texaco Joint Venture (Boulder, CO., U.S.A.) for providing the xanthan variant polymers, Dr. V.J. Morris (AFRC, Norwich, U.K.) for the acetan sample, the U.S. National Science Foundation for financial support (DMB-8916477), and Deb Zerth for word processing.

REFERENCES

1. Janson, P.E.; Keene, L.; Lindberg, B. *Carb. Res.*, **1975**, *45*, 275.
2. Morris, E.R.; Rees, D.A.; Young, G.; Walkinshaw, M.D.; Darke, A. *J. Mol. Biol.*, **1977**, *110*, 1.
3. Okuyama, K.; Arnott, S.; Moorhouse, R.; Walkinshaw, M.D.; Atkins, E.D.T.; Wolf-Ullish, C. *ACS Symp. Ser.*, **1980**, *141*, 411.
4. Betlach, M.R.; Capage, M.A.; Doherty, D.H.; Hassler, R.A.; Henderson, N.M.; Vanderslice, R.W.; Marrelli, J.D.; Ward, M.B. In: Yalpai, M. (Ed.) *Industrial Polysaccharides*, Elsevier, Amsterdam, 1987, 35.
5. Morris, V.J.; Brownsey, G.J.; Cairns, P.; Chilvers, G.R.; Miles, M.J. *Int. J. Biol. Macromol.*, **1989**, *11*, 326.
6. Glicksman, M. In: Glicksman, M. (Ed.) *Food Hydrocolloids*, Vol. 3, CRC press, Boca Raton, 1983, 191.
7. York, W.S.; van Halbeck, H.; Darvill, A.G.; Albersheim, P. *Carbohydr. Res.*, **1990**, *200*, 9.
8. Reid, J.S.G.; Edwards, M.; Dea, I.C.M. In: Phillips, G.O.; Wedlock, D.J.; Williams, P.A. (Eds.) *Gums and Stabilisers for the Food Industry 4*, IRL Press, Oxford, 1988, 391.
9. Millane, R.P.; Arnott, S. *J. Macromol. Sci. Phys.*, **1985**, *B24*, 193.
10. Smith, P.J.C.; Arnott, S. *Acta. Cryst.*, **1978**, *A34*, 3.
11. Millane, R.P.; Wang, B. In: Phillips, G.O.; Wedlock, D.J.; Williams, P.A.; (Eds.), in *Gums and Stabilisers for the Food Industry 6*, Oxford Univ. Press, Oxford, 1992, 541.
12. Millane, R.P.; Narasaiah, T.V. in Phillips, G.O.; Wedlock, D.J.; Williams, P.A.; (Eds.), In: *Gums and Stabilisers for the Food Industry 6*, Oxford Univ. Press, Oxford, 1992, 535.
13. Wang, B. M.S. Thesis, 1991, Purdue University, West Lafayette, Indiana, USA, 1991.
14. Gidley, M.J.; Lillford, P.J.; Rowlands, D.W.; Lang, P.; Denti, M.; Crescenzi, V.; Edwards, M.; Fanutti, C.; Reid, J.S.G. *Carbohydr. Res.*, **1991**, *214*, 299.
15. Ogawa, K.; Hayashi, T.; Okamara, K. *Int. J. Biol. Macromol.*, **1990**, *12*, 218.

15

Structures of mannan II and konjac mannan in polycrystalline fibers

T.L. Hendrixson and R.P. Millane - Whistler Center for Carbohydrate Research, Purdue University, West Lafayette, Indiana 47907-1160, USA.

ABSTRACT

The structure of mannan II has been determined by X-ray fiber diffraction analysis. The polysaccharide crystallizes in an orthorhombic unit cell with space group I222. The molecules form 2-fold ribbon structures and pack antiparallel to form sheets, with adjacent sheets being antiparallel. The structure is stabilized by both intra- and inter-molecular hydrogen bonds and water bridges. X-ray diffraction patterns from native konjac mannan show that there is isomorphous replacement of mannose by glucose in the mannan II crystal structure.

INTRODUCTION

Polysaccharides of the mannan family are abundant in nature [1,2]. Mannan, a homopolymer of (1→4) linked β-D-mannose residues, has been found in the endosperm of coffee beans and ivory nuts, and the cell walls of some types of algae, and as storage carbohydrates in bulbs. Mannan has two distinct crystalline forms, mannan I and mannan II, with the latter form being hydrated [1]. The crystal structure of mannan I has been previously determined [3], although that of mannan II has not.

Glucomannans, in which some of the mannose residues are replaced by D-glucose, are found in softwoods and some roots, tubers and bulbs. Konjac mannan is a high molecular weight glucomannan from the tubers of the *Amorphophallus konjac* plant, and is a linear random co-polymer of (1→4) linked β-D-mannose and β-D-glucose. It has a mannose/glucose ratio of 1.6 [4], and approximately one in fifteen of the sugar units is acetylated, although the exact position of the acetylation is unknown [5,6]. It forms strong, elastic, heat-stable gels when heated in the presence of mild alkali. Although used for centuries as a foodstuff in the Orient, it has seen little application in the West.

The crystal structure of native mannan is of interest because of its structural similarity to cellulose and other more complex polysaccharides such as glucomannans and galactomannans. The crystal structure of mannan II, determined by X-ray fiber diffraction, is presented here. The crystal structure of konjac mannan is of interest as it serves as a model for the junction zones in konjac mannan gels, and for interactions with other polysaccharides. Diffraction patterns from konjac mannan indicate that it packs isomorphously with mannan II, and the structure of the latter is used to examine the steric implications of acetylation in konjac mannan. The details of these structures are presented elsewhere [7,8].

EXPERIMENTAL

The mannan II specimen was obtained from the inner wall of *Cympolia barbata* and dried under tension. The X-ray diffraction pattern (Fig. 1a) was taken at 98% relative humidity using CuKα radiation and a flat-plate pinhole camera [1,7]. Konjac mannan was dispersed in water, heated to 95°C, cooled to room temperature and filtered. A drop of solution (~1% w/w) was cast onto a Teflon block, partially dried and the resulting films were cut into strips and stretched (~100%) at 98% relative humidity, and X-ray diffraction patterns (Fig. 1b) obtained as described above [8]. X-ray diffraction patterns were digitized using an Optronics Photoscan P-1000 rotating drum microdensitometer. The lattice spacings and diffracted intensities were measured using standard methods [9]. Molecular and crystal structure models were constructed and refined using the linked-atom least-squares (LALS) technique [10].

MANNAN II

A diffraction pattern from mannan II is shown in Figure 1a. The reflexions were indexed using an orthorhombic unit cell. As shown in Table 1, the unit cell dimensions are quite different from those for mannan I. The absence of reflexions having $h + k + l = 2n + 1$ implies a body-centered lattice. The absence of meridional reflexions on the odd-numbered layer lines indicates 2_1 helix symmetry. The only orthorhombic space groups which will accommodate such a polar molecule are I222 and I$2_1 2_1 2_1$.

Analysis of the Patterson function shows that four chains pass through the unit cell, oriented with their helix axes parallel to the c axis, at $(u,v) = (0,0)$, $(0,\frac{1}{2})$, $(\frac{1}{2},0)$, $(\frac{1}{2},\frac{1}{2})$. Based upon this, and the molecular symmetry, the helix axes were placed on the 2_1-screw axes. In the space group I222, there is one residue per asymmetric unit and adjacent molecules are oriented antiparallel to each other. In the space group I$2_1 2_1 2_1$, there are two independent residues per asymmetric unit, and there must be a random distribution of up- and down-pointing chains at each site to satisfy the space group constraints.

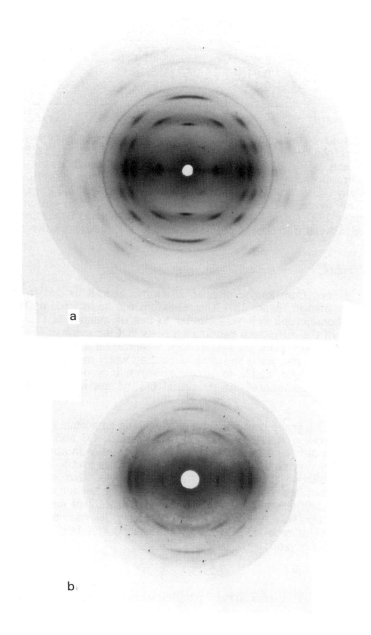

Figure 1.   X-ray fiber diffraction patterns from (a) mannan II and (b) konjac mannan

Table 1. Unit cell constants (Å) for mannan I, mannan II and konjac
mannan.

	a	b	c
Mannan I	8.92	7.21	10.3
Mannan II	8.91	16.46	10.3
Konjac mannan	9.14	16.46	10.3

Molecular models were constructed using standard pyranose ring geometries
[11], and the bond angles at the glycosidic oxygens were fixed at 116.5°.
The orientation and position of the chains, and the conformation of the
hydroxymethyl group, were initially sterically refined, and then co-refined
against the X-ray data. Models in the space group $I2_12_12_1$ gave
unacceptable X-ray agreement and further refinements were carried out only
for the space group I222. The location of a water molecule was determined
using electron density difference maps. The refined structure is shown in
Figure 2.

KONJAC MANNAN

The diffraction pattern from konjac mannan is similar overall to that from
mannan II and is shown in Figure 1b. The reflexions can be indexed on the
basis of an orthorhombic unit cell very similar to that of mannan II (Table
1). The similarity of the diffraction patterns suggests that the mannan II and
konjac mannan structures pack isomorphously. Steric refinements were
performed by taking the mannan II structure (excluding water molecules) and
adding an equatorial oxygen at C2 to simulate the random substitution of
mannose by glucose. Addition of the equatorial O2 did not introduce any
steric hinderances.

The effect of the acetate group was examined by attaching an acetate group at
each available hydroxyl group (O2 (equatorial), O2 (axial), O3 and O6) on
each sugar residue in turn, and then optimizing the structure to minimize the
steric compression by varying the conformation of the acetate group, keeping
the glucomannan backbone fixed. The refinements showed that sterically
acceptable crystal structures could not be generated with acetate substitutions
in three of the four positions. For the remaining position (O2 of mannose),
although the structure was not totally unacceptable, it was sterically
compressed with some non-bonded contacts more than 0.4 Å less than the
sum of the van der Waals radii.

DISCUSSION

The mannan chains pack approximately parallel to the b cell edge in the
mannan II structure, with the molecules within a sheet being antiparallel and
adjacent sheets also being antiparallel (Figure 2). The conformation angles
at the (1→4) linkages are comparable to those in analogous glycans. The
structure is stabilized by both intramolecular and intermolecular hydrogen
bonds. The intramolecular O3-O5 hydrogen bond is identical to that seen in

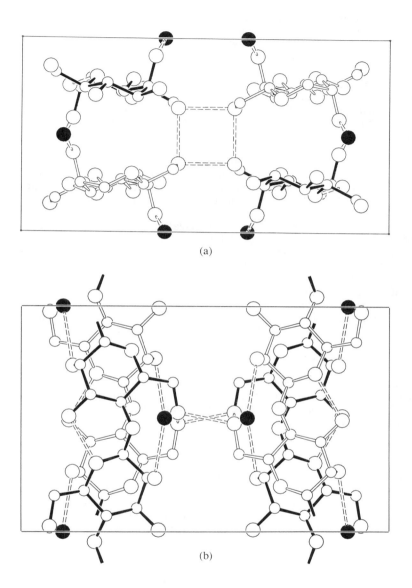

(a)

(b)

Figure 2. Unit cell packing of mannan II. In (a), the **a** axis is vertical, the **b** axis is horizontal and the **c** axis is perpendicular to the page. In (b), the **c** axis is vertical, the **b** axis is horizontal and the **a** axis is perpendicular to the page.   The water molecules are indicated by filled circles.   Up-pointing helices are indicated by filled bonds, down-pointing helices by open bonds and hydrogen bonds by dashed lines.

cellulose and mannan I.   Molecules within a sheet are stabilized by intermolecular hydrogen bonds through the hydroxymethyl groups.   There are also intersheet hydrogen bonds, via water bridges, of the form $O2$-O_{water}-$O2$.

Native konjac mannan packs isomorphously with mannan II in the solid state, showing that the replacement of mannose by glucose does not disrupt the molecular associations adopted by hydrated mannan.   Preliminary analysis does not indicate any interactions involving $O2$ of glucose.   Since the degree of substitution of the acetate groups is low, and since the most sterically accessible position for acetate substitution produces a structure which is still sterically hindered, the most plausible explanation is that the diffraction pattern is due to crystalline regions in the specimen that incorporate only unsubstituted segments of the glucomannan backbone.   If the acetate substituents are randomly distributed on one in every 15 sugars, then the average length of the unsubstituted stretches is 14 sugars.   If the packing of the chains is disrupted by the presence of an acetate group, then the average length of the crystalline regions would be about 80 Å.   Estimates of the crystallite dimensions from the shapes of the reflexions on the diffraction pattern give a length of ~50 Å, which is consistent with the above proposal. This suggests that gel formation may involve associations of acetate-free stretches of the glucomannan in the junction zones.   Presumably then, gel formation occurs on deacetylation since the acetate-free stretches are then long enough for the associations in the junction zones to be energetically favorable [12].

ACKNOWLEDGEMENTS

We are grateful to the U.S. National Science Foundation (DMB-8916477 to RPM) and the Industrial Consortium of the Whistler Center for financial support, Deb Zerth for word processing, and Robert Werberig for photography.

REFERENCES

1. Frei, E.; Preston, R.D. *Proc. Roy. Soc. London Ser. B.* **1968**, *169*, 124.
2. Meier, H.; Reid, J.S.G. *Encyl. Plant Physiol. New Ser.* **1982**, *13A*, 418.
3. Nieduszynsk, I.; Marchessault, R.H. *Canad. J. Chem.* **1972**, *50*, 2130.
4. Kato, K.; Matsuda, K. *Agr. Biol. Chem.* **1969**, 33, 1446.
5. Chanzy, H.D.; Grosrenaud, A.; Joseleau, J.P.; Dube, M.; Marchessault, R.H. *Biopolymers* **1982**, *21*, 301.
6. Gidley, M.J.; McArthur, A.J.; Underwood, D.R. *Food Hydrocolloids* **1991**, *5*, 129.
7. Millane, R.P.; Hendrixson, T.L.; Arnott, S.; Okuyama, K.; Preston, R.D. *Carbohydr. Res.* submitted.
8. Hendrixson, T.L.; Millane, R.P.; Morris, V.J.; Cairns, P. *Carbohydr. Polym.* submitted.
9. Millane, R.P.; Arnott, S. *J. Macromol. Sci. Phys.* **1985**, *B24*, 193.
10. Smith, P.J.C.; Arnott, S. *Acta Crystallogr.* **1968**, *A34*, 3.
11. Arnott, S.; Scott, W.E. *J. Chem. Soc., Perkins Trans.* **1972**, 2, 324.
12. Millane, R.P.; Hendrixson, T.L.; Morris, V.J.; Cairns, P. in Phillips, G.O.; Wedlock, D.J.; Williams, P.A. (Eds.) *Gums and Stabilisers for the Food Industry 6*, Oxford Univ. Press, Oxford, 1992, 531.

16

TEM studies on cellulose solutions

Dietrich Fengel and Alexander Maurer - Institute for Wood Research,
University of Münich, D-8000 Munchen 40, Winzerestrasse 45, Germany.

ABSTRACT

Among the preparation methods for the visualization of individual cellulose molecules
and of processes in the molecular range freeze-etching is obviously the best one.
This method is applicable with various kinds of cellulose solvents. A molecular
resolution in the TEM is achieved by choosing suitable parameters during the
preparation. Some preliminary results are presented concerning the process of
cellulose dissolution in trifluoroacetic acid.

INTRODUCTION

Technical progress in electron microscopy has made the visualization of molecules
and atoms possible. Some problems still remain for organic substances concerning
the staining in molecular dimensions. As to cellulose, the dense packing of molecules
within the fibrillar order system is a further problem.

Separation of individual cellulose chains can be achieved by splitting the hydrogen
bonds by intensive mechanical treatment or by dissolving in suitable solvent.
Endeavours by the OH groups to form new hydrogen bonds require that the individual cellulose molecules be conserved.

The observation of individual cellulose molecules and of their behaviour during
dissolution is seen as a useful method for obtaining information on the order system
of cellulose within fibrils and crystallites.

APPROACH TO THE VISUALIZATION OF MOLECULES

For many years we have been studying specimens deriving from mechanically disintegrated suspension of cellulose [1]. Specimens are obtained by atomizing the suspension onto supporting film where the micro-droplets dry in air. Although very fine elements with diameters of 2 nm and less can be observed one cannot be sure whether individual molecules exist or whether the finest structures are the remaining aggregates of a few cellulose chains or even of reaggregated chains.

Another way is the use of suitable solvents such as trifluoroacetic acid (TFA) which can be very easily removed by evaporation or freeze-drying. After this procedure - starting from highly diluted solutions - individual molecules were observed in the transmission electron microscope (TEM) [2]. These molecules are enveloped by solvate coatings which show a positive staining with uranyl acetate (Fig. 1). These molecules are not, however, in condition of solution as the entire volume of a solution droplet collapses during freeze-drying. Thus rearrangement structures are also visible. Structures very similar to these were observed in thin platelets which build up the "macrocrystals" of cellulose which can be grown from TFA solution [3].

10 nm

Fig. 1   Cellulose molecules freeze-dryed from highly diluted solution (10^{-4} wt.%) in TFA

FREEZE-ETCHING OF CELLULOSE SOLUTIONS

The method of freeze-etching was developed about 30 years ago [4, 5] and is very well suited for cell suspensions. Only few studies however, were performed with solutions of macromolecules [6, 7]. During the usual preparation procedure a certain plane within a small drop is revealed by cutting and the particles (cells or molecules) remain in their position within the frozen solution. There are, however, two main problems which finally obstruct the observation of individual molecules in their original position within the solution.

The first problem is the freezing velocity in the solution droplet. There is a slowing down in freezing velocity from surface to centre related to the freezing medium. Using pentane and liquid nitrogen the formation of solvent microcrystals begins at a very short distance from the surface. The arrangement of the dissolved molecules is greatly influenced by these crystals. Fig. 2 shows that an undisturbed arrangement of disolved molecules exists at the surface only.

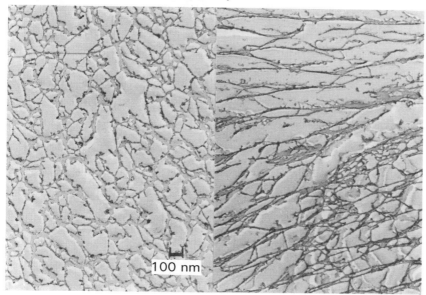

Fig. 2   Methyl cellulose in aqueous solution (10^{-3} wt.%), freeze-etched, surface preparation (left), inner plane (right)

The second problem arises from the fact that the method of freeze-etching includes the preparation of replicas by shadowing and coating with metal and carbon. For specimens of this kind the best resolution to be expected is 3 - 4 nm [8].

However, in the case of cellulose solutions we observed structural details of about 1 nm in diameter, i.e. in the dimensions of the molecular chains (Fig. 3). If these structures observed in the TEM represent individual molecules, compounds with molecular structure different from those of cellulose should also be made visible. Therefore samples of starch, deoxyribonucleic acid (DNA) and ferritin were prepared in the same manner. Starch and DNA show helical structures (Fig. 4), while tetragonal and hexagonal particles can be identified in the ferritin specimen.

An explanation for this "impossible" resolution is seen in a decoration effect caused by the short evaporation time and the small angle for metal shadowing we applied (Fig. 5). As far as we know this technique was applied to cellulose solutions for the first time in our laboratory and some of the preliminary results are presented here. This technique permits the electron microscopic observation of the dissolving process and the behaviour of cellulose in various solvents.

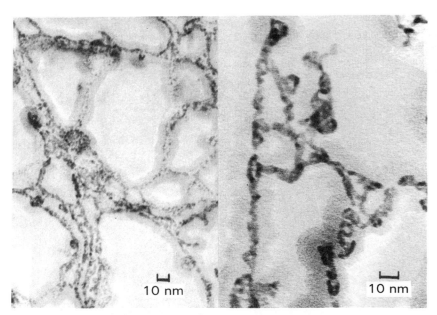

Fig. 3. Cellulose in TFA solution (10^{-5}wt.%), freeze-etched

Fig. 4. DNA in aqueous solution (10^{-4}wt.%), freeze-etched

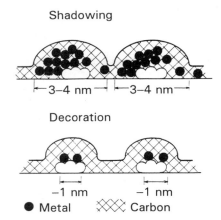

Fig. 5. Comparison of the effects of normal shadowing and decoration on the resolution in the TEM

Starting from highly crystalline cellulose such as cotton linters dissolution in TFA occurs very slowly. Fibrillar aggregates are still observed in concentrations of 10^{-4} wt. %. These aggregates are separated laterally, forming an oriented network (Fig. 6). Further phenomena are loops at the end of the dissolving network. This observation is interpreted by the mobility of the free ends of individual molecules and their endeavour at reaggregating or coiling.

Less crystalline cellulose such as cellulose precipitated from TFA solution forms more random networks during dissolution (Fig. 7). The formation of coils begins at the periphery of the network prior to complete separation of individual molecules.

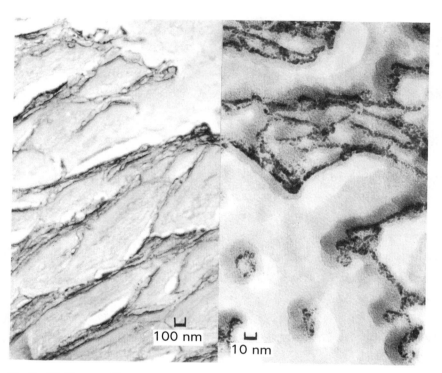

100 nm

10 nm

Fig. 6   Highly crystalline cellulose (cotton linters) in TFA solution (10^{-4} wt.%), freeze-etched

Fig. 7   Re-dissolved cellulose in TFA solution (10^{-4} wt.%), freeze-etched

Further examples for the application of the freeze-etching method are solutions of cellulose in other solvents such as alkali (Fig. 8) or metal complexes or of cellulose derivatives such as methyl cellulose in water (Fig. 2).

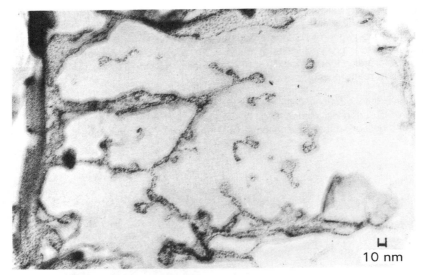

Fig. 8.   Cellulose in 5 % NaOH solution (10^{-4}wt.%)

REFERENCES

1. Fengel, D. *Holzforschung* **1978**, *32*, 37.

2. Stoll, M.; Fengel, D. *Holzforschung* **1989**, *43*, 7.

3. Fengel, D.; Stoll, M. *Wood Sci. Technol.* **1989**, *23*, 85.

4. Steere, R.L. *J.Biophys.Biochem.Cytol.* **1957**, *3*, 45.

5. Moore, H.; Mühlethaler, K. *J.Cell Biol.* **1963**, *17*, 609.

6. Bachmann, L.; Weinkauf, S.; Baumeister, W.; Wildhaber, I.; Bacher, A.
   *J.Mol.Biol.***1989**, *207*, 575.

7. Amend, T.; Belitz, H.D.; Kurthen, C. *Z.Lebensm.Unters.Forsch.* **1990**, *190*, 217.

8. Niedermeyer, W. *Biol.uns.Zeit* **1977**, *7*, 178.

17

Lyotropic liquid crystalline cellulose derivatives – phase behaviour, structures and properties

P. Zugenmaier - Institute of Physical Chemistry, Technical University of Clausthal, W-3392 Clausthal-Zellerfeld, Germany.

ABSTRACT

A newly discovered liquid crystalline phase of a cellulose derivative/solvent system has been characterized by X-ray scattering and DSC experiments. It was found that the cellulosic molecules form parallel arrangements on a local scale, a desired prerequisite for novel fibre forming media. A similar structural arrangement was also achieved by the lyotropic liquid crystalline state of a mixed cellulose derivative.

The interaction of polymer and solvent occurs on a very local scale for the formation of twisted liquid crystalline phases and influences the phase behaviour and the optical spectra considerably.

INTRODUCTION

Lyotropic liquid crystalline cellulose derivatives represent an interesting class of material [1,2]. They exhibit extraordinary optical properties and are regarded as possible media for drawing unusual fibres. They seem to play an important role for producing the stationary phase material coated on silica for enantiomeric separations in chromatography [3] and represent a useful tool for a study of polymer solvent interaction. However, an underlying theory to describe the observed phenomena is not available at the present.

In our group we have recently specialized on investigations mostly on cellulose tricarbanilate (CTC) and closely related cellulose derivatives, as polymeric compounds, and on solvents with high evaporation temperatures and pressures, such as ethylene glycols, to ensure proper experimental conditions. In this report we focus on effects related to polymer solvent interactions with regard to highly ordered unidirectional molecular arrangements which are realized by a phase transition into a newly discovered columnar phase or by a mixed substitution on one cellulose chain, and we discuss the deviation of liquid crystalline cellulose derivative/solvent systems from predicted ORD-spectra.

EXPERIMENTAL

Cellulose triphenyl carbamate and cellulose tri-3-chlorophenyl carbamate were synthesized from commercially available Avicel cellulose according to published procedures [4,5]. The degree of substitution (DS) was determined by infrared and elemental analysis and was higher than 2.9. The preparation of highly concentrated, liquid crystalline solutions and the necessary experiments for a determination of the pitch of the helicoidal cholesteric structures as well as for an evaluation of phase diagrams were performed as described in preceeding papers by polarization microscopic observations, ORD-, UV-VIS-spectroscopy, refractive index and DSC measurements and X-ray scattering analysis [5,6].

RESULTS AND DISCUSSION

Phase behaviour of CTC/DEME.

In a recent paper the phase behaviour of cellulose tricarbanilate (CTC)/diethylene glycol monoethyl ether (DEME) was investigated and a possible "nematic" phase introduced, detected by polarization microscopic observations and birefringence [5]. Further studies now led to the conclusion that the discovered phase deviates from a nematic state and resembles the one termed columnar phase [7].

A sample of 1.1 g/ml CTC/DEME in the columnar phase was

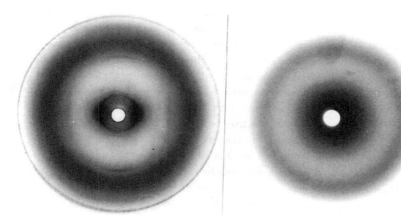

Figure 1. Fibre X-ray pattern for
the columnar phase of CTC/
DEME (1.1 g/ml) at room temperature; fibre axis vertical.

Figure 2. Debye-Scherrer pattern
for the cholesteric phase of CTC/
DEME (1.1 g/ml) at 65 C.

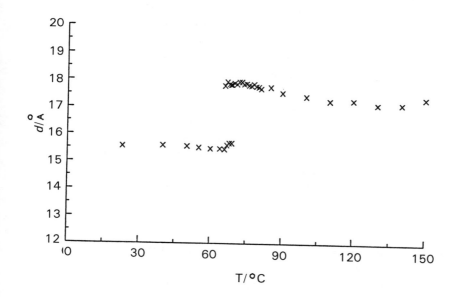

Figure 3. d-spacings of CTC/DEME (1.1 g/ml) as a function of
temperature T.

oriented by shear with a rod inside a capillary and the orientation stabilized by surface effects [8]. An X-ray pattern as represented in Fig. 1 was obtained at room temperature. The X-ray pattern of the same sample at higher temperature in the cholesteric phase resembles the one in the cholesteric phase at lower concentration (0.7 g/ml CTC/DEME) at room temperature (Fig. 2). The change of d-spacings of the innermost reflection with temperature was measured with a Kratky camera by a position sensitive counter and is depicted in Fig. 3. At the columnar-cholesteric phase transition the d-value jumps by 2.5 Å and remains at this high value without any recognizable change when the sample reaches the isotropic state. Since this reflection represents approximately the diameter of the CTC molecule, it can be concluded that the molecules pack more tightly in the columnar phase than in the cholesteric one and, surprisingly, that the clusters, causing this reflection, are still present in the isotropic state.

The fibre diagram of Fig. 1 reveals an uniaxial axis parallel to the main direction of the aligned and ordered CTC molecules with a monomeric repeat of 5 Å indicated by a sharp meridional reflection. Two weak reflections on the equator together with the strong innermost reflection suggest a hexagonal packing within the columnar phase. The X-ray pattern of Fig. 1 resembles more that of a smetic phase than the one of a nematic phase and shows similarities with a pattern of a columnar structure.

The cholesteric phase does not exhibit a fibre X-ray pattern, since the uniaxial optical axis of the helicoidal structure lies perpendicular to the CTC molecular axes, and a parallel alignment of CTC molecules on a molecular scale cannot be achieved. The Debye-Scherrer pattern of Fig. 2 shows a broader innermost reflection as a ring and a halo at the same d-value as in Fig. 1, which originates from the van der Waals distances of the interacting atoms.

Fig. 4 depicts DSC-curves upon cooling for CTC/DEME at various concentrations with changes in enthalpy. A classification of the two transitions (curve c) according to a method of Navard and Haudin [9] suggests a phase transition of first order.

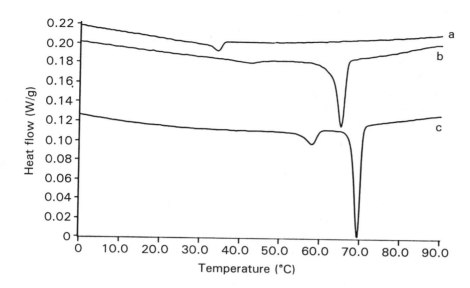

Figure 4. DSC-diagrams on cooling (10 K/min) of CTC/DEME; a: 0.7 g/ml, b: 1.0 g/ml; c: 1.15 g/ml.

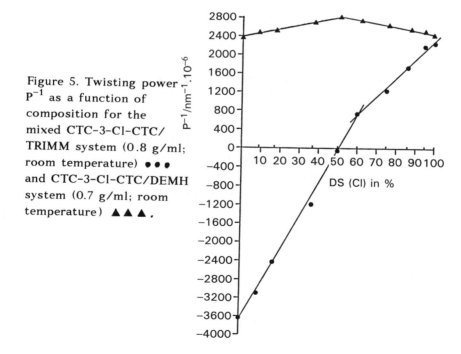

Figure 5. Twisting power P^{-1} as a function of composition for the mixed CTC-3-Cl-CTC/ TRIMM system (0.8 g/ml; room temperature) ●●● and CTC-3-Cl-CTC/DEMH system (0.7 g/ml; room temperature) ▲▲▲.

Quasi-nematic phase of a mixed derivative [10]

The parallel arrangement of cellulose derivative molecules on a local scale is one of the goals for application oriented research, since it might lead to novel cellulose fibre forming processes. It was found that such an arrangement was achieved by a compensation of twisting angle by mixing a right- and left-handed cholesteric structure of cellulose derivative/solvent systems as CTC/DEME and CTC/methyl propyl ketone (MPK). However, a mixture of CTC/triethylene glycol monomethyl ether (TRIMM) forming a left-handed helicoidal structure and 3-Cl-CTC (cellulose tri-3-chlorophenyl carbamate) / TRIMM with a right-handed structure led to phase separation in the desired composition range. Therefore, an attempt was undertaken, if a mixed derivative with a random substitution of phenyl carbamate and 3-chlorophenyl carbamate groups at the same cellulose molecule would accomplish the aim of a compensated cholesteric structure with zero twisting angle. Fig. 5 depicts the result of this study where P^{-1} represents the twisting power which is zero at the quasi-nematic state (pitch $P \to \infty$), negative for left-handed and positive for right-handed twisting. The state $P^{-1} = 0$ is achieved when 50 % of the substituents on a cellulose molecule are 3-chlorophenyl carbamate groups (DS (Cl) = 50 %). However, quite a different pattern for the twisting power is obtained (Fig. 5) for the same mixed cellulose derivatives in another solvent, diethylene glycol monohexyl ether (DEMH). Right-handed twisting occurs for both pure CTC/DEMH and 3-Cl-CTC/DEMH systems in the cholesteric phase. A more or less straight line to connect the two extremes would be expected and was found in a mixed solvent system with CTC [11]. Instead the graph in Fig. 5 shows a sharp maximum at DS (Cl) = 50 %. This investigation on the mixed cellulose derivative systems suggest that the twisting of the nematic sheets which form the cholesteric structure occurs through polymer solvent interaction on a very local scale and can easily be influenced by adding a few other substituents.

ORD-spectra [12]

Polymer solvent interaction influences strongly the phase behaviour of CTC/solvent systems, especially the biphasic region [6] and the ORD-spectra as shown here. Narrow molecular weight fractions were used in this study to avoid problems with the wide molecular weight distribution normally present in liquid crystalline cellulose derivative/solvent systems.

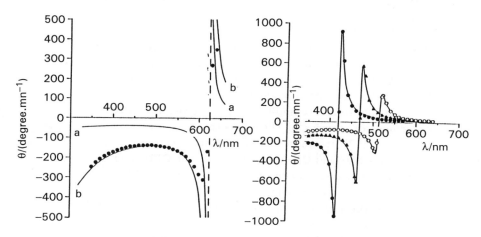

Figure 6. ORD-spectrum of CTC/
DEME (0.8 g/ml), degree of polymerization DP=125, T=293 K
experimental data • • •
a: according to equation of de Vries
b: corrected curve with $\exp(\lambda o /\lambda)$,

Figure 7. ORD-spectra of CTC/
TRIME (0.8 g/ml), DP=125,
at various temperatures
● 293 K, ▲ 303 K, o 313 K
(TRIME (triethylene glycol
monoethyl ether)).

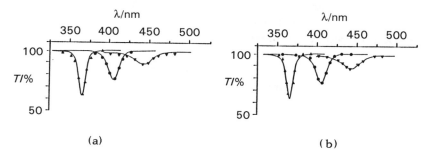

(a)   (b)

Figure 8. UV-VIS-spectra for CTC/TEMM (0.8 g/ml), DP=125 at
various temperatures ▲ 293 K, ● 303 K, ▼ 313 K;
solid line: experimental curves,
points: theoretically calculated ;
(a) Lorentz curves, (b) Gauss curves (TEMM (tetraethylene glycol
monomethyl ether))

Fig 6 compares the experimental ORD-spectrum (Θ rotatory power, λ wavelength) with the one calculated according to the equation derived by de Vries [13]. Deviations between the two curves are expected at the reflection wavelength λo (singularity) but the two curves should coincide at smaller wavelengths as observed for other systems. A rather good agreement (curve b) apart from the singularity is achieved by the introduction of an empirical factor exp (λo/λ). If, in addition, a damping of the curves is added, an excellent agreement of a series of ORD-spectra at various temperatures is obtained (Fig. 7). It can be assumed that a certain relationship holds between ORD- and UV-VIS-spectra as for ORD- and CD-spectra. An attempt was undertaken to represent the UV-VIS-spectra by either a Lorentz or Gauss curve and the ORD-spectra by the same based curves through the Kramers-Kronig relationship [12]. The damping factors (β half width of the UV-VIS curve and u half of the width at 1/e of the UV-VIS curve) are available by a fit of the experimental data by either a Lorenz or Gauss curve and should then be the same for ORD- and UV-VIS-spectra. Table 1 shows a comparison of β (Lorentz curve) and u (Gauss curve). An acceptable agreement is obtained.

It should be noted that the fit of UV-VIS-spectra by a Gauss curve deviates at the low tail ends to smaller values, a Lorentz curve to larger values as shown in Fig. 8. An explanation for the introduced factor exp (λo/λ), nor an interpretation of the damping values can be provided at the present.

Table 1. Constants β and u determined from ORD- and UV-VIS-spectra for CTC/TEMM (tetraethylene glycol monomethyl ether)

T/K	β/nm		u /nm	
	ORD	UV-VIS	ORD	UV-VIS
293	9	9.5	6	5.8
303	10	15.0	7	9.3
313	25	24.0	15	15.0

REFERENCES

1. Gray, D.G. *Appl. Polym. Sci., Appl. Polym. Symp.* **1983,** 37,179.

2. Gilbert, R.D. *Mater. Sci. Monogr.* **1986,** 36, 97.

3. Okomoto, Y; Kawashina, M; Hatada, K. *J. Am. Chem. Soc.* **1984,** 106, 5357.

4.   Hearon, W.M.; Hiatt, G.D.; Fordyce, C.R. *J. Am. Chem. Soc.* **1943,** 65, 829.

5.   Siekmeyer, M.; Zugenmaier, P. *Makromol. Chem.* **1990,** 191, 1177.

6.   Haurand, P.; Zugenmaier, P. *Polymer* **1991,** 32, 3026.

7.   Yamagishi, T.; Fukuda, T.; Miyamoto, T.; Yakoh, Y.; Takashina, Y.; Watanabe, J. *Liquid Crystals* **1991,** 10, 467.

8.   Hildebrandt, F.-I. *Diploma Thesis, Institut für Physikalische Chemie der TU Clausthal, Clausthal-Zellerfeld,* **1991.**

9.   Navard, P.; Haudin, J.M. *J. Therm. Anal.* **1984,** 29, 405 and 415.

10. San-Torcuato, A. *Diploma Thesis, Institut für Physikalische Chemie der TU Clausthal, Clausthal-Zellerfeld,* **1989.** Zugenmaier, P. *Das Papier* **1989,** 43, 658.

11. Zugenmaier, P. in Young, R.A.; Rowell, R.M. (Eds.) *Cellulose – Structure, Modification and Hydrolysis,* J. Wiley & Sons **1986,** 221.

12. Siekmeyer, M. *Dissertation, TU Clausthal, Clausthal-Zellerfeld,* **1989.**

13. De Vries, Hl. *Acta Cryst.* **1951,** 4, 219.

18

Cellulose-based chiral nematic structures

J.-F. Revol, J. Giasson, J.-X. Guo, S.J. Hanley, B. Harkness, R.H. Marchessault and D.G. Gray - Pulp and Paper Research Centre, McGill University, 3420 University St., Montreal, Quebec, H3A 2A7, Canada.

ABSTRACT

The chiral nematic (cholesteric) self-ordering of cellulose-derived materials is now recognized as a widespread phenomenon. A wide range of techniques may be used to elucidate the helicoidal structure of chiral nematic cellulosics in the liquid crystalline phase, and of the films, gels and solids prepared from such phases. Measurements of induced circular dichroism (ICD) of substituent chromophores in new specifically-substituted cellulose derivatives, and from dyes and chromophores incorporated in lyotropic and thermotropic phases gives information on local chirality. The helicoidal chiral nematic structure has recently been observed by transmission electron, optical and atomic force microscopies on samples prepared from aqueous cellulosic gels. Remarkably, dilute aqueous suspensions of cellulose crystallites prepared by acid degradation also show chiral nematic order; the order is preserved in dry films prepared from the suspensions. The structure of some of these samples prepared *in vitro* shows a marked resemblance to structures observed *in vivo* by others.

CHIRAL NEMATIC STRUCTURE

Chiral objects lack reflection symmetry; like right and left hands or enantiomeric carbon atoms, chiral objects are mirror images of each other [1]. In this paper, we are concerned with a particular helicoidal orientational order displayed by chiral nematic liquid crystalline phases. (The term 'chiral nematic' is used instead of the more general term 'cholesteric' because cholesteryl esters display other distinct 'blue' phases [2] just below their clearing temperature)

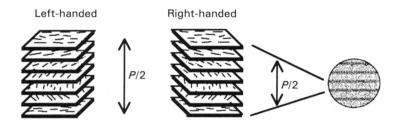

Figure 1. *Right and left-handed chiral nematic structures consisting of short rods. When viewed normal to the helicoidal axis, the structure has the microscopic fingerprint pattern shown on the right. P is the helicoidal pitch.*

Right- and left-handed helicoidal arrangements of small rod-like species are sketched in Figure 1. Two points should be emphasized regarding chiral nematic liquid crystals:

(i)   The liquid crystalline phase is <u>fluid</u>.

(ii)   The layers in the sketch are not real and the orientational order is imperfect, with the rods showing a local distribution of orientation.

Derivatives of cellulose form chiral nematic phases, both in solution (lyotropic liquid crystals) [3] and in the absence of solvent (thermotropic liquid crystals) [4]. The helicoidal structure of these phases is inferred from optical and electron microscopic techniques.

Evidence for the preservation of the helicoidal organization of the liquid crystalline phase of cellulose and its derivatives in films [5] and crosslinked gels [6] has been presented. The analogy between the molecular orientation in the liquid crystalline state and the orientation of microfibrils in many biological membrane structures has been pointed out by Bouligand [7]. All of these systems show chiral nematic structure; in the case of lyotropic and thermotropic liquid crystalline phases, and of polymer glasses, films and gels derived from these phases, the oriented elements of the chiral nematic structure are the molecular chain segments. In the case of biological membranes, the chiral nematic structure is composed of oriented crystalline microfibrils embedded in a matrix. Examples include chitin fibrils in insect and crustacean cuticle [7] collagen in skeletal tissue [8] and cellulose microfibrils in certain plant cell walls [9,10].

HANDEDNESS AND PITCH OF CHIRAL NEMATIC STRUCTURES

A wide range of cellulose derivatives form chiral nematic liquid crystalline phases, both left- and right-handed, and with pitches ranging from $\sim 0.1\ \mu m$ up to infinity (for compensated nematic phases). In some systems, the pitch increases with increasing temperature; in other cases, the pitch decreases with temperature. Recently significant progress has been made in the theory of chiral nematic phases of rod-like molecules [11,12] but it is still not possible to relate the handedness and pitch of cellulosic liquid crystals to the molecular properties of the cellulose derivative, the amount and nature of solvent (in the case of lyotropics) or to the temperature. In view of the unexpected sensitivity of pitch and handedness to minor changes in substitution [13], we have been preparing a series of cellulose-based polymers with different substituents at specific positions on the sugar rings. The long-term aim is to see if we can relate the chiral properties of the individual chains (measured in dilute solution) to the pitch and handedness of the liquid crystalline phases. A recent example of this approach is the synthesis and

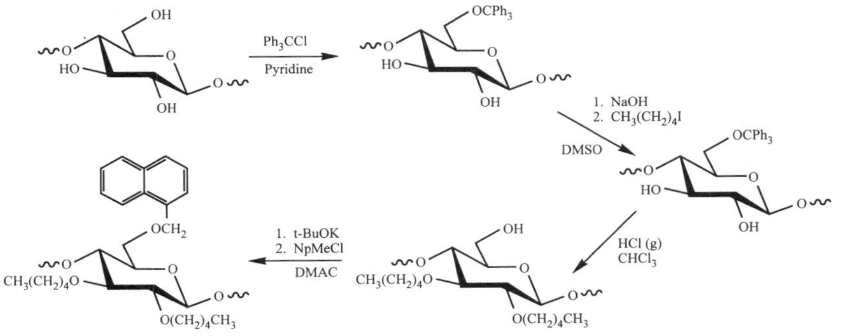

Scheme 1. *Preparation of 6-O-α(1-naphthylmethyl)-2,3-di-O-pentylcellulose.*

examination of the chiroptical properties of 6-*O*-α(1-naphthylmethyl)-2,3-di-*O*-pentyl cellulose [14]. Here, the naphthylmethyl group at the 6-position provides an accessible chromophore for measurements of circular dichroism (CD) and the pentyl ethers at the 2 and 3 position provide the flexible side-chains that facilitate liquid crystalline formation. The polymer forms a thermotropic liquid crystalline phase in the temperature range $\sim 50\text{-}105\,^{\circ}C$.

This phase displays iridescent colours, and has a right-handed chiral nematic structure with pitch values of the order of visible light. In dilute solution, the CD signals from the naphthyl chromophore are weak, presumably due to the four bonds separating the chromophore from the chiral centre on the cellulose chain, and the shape of the CD peaks provided some (weak) evidence for a helical chain configuration in dilute solution. Two distinct sets of CD signals are observed for the thermotropic liquid crystalline phase. (i) The reflection of circularly polarized light by the chiral nematic structure gives apparent CD peaks whose sign and wavelength are related to the handedness and pitch of the liquid crystalline phase. (ii) Circularly polarized UV light is absorbed by chromophores in a chiral environment.

Thin layers of (naphthylmethyl)(pentyl)cellulose liquid crystalline phase showed a marked CD peak at 280 nm resulting from the naphthyl chromophores; curiously, the sign of the CD peak was reversed compared to that in dilute solution, and the magnitude was larger. Although there remains a possibility that the CD signals from the second source (ii) are an artefact due to macroscopic birefringence effects, it seems from this and other evidence [15] that the sign of CD peaks for the chromophores is more strongly affected by the helicoidal chiral nematic environment than by the proximity of chiral centres on the molecular chains. This interpretation is supported by the reversal in sign of the induced CD signal for a dye (acridine orange) dissolved in a chiral nematic mesophase of (acetyl)(ethyl)cellulose in m-cresol. Changing the acetyl DS while holding the polymer concentration at 42 wt% causes a reversal in handedness of the chiral nematic phase, and this is reflected in a change in sign of the induced CD peak [15].

CHIRAL NEMATIC FILMS AND GELS OF CELLULOSE DERIVATIVES

Chiral nematic liquid crystalline phases are fluid; the chiral nematic order is readily destroyed by shear, reforming when the flow ceases. The structure is of course also destroyed by dilution below the critical concentration or heating above the critical temperature for mesophase formation. The chiral nematic order of liquid crystalline phases can be trapped in cellulosic solids by a variety of techniques, the simplest being casting films from a mesophase solution [16]. The chiral nematic order may be inferred from transmission electron microscope (TEM) images of thin cross-section of solvent-cast films. The periodicity corresponding to half the chiral nematic pitch was clearly visible, along with some of the disclinations associated with discontinuities in the chiral nematic structure [17]. The observed periodicity is apparently due to a periodic variation in thickness of the samples resulting from an artefact in microtomy [7]. We examined microtomed cross-sections of some solvent-cast films of (trityl)(hexyl)cellulose by TEM and also by atomic force microscopy (AFM), a technique that measures the surface profile of materials with a resolution that can approach atomic dimensions.

The AFM micrographs show a pattern similar to that observed by TEM, confirming that chiral nematic structure is reflected in the surface topology. (It is not yet possible to approach molecular resolution on these orientationally ordered samples. However, AFM scans of the surface of highly crystalline valonia cellulose microfibrils showed periodicities along the fibril axis of 1.07 nm and 0.54 nm, presumably corresponding to the fiber repeat and glucose unit length, respectively [18]. One advantage of the AFM technique is that high vacuum is not required and samples can be examined under ambient conditions of temperature, pressure and humidity.)

It is of interest to try and trap the chiral nematic order in a solvent-swollen gel by crosslinking the orientationally ordered chains. Anisotropic gels have been prepared by crosslinking lyotropic (hydroxypropyl)cellulose liquid crystalline solutions [19,20], but unequivocal evidence of chiral nematic order was not presented. Recently, we managed to preserve the chiral nematic order of (hydroxypropyl)cellulose solutions and films by irradiation cross-linking [6]. The gels swell but do not dissolve in water. This is, we think, the first preparation of an aqueous gel with chiral nematic properties.

CHIRAL NEMATIC SUSPENSIONS OF CELLULOSE CRYSTALLITES

The chiral nematic arrangement of cellulose molecular chains in liquid crystalline solutions, melts, films and gels is thus well established. However, as mentioned above, the species displaying chiral nematic order in natural composites are polymeric microfibrils rather than individual polymer molecules. Marchessault et al. reported the formation of a "liquid crystalline system" in stable suspensions (> 13% by weight) of cellulose and chitin microcrystallites prepared by acid hydrolysis [21]. A naturally occurring fluid state incorporating chiral nematic ordering of a glucuronoxylan-stabilized suspension of cellulose microfibrils has been observed in quince seed mucilage [22,23]. This phenomenon has been reproduced by Revol et al. [24] starting from cellulose fibers from a variety of sources. For example, treatment of a low-yield bleached kraft wood pulp with 65% sulfuric acid at 70°C for 30 minutes, followed by washing to neutrality and ultrasonic dispersion, gave a stable colloidal suspension of cellulose microcrystallites with lengths ~ 100 nm and widths of ~ 5 nm. At concentrations of ~ 3%, the suspensions separated on standing into an upper isotropic phase and a lower anisotropic phase. The anisotropic phase showed the finger-print lines characteristic of a chiral nematic phase. Furthermore, the chiral nematic structure of the suspensions was preserved on careful drying, to give a film of 'paper' in which the cellulose crystallites are arranged in chiral nematic order. Transmission electron microscope images of oblique sections through this film (Figure 2) show the arced pattern characteristic of a chiral nematic arrangement of microfibrils (Figure 3) [7]. The image in Figure 2 is virtually identical to electron microscopic images of sections of certain plant cell walls [10] and other biological composites.

The formation of the ordered phase in these suspensions is attributed to the well-known entropically driven self-orientation of rod-like species above some critical concentration [24], to give nematic order [25]. More surprising is the observation that the suspensions display chiral nematic order; the source of the chiral interactions between the crystallites is not obvious, as the distance

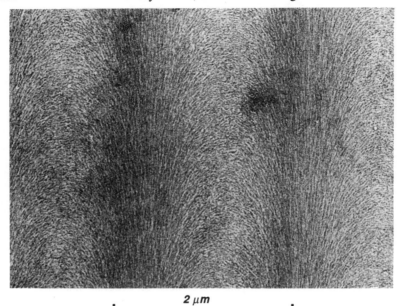

Figure 2. *Transmission electron micrograph of an ultrathin cross-section of a solid film prepared by evaporation of water from a chiral nematic suspension of cellulose crystallites. The characteristic arced pattern with periodicity of 2 μm is visible.*

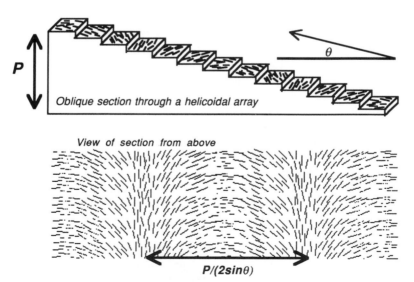

Figure 3. *Sketch of arcs formed by oblique section through a helicoidal arrangement of rods.*

separating the crystallites in the dilute suspensions must be much larger than that separating cellulose molecules in the liquid crystalline phases of cellulose derivatives. Nevertheless, it is clear from these observations that purely physical factors can contribute to the chiral nematic order in natural composites.

CONCLUSIONS

The chiral properties of cellulose are expressed at several levels, as summarized in Figure 4. The relationships between the chirality at different levels of structure are not at all understood, and exploitation of the chiral properties of cellulose is at a very early stage.

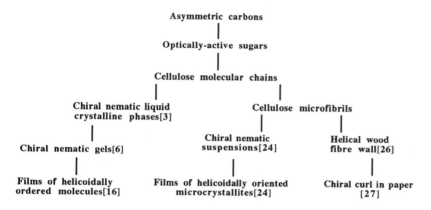

Figure 4. Interrelation between some of the chiral structures observed for cellulosic materials in the laboratory (references in parentheses).

ACKNOWLEDGEMENTS

The support of the Pulp and Paper Research Institute of Canada and the Natural Sciences and Engineering Research Council of Canada is gratefully acknowledged.

REFERENCES

1.   For a general introduction to chirality in nature, see, Hegstrom, R.A.; Kondepudi, D.K. *Scientific American*, January 1990, p. 108.
2.   Crooker, P.A. *Liquid Crystals* **1989**, *5*, 751.
3.   Werbowyj, R.S.; Gray, D.G. *Mol. Cryst. Liq. Cryst. (Letters)* **1976**, *34*, 97.

4. Tseng, S.-L.; Valente, A.; Gray, D.G. *Macromolecules* **1981**, *14*, 715.

5. Bhadani, S.N.; Gray, D.G. *Mol. Cryst. Liq. Cryst. (Letters)* **1984**, *102*, 255.

6. Giasson, J.; Revol, J.-F.; Gray, D.G.; St-Pierre, J. *Macromolecules* **1991** *24*, 1694.

7. Bouligand, Y. *Tissue and Cell.* **1972**, *4*, 189.

8. Bouligand, Y.; Giraud-Guille, M.-M. in Bairati, A.; Garrone, R. (Eds.) *Biology of Invertebrate and Lower Vertebrate Collagens*, Plenum Press, London, 1985, pp. 115-134.

9. Neville, A.C.; Gubb, D.C.; Crawford, R.M. *Protoplasma* **1976**, *90*, 307.

10. Roland, J.-C.; Reis, D.; Vian, B.; Roy, S. *Biology of the Cell* **1989**, *67*, 209.

11. Varichon, L.; ten Bosch, A.; Sixou, P. *Liquid Crystals* **1991**, *9*, 701.

12. Osipov, M.A. in Shibaev, V.; Lam, L. (Eds.) *Liquid Crystal and Mesomorphic Polymers*, Springer, New York, 1991, in press.

13. Guo, J.-X.; Gray, D.G. *Macromolecules* **1989**, *22*, 2086.

14. Harkness, B.R.; Gray, D.G. *Macromolecules* **1991**, *24*, 1800.

15. Guo, J.-X.; Gray, D.G. to be published.

16. Charlet, G.; Gray, D.G. *Macromolecules* **1987**, *20*, 33.

17. Giasson, J.; Revol, J.-F.; Ritcey, A.M.; Gray, D.G. *Biopolymers* **1988**, *21*, 1999.

18. Hanley, S.J.; Giasson, J.; Revol, J.-F.; Gray, D.G. to be published.

19. Suto, S.; Tashiro, H. *Polymer* **1989**, *30*, 2063.

20. Song, Q.C.; Litt, M.H.; Manas-Zloczower, I. *Polym. Mater. Sci. Eng.* **1990**, *63*, 445.

21. Marchessault, R.H.; Morehead, F.F.; Walter, N.M. *Nature* **1959**, *184*, 632.

22. Willison, J.H.M.; Abeyesekera, R.M. *J. Polym. Sci., Polym. Letters* **1988**, *26*, 71.

23. Reis, D.; Vian, B.; Chanzy, H.; Roland, J.-G. *Biology of the Cell* in press.

24. Revol, J.-F.; Bradford, H.; Giasson, J.; Marchessault, R.H.; Gray, D.G. to be published.

25. Onsager, L. *Ann. New York Acad. Sci.* **1949**, *51*, 627.

26. See for example Meylan, B.A.; Butterfield, B.G. *Wood Sci. Technol.* **1978**, *12* 219.

27. Gray, D.G. *J. Pulp Paper Sci.* **1989**, *15*, J105.

PART 2

19

Highly advanced utilization and functionalization of cellulose from the aspects of both chemistry and industry

Iwao Toyoshima - Daicel Chemical Industries Ltd., 3-8-1, Kasumigaseki, Chiyoda-ku, Tokyo 100, Japan.

ABSTRACT

A perspective summary is presented on the macroscopic trends in the world cellulose and derivatives industry during the past half century. Based on a long-term personal involvement in this field some examples of R&D activities at Daicel for the advanced technologies in cellulose acetate production and the new highly functionalized cellulose materials are reported. From industrial and business point of view, focusing on special difficulties for cellulose materials, problems to be solved for future development of this field and key factors for success in R&D are examined. Lastly my personal views and expectations on the future of cellulose industry are presented, considering the environment of the earth and desirable innovative technologies and new resources.

INTRODUCTION

Cellulose is a material that has aroused interest in the academic and industrial fields for a long time. It has been a popular topic to discuss occasionally over the years, and it is now watched with interest again these days. Its usefulness derives from being one of the renewable non-fossil resources.

I have been working for Daicel Chemical Industries for many years in charge of cellulose-related divisions, such as planning, research and development, production and management. I have some knowledge and experience with nitrocellulose, celluloid, cellulose acetates, acetate fiber, acetate plastics, cellulosic films, CMC, cigarette filters, membranes, specialty paper, chiral resolution columns, HPCE and so on. Having worked with cellulose all these years, my sincere wish is to see the cellulose industry continue to prosper.

From my affection to cellulose, I would like to share some of my studies and experiences on functionalization and the utilization of cellulose, as well as my views about the future of cellulose.

CELLULOSE INDUSTRIES IN THE WORLD

Generally speaking, a material that is able to take a dominant position in the industry must have a stable supply source, suitable processing technologies and cost advantages over competitive materials. The history of the cellulose industry during the last half of the 20th century is reviewed briefly.

I joined Daicel soon after graduation from university in 1953, and I began to work on cellulose. At that time, there was a book I read thoroughly, titled "Cellulose and Cellulose Derivatives"(ed. E.Ott, H.M.Spurlin, M.W.Graffline, Interscience, N.Y.,1954). I noted a tree-chart of the world's cellulose industry prepared by Emil Ott and H.G. Tennent. I kept looking at the figures and hoped to do something myself to influence them. Figure 1 shows the amount of products derived from wood and cotton using the same tree chart.

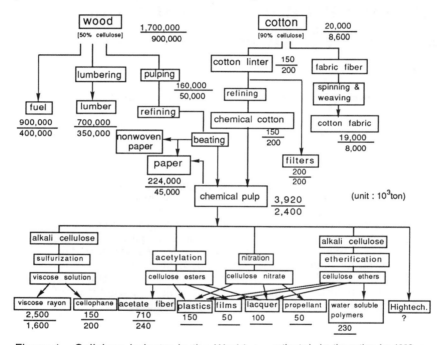

Figure 1. Cellulose Industry in the World   estimated by the author for 1990 / estimated by Ott & Tennent for 1950

The figures above bars in Fig.1 are those estimated for 1990[1] and those below are by Ott and Tennent in 1950.   The upper half refers to products involving no chemical reactions, and the bottom half refers to products with chemical reactions including derivatives. These figures indicate a rough estimate and are not intended to be accurate.

As to the scale of the industry[2], about 1.7 billion tons of wood was produced annually in 1990. This is a fairly large quantity compared with the production of steel (0.8 billion tons). In the same year, 19 million tons of cotton fabric was produced; this is also much larger than polyester fiber (8 million tons).

Then 224 million tons of paper is compared with 80 million tons of synthetic polymers. However, only *ca.* 4 million tons of chemical pulp, from which a variety of derivatives are produced, are manufactured a year, that is less than one tenth of the 60 million tons of ethylene. Furthermore, only 600 thousand tons of pulp for acetate was produced in 1990, only 15% of the amount of chemical pulp.

Table 1. Growth Pattern in Cellulose Industry

Products in Growth	ratio	Products on Decline	ratio
Wood	1.88	Chemical Pulp	1.63
Paper	4.98	Cellophane	0.75
Cotton	2.33	Viscose	1.56
Cellulose Acetate	2.96	Lacquer	??
WSP	??	Cellulose Plastics	??
New High Functional Products	??	(ratio:1990/1950)	

Table 1 shows the comparison of production levels both in 1950 and 1990, indicating the growth ratios. It is realized that some products began to decline after their peaks in those 40 years. The Table shows growth patterns of the products roughly classified, from which we can see which products have contributed to the growth of cellulose industry. Declines in viscose rayon and cellophane on a large scale resulted in a slowdown of the whole cellulose industry, due to competition from synthetic fibers and plastics. The cellulose industry has so far failed to create a big new product to cover the loss.

After all the total growth rate of cellulose industry was considerably low compared with that of other industries, specially in the field of products chemically modified. This over-all slowdown had a variety of influences on the industrial and academic fields, namely, a decrease in the number of researchers and companies interested in cellulose, giving way to the petrochemicals and synthetic polymers.

EXAMPLES FOR R&D OF HIGHLY FUNCTIONALIZED CELLULOSE

Some R&D activities at Daicel in the past are described in this chapter as for examples. Daicel Chemical Industries[3] has been doing business in cellulosics since 1907. Daicel's total sales in 1990 were *ca.* 170 billion yen, 30% of which came from cellulose derivatives.

Two definite targets have been pursued in R&D in this field;

(1) to cut down the cost of production to be competitive with improved profitability, and

(2) to develop highly technical and functionalized new products to acquire new markets with enlarged sales and higher profits.

A NEW PROCESS FOR CELLULOSE ACETATE

As an example for (1) above, a new process for cellulose acetate production is described in some detail. Industrial processes for cellulose acetate are thought to be fully established, but there are essential problems that should be solved in order to compete with other materials. These problems derive from the nature of cellulose itself used as a starting material. The first oil crisis revealed these problems and urged us to improve Daicel's cellulose acetate process on the following points:

(1) A highly pure alpha-cellulose is necessary as a starting material.
(2) The process requires a lot of utilities and energy, for example: A large amount of sulfuric acid is used as a catalyst, for which an equivalent amount of neutralizer is needed. Lots of electricity are required for refrigeration, which is necessary to control the exothermic acetylation reaction. Lots of utilities and energy are needed to recover acetic acid used as a solvent.
(3) Temperatures for acetylation must be kept low to get a desired degree of polymerization (DP). This requirement prevented us from cutting down the reaction time to raise productivity.

Thus, the targets set for our process innovation are summarized as follows:
1: To employ a low grade pulp with less purity as a starting material.
2: To change the process to save resources and energy, and
3: To revise the process for greater productivity.

It took 10 years to achieve the targets starting from a fundamental study. The achievements of this project are summarized here:

(1) A low grade pulp with 92% alpha cellulose was made available, which allowed extension of the range of usable materials markedly.
(2) Catalysts and neutralizing reagents were reduced by ca.70%.
(3) Consumption of electricity was reduced by ca. 17%.
(4) Steam was reduced by ca.17%.
(5) The temperature of acetylation and hydrolysis was raised, and productivity was improved by ca. 20%.
(6) Cellulose acetate produced by the new process is smoothly filtered and spun in the spinning process, because dope properties were improved.

Figure 2 shows a flow sheet of Daicel's new cellulose acetate process ("H-H process")[4].

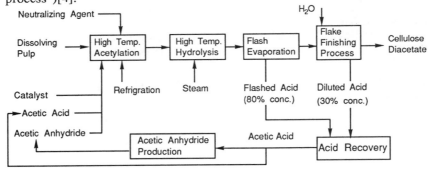

Figure 2. Flow Sheet of Daicel "H - H Process"

The characteristic features of the "H-H process" are a combination of high temperature acetylation, high temperature hydrolysis, flash evaporation and new flake finishing. Compared to the conventional process, the high temperature acetylation requires only a half or one third of the catalyst at a reaction temperature of about 10 to 20 degrees higher without any cooling. Under these conditions, the reaction time is reduced considerably.   In the high temperature hydrolysis, cellulose triacetate is subjected to hydrolysis at 130 to 160℃ under slightly increased pressure. Upon completion of hydrolysis, to attain the desired degree of substitution, the reaction mixture is concentrated by flash evaporation, that removes acetic acid-water mixture(80% conc.) allowing to proceed directly to the acetic acid recovery process.   Consumption of steam is substantially reduced by this change.

Notably the high temperature acetylation could never be achieved without being followed by high temperature hydrolysis. This stems from the fact that the hydrophobicity of the acetylation product is such that higher temperatures are required for dissolution for smooth hydrolysis. In "H-H process" depolymerization is suppressed by a reduction in the catalyst concentration. Even with the reduced catalyst concentrations, however, hydrophobic cellulose triacetate is formed, which is poorly soluble in a mixture of acetic acid and water, and is reluctantly hydrolyzable. Use of low-alpha pulp with an increased content of hemicellulose poses another problem of formation of hemicellulose acetates, which are insoluble in solvents, such as acetone, and form aggregates. This results in inferior filterability, lower transparency, unusual viscosity behaviors of cellulose diacetate solutions. Deterioration of filterability and spinnability was encountered in the spinning process of cellulose acetate fiber.

It was found possible to overcome these difficulties by employing high temperature hydrolysis. High temperature hydrolysis, which had never been employed on an industrial scale, somehow changes the structure of hemicellulose acetate and reduces cellulose acetate coagulation in the acetone solution. High temperature hydrolysis also converts cellulose triacetate to cellulose diacetate at a reasonable rate. Properties of the acetone solution are fairly good, with a smaller amount of insoluble gels, good filterability and transparency. The solution does not show false viscosity effects found in cellulose diacetate produced from cotton linter.

Table 2 shows a smaller gel content in cellulose diacetate produced in the "H-H process" compared with the conventional process.

Table 2. Comparison of Weight Fraction for Cellulose Diacetate and its Fractions as Prepared by Conventional and New Processes

		Diacetate	CDA	Gel I	Gel II
Conventional	wt.%	100	95.4	4.4	0.2
New	wt.%	100	99.1	0.6	0.3

pulp;soft wood, sulfate, α =94.2%

Figure 3(next page) also shows improvement of the false viscosity effects by the high temperature hydrolysis.

Daicel was awarded for "H-H process" the prizes of the Japan Chemical Industry Association and the Japan Institute of Invention and Innovation for advanced technologies.

APPLICATIONS TO SEPARATION TECHNOLOGY

In the 1960s, Daicel started a research into the reverse osmotic membrane for a new application. Then various uses of cellulosic membranes were developed, and eventually, synthetic ones were supplied to the food processing, health care, electronic, and waste water disposal treatment industries.

Cellulose derivatives are also important in chromatographic separation for

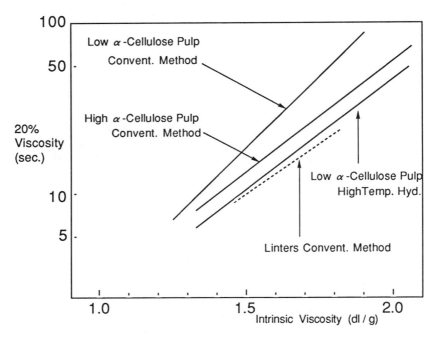

Figure 3. Effects of High Temperature Hydrolysis on Intrinsic Viscosity

further refined applications.   Conventionally cellulose has been used as a support for ligands of functional groups. The required properties for supports are: low adsorbability, high mechanical strength, appropriate pore size, pore volume and particle size, spherical particle shape, and easy introduction of ligands.

As a good example in this category, Daicel has marketed an adsorbent for removal of pyrogens in the injection solution named "PyroSep"[5,6] in cooperation with Tanabe Seiyaku Co., Ltd.   "PyroSep" are structurally characterized as a chemically modified cellulose which incorporates imidazole rings or primary amino groups as active ligands bonded to porous spherical cellulose beads.

Another important application developed by Daicel of cellulose-based materials is in the separation of optical isomers. It is known that microcrystalline cellulose triacetate is useful for recognizing chirality of optical isomers. It was found in Daicel that various acyl and carbamoyl cellulose derivatives display better optical resolution capability.

Daicel has also developed a microfibrilated cellulose(MFC)[7] in technical cooperation with ITT-Rayonier Inc. MFC is prepared from pulp or cotton linter being homogenized in water, and cellulose fibers are split along the fiber axis, hammered with severe impact and shear forces, to give a finely divided dispersion. Daicel produces and distributes MFC exclusively. MFC has found a unique use as the best filter aid for delicate filtration of Japanese 'sake'. Daicel has also developed applications as a filter aid to the food processing and other industries.

DIFFICULTIES IN THE CELLULOSE INDUSTRY

During the research studies described here, various difficulties, some of which are seldom recognized in the academic field, were in the way. These difficulties in using cellulose as an industrial raw material, should be overcome in order to attain further development of the cellulose industry.
They are:
(1) The starting materials are usually heterogeneous and contain various impurities.
(2) Reproducibility of experimental data and application in the actual plant are not easily attainable.
(3) Reactions are usually conducted in the heterogeneous phase, which makes it very difficult to control.
(4) Purification is hard and annoying.
(5) Disposal of effluents and wastes is troublesome.
(6) Processes and product quality are hardly controllable.
(7) Huge and heavy installations are required.
What is required for cellulose to survive in tough competition with the rivals? Apart from highly functionalized products, the production cost must be reduced to about half the present level. In many cases, a raw material, that is a chemical pulp, is already expensive. This is attributed to the fact that only 50% of wood is usable, collection costs and various costs for the pulping process are high, and so on. These problems should be solved systematically and integratively. Much attention has been paid to cellulose lately as one of the non-fossil, renewable and biodegradable natural resources, but high prices of the cellulosic products would prevent the business from lasting long.
Once a study was undertaken on the feasibility of producing chemicals, which are at present made from petroleum, by utilizing the cellulosic resources. As a conclusion collection costs for cellulosic resources are much higher than those for petroleum and coal, because cellulosic resources are spread dispersedly all over the world. Another result is that cellulosic resources are not competitive with fossil resources, unless three major components of wood, namely, cellulose, hemicellulose and lignin, are utilized effectively.
Some of these problems have been solved and some were left untouched because they were not profitable, but hopefully all these problems are solved with new technologies.

KEY FACTORS IN R&D OF CELLULOSE

Next, there will be given a personal view on how to succeed in R&D for business as is summarized in Figure 4(next page). Success in the research work is not sufficient for a raw material to grow and get a predominant position in the industry. Based on the developmental works described here, some important factors of success are considered. This kind of argument should not be limited to cellulose only, and we have to consider problems characteristic of cellulose separately from common ones.
Important factors in success are summarized as follows as characteristic problems of cellulose.
(1) It is necessary to have superiority in ideas and technology in order to make materials highly functional and to satisfy the market need. This is essential in planning and proceeding with the research work.
(2) It is necessary to have some distinctions in manufacturing and engineering technologies that can make the products highly competitive in cost and quality.

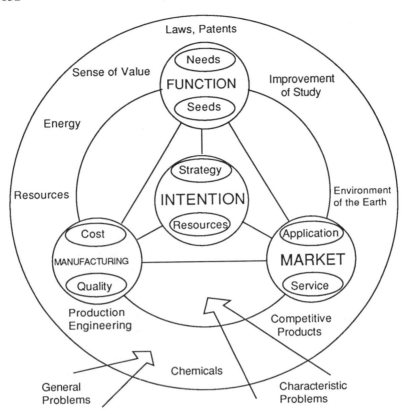

Figure 4. Key Factors in R & D for Success in Business

This might apply to research and improvement in the pilot plant and actual plant. (3) The third point is to have full technical information on a product to meet customers' requests for cost, quality, and application. This is for development. (4) The last necessity is a strong will and the long-term strategic vision of the top board for the project or the business. This is essential in all stages of R&D.

I would like to mention some general and technological factors that may influence the characteristic problems in cellulose chemistry. For example, changes in people's sense of value would affect the market, requiring new products that could expel the conventional ones. Technical progress in other fields or an appearance of a new economical chemical would also affect the manufacturing technologies in the cellulose industry by reducing the production cost. In order to take advantage of these effects, continuous monitoring of progress in other fields and tight cooperation with researchers of other technologies is necessary.

The value of research is not ultimate for the business, and should not be judged by the success or failure of products derived from a given project. Conversely speaking we cannot expect success in business, where we have to handle a very unique material like cellulose, even if we have strengthened certain application research since the business seemed interesting at the first glance. This means that it is important to continue fundamental research work from several points of view and accumulate know-how in handling cellulosics. At the same time, it is also necessary to create novel R&D themes in a long term strategic plan, which should be derived from general trends in the world.

PERSONAL VIEW ON THE FUTURE OF CELLULOSE INDUSTRY

Partly as a summary I would like to make a forecast about the future of the cellulose industry.

Chemically less-modified cellulose products such as lumber, pulp, cotton fabric, paper and so on, have kept growing, but others are not growing so well. The demand for paper should grow steadily in the future, but saving resources by recycling materials must be seriously considered. No new principal products have appeared yet that could cover losses caused by the reduction of viscose. It is a matter of concern that some cellulose businesses will be reduced or abandoned, and the position of cellulose in the whole industry would fall to be a low-value industry, if no effective action is taken.

Although many companies have lost interest in cellulose, Daicel has continuously made efforts on two items in its R&D. One is to reduce costs by overcoming difficulties resulting from the complicated structure of cellulose. The other is to develop new highly functionalized specialties taking advantages of the characteristics of cellulose. In the latter, there may be great interest in the technology, but there will be a lot of obstacles to overcome before succeeding in business. We could enjoy a successful business in the specialty field working hard with such effort. However, for some time to come, development of such highly technological functionalized products, which might be small in production scale, shall continue. We, not expecting the appearance of new large-scale products at once but through untiring efforts, should seek the coordination of chemistry and engineering in other fields than that of cellulose, and be prepared for the forthcoming generation.

As mentioned earlier, the cellulose industry would not grow further unless much cheaper raw materials were made available, but, we cannot expect to discover such raw materials easily, because forests have been devastated by reckless deforestation and acid rain. In other words, the appearance of a new principal cellulosic product would accelerate the destruction of forest resources. We have an important mission to make good use of the remaining cellulose resources carefully.

As a conclusion, my expectations in the long term are as follows:

(1) Improvement of productivity in forestry by thorough rationalization to make it a more attractive industry.

(2) Development of a plant for effective photosynthesis of useful substrates that are easily handled as industrial raw materials.

(3) Large scale production of cellulosic resources such as bacterial cellulose.

As the result of these actions, increasing the number of plants and improvement of photosynthesis might consume a lot of carbon dioxide, which will prevent the greenhouse effect. These investigations will be helpful for mass production of a raw material for the industry in the future, when a utilization technology of plants as non-fossil carbon sources will be developed. What new technologies can we expect? Based on the trend of recent research of cellulose[8], some innovative items are shown as follows, which may be able to overcome obstacles to further the development of the cellulose industry. The first one regards techniques for hyperfunctionalization. The method of using derivatives will also be important in the future, and practical development is anticipated on:

(1) New reaction systems using newly developed solvents,
(2) Control of substituting group distribution,
(3) Relation between structures and physical properties,
(4) Functional gels, membranes, liquid crystals, and other functional materials.
Furthermore, hyperfunctionalization by means of physical processing that makes good use of the natural structure of cellulose, and technologies availing biodegradability of cellulose, are also worth notice.
The second one regards techniques for hyperutilization of materials. Those to obtain raw materials from not only lumber but something else extending selectivity, and those related to recycling and new pulp production, or purification of material are particularly important.
Further, efficient use of lignin, composite materials and gasification techniques are listed, for the purpose of making use of all the components of plants.
The third one is the techniques related to engineering to overcome all industrial difficulties, details of which have already been explained previously.
Our dream of a new principal cellulosic product will come true upon the realization of these technologies and solutions to the problems related to the resources mentioned before.

CONCLUSION

I think I have to conclude that the future of the cellulose industry will not be bright unless breakthroughs are made in functionalization, competitive power, and the supply of raw materials. I hope that cellulose will have a strong impact on the structure of the whole industry, and can contribute to the preservation of the global environment. Further, I believe that contribution of science in various fields and the consent of the people on a global basis are required for cellulose technologies grow into a great industry beneficial for mankind.

ACKNOWLEDGMENT

The author gratefully acknowledges the untiring enthusiasm and invaluable collaboration of colleagues who have contributed their best at the various stages of the studies described here over 38 long years since 1953.

REFERENCES

[1] SRI, Chemical Economics Handbook, Industry Overview, 1991 Textile Organon, June, 1981.
    USDA, Foreign Agriculture Circular, Tobacco, December, 1975.
[2] "Sekai Kagaku Kogyo Hakusho"(White Paper of the World Chemical Industry 1991", Kagaku Keizai(Chemical Economy), March, 1991.
"Nippon Kogyo Nenkan 1990"(Japanese Industry Yearbook 1990), Nippon Kogyo Shimbun Sha, March, 1990.
[3] Hiroshi Inagaki, Kyoto University, "Historical Survey of Cellulose Research in Japan", 1988.
[4] USP-4,439,605, USP-4,504,355, BP-2,111,059, JP-1,293,645, JP-1,473,990.
[5] S. Minobe, T. Tosa, I. Shibata: J. Chromatography, 248, 401 (1982).
[6] Japanese Patent Toku-Kai-Sho 57-183712.
[7] USP-4,481,076.
[8] "Cellulose 1988 Japan": J. Fiber Science and Tech., Japan, vol. 45, 1989.

20

The application of FTIR spectroscopy in cellulose research

Dietrich Fengel - Institute for Wood Research, University of Munich, D-8000 München 40, Winzerestrasse 45, Germany.

ABSTRACT

Deconvolution of the IR spectra of cellulose and cellulose derivatives in the range of the OH stretching vibrations gives detailed evidence on crystallinity, crystal modification and degree of substitution. Cellulose nitrate is used as an example to demonstrate the possibility of quantifying these effects. This method of spectra analysis is applicable for IR spectra obtained from KBr embedded samples as well as from micro samples embedded in nujol.

INTRODUCTION

For the past 40 years infrared spectroscopy has proved a very useful method in the field of cellulose and wood chemistry [1-3]. The past decade has seen the introduction of a new generation of instruments, the Fourier-transformation spectrometers, which offer new methods for measurement and evaluation.

Among the various new possibilities are micro-spectroscopy of small samples such as individual fibres and the improvement of spectra by mathematical means, e.g. by deconvolution.

This paper gives a survey on the use of these methods which lead to a more sophisticated study of cellulose and cellulose derivatives particularly in the range of the OH stretching vibrations. More detailed results were published elsewhere [4-7].

METHODS

For these studies the conventional KBr method (cellulose fibres) and the micro-IR method (cellulose nitrate) were applied. In the latter the samples were embedded in nujol and survey spectra were produced using a circular aperture in the microscope. The spectra of the KBr embedded samples result from measurements involving 64 scans at a spectral resolution of 8 cm^{-1}. The micro samples were measured with 256 scans at the same resolution.

The deconvolution of the spectra in the range 3200 - 3700 cm^{-1} was carried out applying a triangular apodization function with a K-factor of 2 and a half-width of 100 cm^{-1} [8].
The instrument used was a Digilab FTS 40 spectrometer equipped with a UMA 300 A microscope.

THE ADVANTAGE OF DECONVOLUTION

The most important groups whose position and bonding is influenced by crystallinity, crystal modification and substitution of cellulose are the OH groups. Their main absorption bands are in the range of 3200 - 3700 cm^{-1} where the resolution obtained in conventional spectra is usually not satisfactory.

Two different cellulose fibres, cotton hairs and viscose fibres give two different IR spectra, particularly in the range 450 - 1500 cm^{-1} (Fig. 1). These variations in spectra can be attributed to differences in both crystallinity and crystal modification. After a short ball-milling of the two cellulose samples no differences can be seen when comparing their IR spectra (Fig. 2).

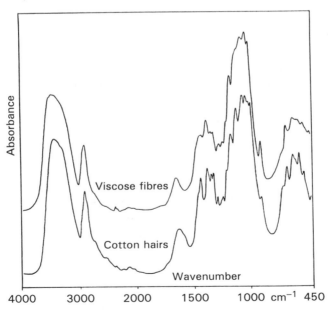

Fig. 1   FTIR spectra of cotton hairs and viscose fibres

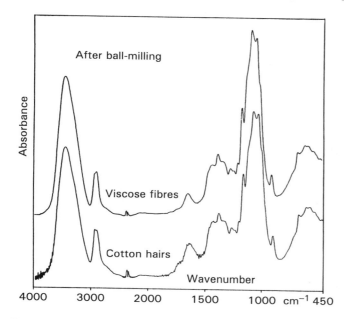

Fig. 2   FTIR spectra of cotton and viscose after 10 minutes ball-milling

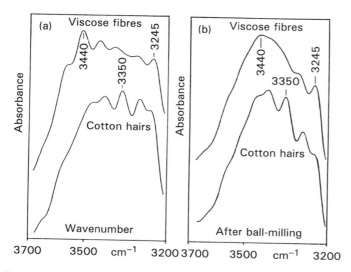

Fig. 3   Deconvoluted FTIR spectra of cotton hairs and viscose fibres
         before (a) and after 10 minutes ball-milling (b)

Detailed information about the variations that take place during ball-milling is revealed by the resolution of the absorption bands hidden in the broad area between 3200 and 3700 cm^{-1} when applying deconvolution.

Cotton hairs and viscose fibres show different positions of absorption bands except for 3245 cm^{-1} (Fig. 3a). Although band intensities are changed after ball-milling the samples can still be attributed to cellulose I and cellulose II according to the band positions (Fig. 3b).

CHANGE IN CRYSTALLINITY

To find out which bands change with variation in crystallinity cotton linters were ball-milled for different periods of time.

Three main phenomena were observed (Fig. 4):

a. The bands at 3350 and 3418 cm^{-1} disappear with increasing milling time (after 20 to 60 minutes);

b. The band at 3466 cm^{-1} is shifted to 3438 cm^{-1}. This shift begins after 20 minutes milling;

c. The amorphous condition of the cellulose is obviously reached after 20 to 60 minutes. After 60 minutes milling no more changes are observed in the IR spectrum.

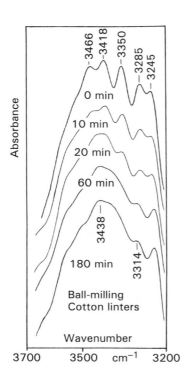

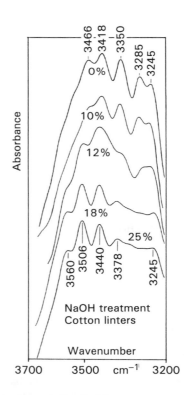

Fig. 4   Deconvoluted FTIR spectra of cotton linters ball-milled for various periods of time

Fig. 5   Deconvoluted FTIR spectra of cotton linters after treatment with NaOH of different concentrations

CHANGE IN CRYSTAL MODIFICATION

The change in the OH group vibrations during the transformation of cellulose I to cellulose II can be traced by studying cellulose after treatment with increasing concentrations of NaOH (Fig. 5). According to the IR spectra the transformation of cellulose I to cellulose II begins with treatment with 12 % NaOH as the following phenomena are observed:

a. The band at 3285 cm^{-1} disappears;
b. The band at 3350 cm^{-1} is reduced in height and shifted to 3378 cm^{-1};
c. The bands at 3418 and 3466 cm^{-1} are shifted to 3441 and 3506 cm^{-1}, respectively;
d. A new band appears at 3560 cm^{-1};
e. There is no further change in the spectrum after treatment with
   25 % NaOH compared to that with 18 % NaOH.

DEGREE OF SUBSTITUTION

The substitution of OH groups during the derivatization of cellulose has a great influence on the vibration behaviour in the IR range of 3200 - 3700 cm^{-1}. This effect is demonstrated using cellulose nitrate as an example.

The introduction of nitrate groups causes a total change in the pattern of bands in this range even at low degrees of substitution (DS) compared to the original cellulose (Fig. 6). New bands appear at 3590 and 3660 cm^{-1}. A closer look at certain bands of the IR spectra of cellulose nitrates reveals a variation in height related to the DS.

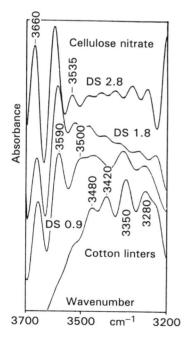

Fig. 6   Deconvoluted FTIR spectra of cotton linters and cellulose nitrates
          of various degrees of substitution (DS)

The best correlation between in relative height of a band and the DS was found for 3500 - 3535 cm^{-1}. The band is shifted from 3500 cm^{-1} at low DS to 3535 cm^{-1} at high DS. The data can be linked by a straight line, the standard deviation amounts to 1 - 5 % depending on the DS (Fig. 7).

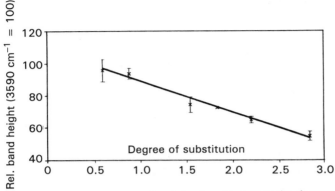

Fig. 7   Correlation between relative band height in the deconvoluted FTIR spectra and the DS of cellulose nitrates

ACKNOWLEDGEMENT

The cooperation of Dr. Martina Ludwig and Mrs. Margarete Przyklenk is greatly appreciated.

REFERENCES

1. Marrinan, H.J.; Mann, J. *J. Polymer Sci.* **1956,** *21,* 301.
2. Liang. C.Y.;Marchessault R.H. *J.Polymer Sci.* **1959,** *37,* 385.
3. Nelson, M.L.; O'Connor, R.T. *J. Appl. Polymer Sci.* **1964,** *8,* 1311.
4. Ludwig, M.; Fengel, D. *Papier* **1990,** *44,* 661.
5. Fengel, D.; Ludwig, M. *Papier* **1991,** *45,* 45.
6. Fengel, D. *Papier* **1991,** *45,* 97.
7. Fengel, D. *Holzforschung,* **1992,** 46 in press.
8. Kauppinen, J.K. Moffatt, D.J.; Mantsch, H.H.; Cameron, D.G.
   Appl.Spectr. **1981,** *35,* 271.

21

The application of ^{13}C solid state NMR to the study of surface reactions of cellulose with succinic anhydride

J.C. Roberts and J. Tatham - Department of Paper Science, UMIST, Manchester, M60 1QD, England.

ABSTRACT

Cellulose pulp fibres, which have been modified by the introduction of carboxyl groups by succinylation with succinic anhydride, have properties which have been shown to depend upon the counterion of the carboxylate group. Substantial swelling of the cell wall and a concomitant increase in tensile strength of handsheets was observed when derivatised pulps were in their sodium salt form. In contrast, very little effect was found when pulps were in their free carboxylic acid form. The reaction has been observed by isotopic (^{13}C) labelling of the anhydride and solid state CP/MAS NMR. The technique has been shown to be very effective for the study of such low-level surface derivatisation reactions.

INTRODUCTION

Cellulosic fibres can be esterified easily by derivatives of succinic anhydride. This is the basis for the control of water penetration in paper (sizing) by alkenyl succinic anhydrides (1-3). These reactions are believed to result in the formation of mono-esters and therefore also to introduce a free carboxyl groups into the fibre :

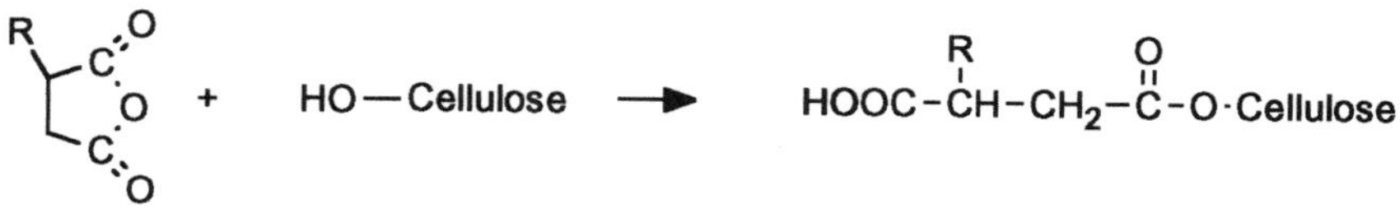

Where R = H (Succinic anhydride) and R = Alkene (Alkenyl succinic anhydride)

The introduction of carboxyl groups into fibres would be expected to assist swelling in water (cf. carboxymethylated pulps). However, this does not happen with alkenyl substituted succinic anhydrides, and it is presumably because the bulky hydrophobic alkenyl group exerts a dominating steric effect. It may be possible to modify selectively the swelling properties of fibres by reaction with various substituted anhydrides.   In this paper, fibre and sheet properties have been modified by treatment with unsubstituted succinic anhydride. The counterion of the carboxyl group has a marked effect on the properties of the pulp and paper and similar effects have been reported for carboxymethylated pulps (4). Levels of reaction are relatively low, and techniques for studying these levels of reaction are limited. ^{13}C substrate labelling and solid state NMR have been used for the study of the reaction.

RESULTS AND DISCUSSION

The method of succinylation was adopted from that of Garves (5) and Matsuda (6). The reaction was carried out in dimethylformamide, which has a high swelling ability for wood (6) and for cellulosic gels (7). It has been suggested that it is an ideal solvent for cellulose esterification, because of its ability to disturb the hydrogen bonding network in cotton and penetrate regions of low order with relative ease (8). Varying amounts of succinic anhydride and different reaction times were employed to vary the extent of substitution. Unreacted succinic anhydride and any succinic acid arising from hydrolysis were removed by extraction in a water/acetonitrile mixture before any fibre properties were investigated. The degree of substitution of treated pulps was determined by weight gain and carboxyl content. Carboxyl contents agreed well with the theoretically calculated values determined from weight gains (figure 1). As the calculation assumes the presence of only one carboxyl group per mole of substituent, this confirms that only monoesterification has occurred.

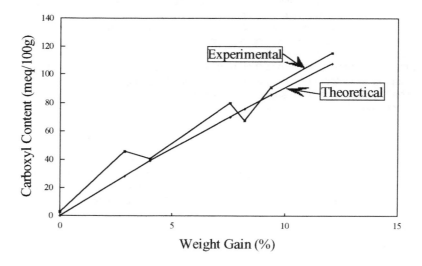

Figure 1 : Experimental and Theoretical Carboxyl Contents for Treated Pulps

It was possible to vary the degree of substitution by varying the amount of succinic anhydride used and also the reaction conditions. As the amount of succinic anhydride increased so the degree of substitution began to level off, despite the fact that there was still an excess of anhydride. This may be due to the increased difficulty in substituting the hydroxyl groups once the more accessible groups have been attacked, or it may be due to the fact that, after this period of time, there has been some hydrolysis of the excess anhydride by equilibrium moisture present in the sheet. Prior to testing, pulps were prepared in their free acid and also in their sodium salt form, as this is known to affect fibre properties.

The water retention value (WRV) of the free-acid form of the original pulp control was 1.2g/g and, on conversion to its sodium salt form, this slightly increased to 1.4g/g. However, the WRV's of the sodium salt forms of the derivatised pulps were found to increase greatly as the degree of substitution increased, whereas those of the pulps in their free acid form showed a small decrease (figure 2).

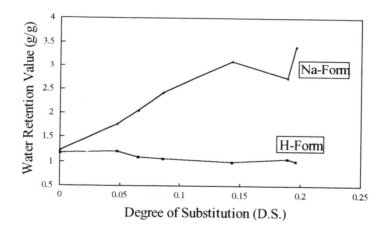

Figure 2 : The effect of Degree of Substitution of Treated pulps on their Water Retention Value

The high swelling values can be explained by reference to early work on the mechanism of swelling gels, first introduced by Proctor (9) to explain the behaviour of gelatin. Swelling was attributed to the presence of ionisable groups attached to the macromelecular network. In the presence of water, a fraction of these groups dissociate releasing counterions which are held within the matrix of the gel to retain electrical neutrality. The gel then swells as a result of the water entering in order to reduce the difference in osmotic pressure between the ionic solution within the gel and the water outside the gel. Proctor further suggested that the water would continue to enter the gel until the osmotic pressure differential equalled the resistance to further expansion brought about by the cohesion forces of the macromelecular network. Scallan (10,11) has adopted this theory to explain the correlation between papermaking properties of pulps and their degree of swelling. He noted that only the free ions contribute to the osmotic pressure difference which existed between the solutions of free ions within pulps and the aqueous phase outside the pulp The difference in swelling between the H form and the Na form can therefore be explained by the fact that the sodium form is much more readily dissociated than the hydrogen form.

The apparent density of paper affects nearly every physical property of the sheet (12) and can be considered to be a measure of bonding. As the inter-fibre forces holding the assembly together increase so the density and strength also increase. The apparent densities of the sheets made from treated fibres in their free acid form showed a slight decrease with degree of substitution. Conversion to the sodium salt form was accompanied by increased swelling and plasticisation of the fibres which in turn led to the formation of denser sheets (Fig.3).

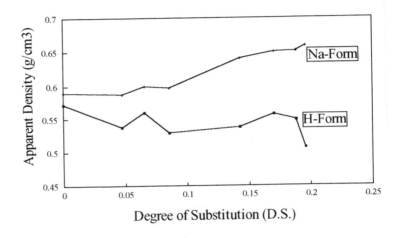

Fig. 3   The effect of Degree of Substitution on the Apparent Density of Handsheets made from Treated Pulps.

The tensile index allows a comparison of samples with different grammages. The introduction of the carboxyl groups in their free acid form causes a slight decline in tensile strength (figure 4). This is probably due to some loss of plasticity of the fibres through the structural change in the cell wall caused by drying and rewetting of the fibres. However, the sodium salt form showed a very substantial increase in strength with increasing degree of substitution. This increase in tensile strength can be attributed to the increased swelling and plasticisation of the modified fibres.

The zero-span tensile strength of treated sheets has been used as a measure of intrinsic fibre strength. The free acid form shows a small but steady decrease in the zero-span tensile index as the degree of substitution increased whereas the sodium salt form shows a slight increase. By combining the short and long span tensile data it is possible, by theoretical treatments of bonding, to determine the bond strength and this increases for pulps in their sodium salt form (figure 5). The increased water absorbancy and plasticisation of the fibres gives rise to better bonding.

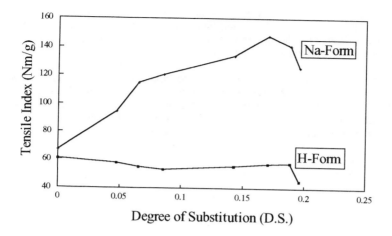

Fig. 4 The effect of Degree of Substitution on the Tensile Index of Handsheets .

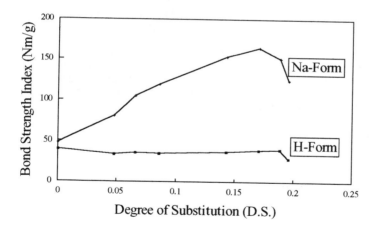

Fig 5 The effect of Degree of Substitution on the Bond Strength of Handsheets made from Treated Pulps

Succinic anhydride might by, virtue of the possibility of some di-ester formation, be expected to cause some cross-linking and hence some wet strength. However, the maximum percentage of tensile strength retained when the fibres were saturated with water was only 7.1%, and this is not very significant.

High resolution NMR was at one time limited to solutions, but techniques have recently been developed which make it possible to obtain well resolved spectra for solids (13-17). The technique has been used to examine the substitution reaction and to demonstrate the general applicability to surface reactions of cellulose. [1,4 - ^{13}C] succinic anhydride was prepared from labelled succinic acid :

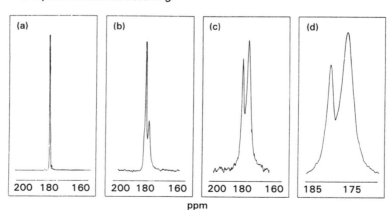

Initial experiments were performed with microcrystalline cellulose, whose solid state spectrum exhibits chemical shifts for all six carbons in the anhydroglucose unit in the range 60-110 ppm. These have been previously assigned (18,19).

Because of the possibility of hydrolysis of succinic anhydride during or after treatment the the solid state CP/MAS spectrum of the [1,4 - ^{13}C] succinic acid was recorded (figure 6a). An identical carboxylic acid carbonyl peak, at 180.1ppm, to that of the unlabelled form was observed and this was shifted downfield by 2.1 ppm from the solution spectrum (20) . Two peaks were observed for the upfield methylene groups at 36.67 ppm and 27.12 ppm (not shown in figure 6a). As the crystal structure of succinic acid shows that there are two molecules per unit cell and that the molecules have a plane of symmetry (21), the only explanation is that there is a ^{13}C - ^{13}C dipolar interaction occurring.

Fig 6   Solid state ^{13}C CP/MAS NMR spectra (carbonyl region) of (a) [1,4 - ^{13}C] succinic acid, (b) and (c) [1,4 - ^{13}C] succinic acid and microcrystalline cellulose (addition methods 1 and 2) and (d) expanded spectrum of (c).

Labelled succinic acid was then added to microcrystalline cellulose from aqueous solution and the cellulose dried in a vacuum dessiccator at room temperature. Two peaks were now observed in the carbonyl region, one at 180.2ppm - the position of the original shift in solid crystalline succinic acid, and another smaller peak at 176.9ppm (Fig.6b). It seems reasonable to assign the peak at 180.2 ppm to succinic acid which has crystallised during evaporation of water. The new peak at 176.9 ppm has been assigned to a non-crystalline form of the acid which may be in an adsorbed state and interacting through its carboxyl groups with the cellulose surface. In an attempt to confirm this idea, succinic acid was added to the cellulose in a different manner such that it was mixed with the air-dry cellulose (moisture content approx. 5%) before water was added. This seems likely to favour the non-crystalline form and to decrease the relative size of the peak at 180.2 ppm. This was indeed found to be the case (figure 6c). Higher resolution (figure 6d) showed that the upfield peak was also broader and supports the contention that it is in a non-crystalline form. Extraction with acetonitrile and water resulted in the removal of the peaks. This shows that there is no significant esterification by the acid under these conditions.

The carbonyl carbons of cyclic anhydrides are shielded by approximately 10 ppm with respect to dicarboxylic acids due to inductive and ring strain effects (22). The solution NMR of unlabelled succinic anhydride in deuterochloroform under ^{1}H broad band decoupling conditions gave a single carbonyl resonance line at 173.1 ppm which is some way upfield from the acid, and is similar to values reported in the literature (173.1ppm in D6-DMSO (20) and 172.9ppm in CDCl$_3$ (23)). The solution NMR spectrum (figure 7a) of the labelled anhydride (20) also shows a single peak in the carbonyl region at 170.6 ppm which is 7.3 ppm upfield from [1,4 - ^{13}C] succinic acid. The intensity of the carbonyl peak is also far greater than the CH$_2$ as would be expected since the carbonyl carbon isotope has been enhanced. The solid state CP/MAS spectra of both the unlabelled and labelled anhydride (Figs 7b and 7c) displayed two peaks for the carbonyl carbons at 174.3 and 176.7 ppm . There is a decline in symmetry of succinic anhydride in the solid state compared to the solution state because the anhydride ring is non-planar in the crystal lattice (24). The chemically equivalent carbonyls are therefore magnetically inequivalent in the solid state and two carbonyl resonances are observed. The solid state spectrum of the anhydride was recorded after a period of several months and three peaks were observed in the carbonyl region (Fig. 7d). It is clear that there is a doublet at 176.7ppm and 174.3ppm which can be assigned to the anhydride and a singlet at 180.1ppm which can be assigned to succinic acid. This demonstrates that the anhydride had undergone partial hydrolysis, as the resonance line at 180.1ppm agrees with the chemical shift of succinic acid.

As mono-esters are reported to be formed between cyclic anhydrides and cellulose (2,6), [1,4 - ^{13}C] methyl hydrogen succinate was prepared as a model compound and its solid state CP/MAS NMR spectrum recorded. The chemical shifts of the carbonyl peaks were observed at 183.3ppm (probably corresponding to the carboxylic acid carbonyl), and at 176.4 ppm (which can probably be assigned to the ester carbonyl). Ester carbonyls are generally found over a wide range of chemical shifts between 160ppm and 180ppm. Examination, in the literature, of the chemical shifts of diesters supports the view that the formation of the monoester takes place.

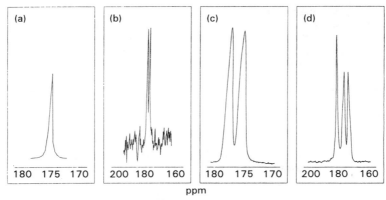

Fig. 7 Solution and solid state ^{13}C CP/MAS NMR spectra (carbonyl region) of (a) [1,4 - [13]C] succinic anhydride (solution spectrum), (b) unlabelled succinic anhydride (solid), (c) [1,4 - [13]C] succinic anhydride (solid) and (d) partly hydrolysed [1,4 - [13]C] succinic anhydride (solid)

Once the chemical shifts for the solid state acid and anhydride had been assigned as well as the spectra for cellulose and succinic acid, the spectra of succinic anhydride and cellulose could be examined. The labelled anhydride (25mg) was added to cellulose (1g) from the solvent chloroform and, after evaporation, the spectrum was recorded. From the expanded spectrum (Fig. 8a) two broad peaks in the carbonyl region at 176.7ppm and 174.4ppm were observed together with the well-resolved cellulose resonance lines. On careful inspection a small shoulder is observed at about 180 ppm, which is probably due to formation of a small amount of crystalline succinic acid arising from hydrolysis of some of the anhydride. This would be expected in view of the moisture content of the microcrystalline cellulose (5%). The broad peak may be due to unreacted anhydride (non crystalline) and/or to monoester formation. Solid state NMR signals of amorphous materials have significant line broadening, and peaks may have linewidths of up to 15ppm.

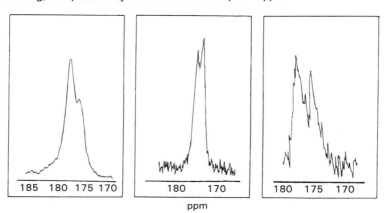

Fig 8. Solid state ^{13}C CP/MAS NMR spectra (carbonyl region) of [1,4 - [13]C] succinic anhydride (a) in the presence of microcrystalline cellulose (before extracion), (b) as for (a) but after extraction, and (c) in the presence of a Lapponia pine pulp (after extraction)

On removal of unreacted compounds by soxhlet extraction a smaller broad peak was observed with no sign of the peak at 180 ppm associated with crystalline succinic acid. An expanded spectrum (Fig. 8b) reveals that this broad peak may be a doublet or a multiplet with resonance lines at 176.4ppm and 174.3ppm, this may therefore be indicative of carbonyls of several types. The region being examined is also around the chemical shift of non-crystalline succinic acid but, as the acid can be successfully removed by extraction this is implausible. It is also unlikely to be due to the anhydride as this should also be removed by extraction. The most likely explanation is therefore that monoester formation has taken place and that the two peaks in the spectrum of the extracted material are the ester and carboxylic acid carbonyls. The carbonyls in different environments may contribute to the multiplicities of the resonance lines. The labelled anhydride (25mg) was also added to Lapponia pine pulp (1g) in an identical manner. The recorded spectrum shows two broad peaks in the carbonyl region positioned at 176.3ppm and 173.4ppm. An expanded spectrum (Fig. 8c) reveals that the peaks are a multiple of signals suggesting that the ester and carboxylic acid carbonyls are in different environments.

EXPERIMENTAL

A fully bleached sulphate pulp (Laponia) was disintegrated by a standard method (25) after soaking in deionised water for four hours. The pulp suspension was concentrated by filtration and then freeze dried to give a manageable, fibrous pulp. Pulps were esterified by adding varying amounts (3-24 g) of succinic anhydride (Aldrich Chemical Co. Ltd.) dissolved in dimethylformamide (350 cm^3) to pulp (24g air dry). In some cases a trimethylamine catalyst (2 cm^3) was added. The mixture was stirred thoroughly with a glass rod and allowed to stand at room temperature for between 2 and 48 hours. The DMF was removed by rotary evaporation and unreacted succinic anhydride and any succinic acid formed by hydrolysis was removed by soxhlet extraction with water/acetonitrile (1:1 v/v) for two hours. The pulp was then thoroughly washed with deionised water and allowed to air dry. Conversion from the free acid to the sodium salt form was achieved by soaking, overnight, at a 1% consistency in 0.1M sodium bicarbonate solution. Pulps were washed to neutrality using deionised water. Degrees of substitution were measured by weight gain and carboxyl content using a modification of the Tappi standard volumetric method (26). Water Retention Value was measured by a modified procedure (27).

Pulps were soaked in deionised water for 24 hours at a 1.2% consistency, before being disintergrated in a Weverk disintergrator. The stock was diluted (to 3.4 g/dm^3) and handsheets formed in accordance with the Tappi standard (28). Sheets were dried under restraint at constant temperature and humidity (50% and 23°C for 24 hours). Long-span Tensile Indices were determined from the mean of five handsheets. Tests were carried out on an Instron Tester with a cross-head speed of 20mm/min (29). The width of the test piece was 15mm and the length under test was 100mm throughout. A soaking attachment known as the Finch device was employed for wet tensile measurements (30). Zero-span Tensile Indeces were determined using a Pulmac short span tensile tester.

[1,4 - ^{13}C] Succinic Anhydride : In a round bottom flask (25 ml) provided with a reflux condensor protected by a cotton wool/calcium chloride drying tube, [1,4 - ^{13}C] succinic acid (MSD isotopes) was weighed (0.25 g, 2.12 mmole). Acetic anhydride

(0.43 g, 4.24 mmole) was added and the mixture refluxed gently on a water bath, with occasional shaking, until a clear solution was obtained (approximately 30 mins). Further refluxing for one hour completed the reaction. The assembly was removed from the water bath and allowed to cool. The [1,4 - ^{13}C] succinic anhydride crystals separated out and were collected on a sintered glass filter (no.1), washed with two portions of ether (1ml) and dried in a vacuum dessiccator. The melting point, infrared and nuclear magnetic resonance spectra were obtained.

Solution ^{13}C NMR spectra were recorded at room temperature using a Bruker AC 300 pulse Fourier Transform Carbon-13 Spectrometer with the ^{13}C channel operating at 75.47 MHz. Succinic acid was dissolved in D_2O. Succinic anhydride was dissolved in deutero chloroform .

Solid State ^{13}C NMR : Compounds were applied to cellulose from solution. Succinic acid (10mg) was dissolved in water (2 cm^3) and microcrystalline cellulose was added (350 mg) with stirring (method 1). Succinic acid was also applied by mixing the acid and the cellulose together before adding the water (method 2). Succinic anhydride (25mg) was dissolved in chloroform (5 cm^3) and microcrystalline cellulose (1.0g) was added. The solvent was allowed to evaporate in a vacuum dessiccator. Unwanted succinic acid or anhydride was extracted in a soxhlet apparatus. High resolution, solid state, magic angle spinning, cross polarisation ^{13}C nuclear magnetic resonance experiments were preformed at room temperature on a Bruker AC 250 spectrometer with the ^{13}C channel frquency operating at 62.9 MHz. The samples (350mg) enclosed in the rotor were spun at the magic angle at a rate of 4-4.5 kHz. ^{13}C magnetisation was generated by spin-lock cross polarisation when required (17).

ACKNOWLEDGEMENT

The authors wish to thank the Department of Wood Science and the Department of Chemistry, University of Wales Bangor for the use of solid state nmr facilities. We would also like to acknowledge the help of Mr E.Lewis in running spectra and of Dr. M. Anderson (UMIST) in interpreting spectra.

REFERENCES

1. Wasser, R.B; Tappi Seminar Notes, Alkaline Papermaking Seminar, 17-20 (1985)

2. McCarthy, W. R; Stratton, R. A; Tappi. J. 12 117-121 (1987)

3. Wan Daud, W. R; PhD Thesis, Paper Science Dept. UMIST, Manchester (1988)

4. Nelson, P. F, Kalkipsakis, C. G, Tappi. 47 (3) 170-176 (1964)

5. Garves, K; Tappi. J. 55(2) 263 (1972)

6. Matsuda, H; Wood Sci. Technol. 21, 75-88 (1987)

7. Chitumbo, K; Brown, W; De Ruvo, A; J. Polymer Sci. Symp. 47, 261-268 (1974)

8. McKelvey, J. B; Berni, R. J; Text. Res. J. 36, 828-37 (1966)

9. Proctor H.R; J.Chem. Soc. 105,313, (1914)

10. Stone, J. E; Scallan, A. M; "Consolidation of the Paper Web" (ed. Bolam, F.)
    Tech. Div. of the British Paper & Board Industry Fed. London, p. 145 (1966)

11. Scallan, A. M; Grignon, J;   Svensk Paperstidn. 82 (2) 40-47 (1979)

12. Emerton, H. W; Fundamentals of the Beating Process, The British Paper and
    Board Ind. Res. Assoc. Kenley, p. 145-173 (1957)

13. Fyfe, C. A; Solid State NMR for Chemists, CFC Press, Canada, (1983)

14. Kalinowski, H; Borger, S; Brain, S; Carbon-13 NMR Spectroscopy, John Wiley,
    England, 79-81 (1988)

15. Haeberten, U; High Resolution NMR in Solids: Selective Averaging, Academic
    Press, New York (1976)

16. Andrew, E. R; Int. Rev. Phys. Chem. 1, 195-224 (1981)

17. Pines, A; Gibby, M. G; Waugh, J. S; J. Chem. Phys. 59, 569 (1973)

18. Atalla R. H; Gast, J. C; Sindorf, D. W Bartuska, V. J; Maciel, G. E; J. Am.
    Chem. Soc. 102 (9) 3249-51 (1980)

19. Earl, W. I; VanderHart, D. L;   J. Amer. Chem. Soc. 102 (9) 3251-52 (1980)

20. Bruker; ^{13}C Data Bank, vol 1

21. Yardley, K; Proc. Roy. Soc. (London) 105A, 451-67 (1924)

22. Williamson, K. L; Hasan, M. U; Clutter, D. R; J. Mag. Res. 30, 367-83 (1978)

23. Silverstein, R. M; Bassler, G. C; Morrill, T. C; Spectrometric Identification of
    Org. Compounds, 4th edition, John Wiley & Sons, New York, USA, 270-1 (1981)

24 Ehrenberg, M; Acta. Cryst. 19, 698-703 (1965)

25. Tappi; T205 om-88, Official test method (1980)

26. Tappi; T237 om-88, Official test method 1983

27. Kropholler, H. W; "Determination of water retention value", Internal Laboratory
    method Dept. of Paper Science, UMIST, Manchester

28. Tappi; T402 om-88, Official test method 1983

29. Tappi; T494 om-88, Official test method 1981

30. British Standard Method, BS   2922 sec.8.2.3 Finch Method 1984

22

The characterisation of cellulose copolymer carbanilates by ^{13}C NMR spectroscopy – solution and solid state studies

J.T. Guthrie and R.D. Hirst* - The Department of Colour Chemistry, The University of Leeds, Leeds LS2 9JT, UK.

ABSTRACT

A series of cellulose-g.co-styrene polymeric assemblies has been developed using radiation induced grafting procedures. The resulting copolymers were converted to their carbanilated equivalent, after extraction of ungrafted cellulose and of poly(styrene) homopolymer. The extracted cellulose and the parent cellulose were also carbanilated. ^{13}C NMR studies were carried out on the various polymeric systems, in the solution state and/or in the solid state, as appropriate. Spectra interpretation has shown that assignments can be made with ease and that the copolymeric products are heterogeneous in character.

INTRODUCTION

Various forms of cellulose (cotton, wood pulp and ramie) have been used in graft copolymerisation studies (1-4) in a variety of physical forms. Processed cellulosics also have been studied. Such copolymers have been evaluated for their physical and chemical properties and also for their compositional and structural features. Solid state characterisation of the underivatised copolymers has been applied because of their lack of suitable solubility. Solution state characterisation has been successfully

* Present address: Scott-Bader Co. Ltd., Wellingborough, Northants NN9 7RL (UK).

applied to cellulose derivative copolymers and to derivatised cellulose copolymers (5-8). Despite this high level of attention, the need remains for fundamental information, coupled with appreciation of the physical and of the chemical attributes of such copolymers. This need must be met if cellulosic copolymers are to take part in the large and the intermediate-scale applications enjoyed by cellulose derivatives. Attention must also be given to compositional factors in that the copolymers and derivatives are heterogeneous.

This heterogeneity shows itself in molar mass related properties, chemical group distribution, structure and morphology, physical and chemical factors, preferential adsorption, solution stability, dispersion stability and in selective solubility. As factors governing heterogeneity become better understood, the design and assembly of optimisied copolymers and derivatives, with respect to particular properties, processes and applications should be facilitated.

In this paper, emphasis is given to the ^{13}C NMR characterisation of a selected cellulose-g.co-styrene assembly and its related components. The system identified is part of a much wider study (6).

EXPERIMENTAL

Materials   Saiccor High Grade Wood Pulp (WP) was obtained from Courtaulds Research, Coventry (UK). To convert this to a suitable form, sheets of the pulp were chopped and then shredded in a Kenwood blender for 30 seconds. The shredded pulp was then soaked in cold water for 72 hours, filtered and dried in an air blown, heated (45°C) atmosphere to yield a fluffy pulp form. This pulp was finally dried to constant mass in a vacuum oven at 40°C. Styrene and tetrahydrofuran (Aldrich, Gillingham, Kent, UK) were purified according to proven procedures (6,9). Other reagents of AR grade were obtained from BDH, Poole, Dorset, UK.

The experimental procedures used in monomer preparation, evacuation procedures, radiation-induced initiation, and post-polymerisation treatments have been fully described elsewhere (6) as have details relating to derivatisation (carbanilation) (9,10).

^{13}C NMR STUDIES OF COPOLYMERS

High resolution, ^{13}C NMR spectra were recorded for cellulose (WP), WP-cellulose tricarbanilate (WPCTC), fully carbanilated 'apparent' copolymers and full carbanilated true copolymers (i.e. after extraction of ungrafted cellulose). Measurements were made in solution (in pyridine and in dioxan) and also in the solid state. The solution state spectra were recorded at the SERC facility of the University of Edinburgh (UK) and at the Shell Research Laboratories (Amsterdam). The solid state spectra were taken at University of Durham (UDIRL) (UK - SERC supported facility) and at Shell Research Laboratories (Amsterdam).

At the University of Edinburgh, the spectra were recorded on a Bruker WH360 spectrometer at 25°C, using pyridine as a solvent, at 90.6 MHz using proton decoupling with no relaxation delay. At Shell, the Spectra were recorded at 100 MHz on a Bruker WN400 unit at 97°C with dioxan as the solvent. Proton decoupling with a WALTZ sequence was applied to remove small, unresolved residual splittings arising from couplings between the carbon and hydrogen nuclei and to avoid overheating of the sample by the radio-frequency field. A 30 degree (4μs) detection pulse with a relaxation delay of 1.5s was used.

At Shell Research Laboratories, the ^{13}C, CP (cross polarisation) MAS proton decoupled spectra were obtained on a XCP200 unit at ambient temperature at 50.3 MHz with CP times of 2ms and spinning speeds of about 3 kHz. At UDIRL, the ^{13}C, CP MAS proton decoupled spectra were obtained on a Varian VXR 300 unit at 75.4 MHz and CP times of 1ms, spinning speeds of 3 to 4ms and a relaxation time of 2s.

RESULTS AND DISCUSSION
C-NMR STUDIES OF CELLULOSE TRICARBANILATE (WPCTC) - IN SOLUTION

The approach to assignments, used in the C-NMR spectra, can be seen by considering the modified anhydroglucose unit shown below as Figure 1. The ^{13}C solution state, NMR spectrum of the WPCTC is shown as Figure 2.

Chemical shift values are not supplied, but assignment is straightforward using comparison with those quoted for simple monosaccharides and for those of ^{13}C solid state spectra of celluloses, where small shifts are seen relative to solution spectra (11-13).

Chemical shift values are supplied for the ^{13}C resonances of the carbanilate portion of the molecule (Table 1). Relative intensities of each of the peaks in this region of the spectrum and for the C_1 carbon of the anhydroglucose unit have been evaluated. Those of the other cellulose carbons could not be evaluated properly due to overlap with the strong solvent resonances.

Assignment of peaks for the carbanilate portion is again straightforward. The carbonyl peak appears as a triple resonance at 152-154 ppm. This extremely high chemical shift value is due to the strong deshielding of

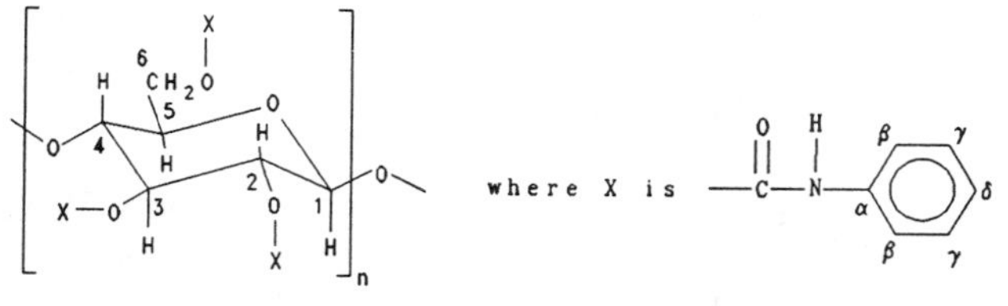

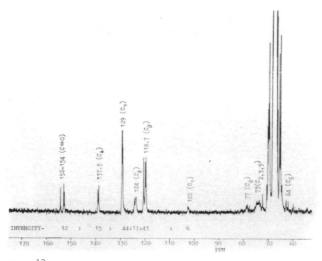

Figure 1.   Modified anhydroglucose unit

this group.   The appearance as a triple resonance indicates that the carbanilate rings are attached to three inequivalent positions, corresponding to substitution at C_2, C_3 and C_6 of the anhydroglucose residue.   All the aromatic carbons show splitting into a maximum of 3 peaks, again indicating inequivalence of substitution position.   Assignment of these peaks has been made by adding-on quoted incremental substituent effects to the value of 128.5 ppm for the chemical shift of a carbon atom in an unsubstituted benzene ring (14).   The experimental and calculated values quoted in Table 1 show excellent agreement.   The appearance of the aromatic carbons at low field is due to deshielding by the aromatic π -electrons.

Consideration of the integrated intensities allows assessment of the use

Figure 2. ^{13}C Solution state spectrum of WPCTC in dioxan of the spectra for quantitative analysis. The triple resonances of the carbonyl group were found to be in the ratio of 1:1:1 as expected for full carbanilation. Comparison of the intensity of the carbonyl group with that of the C_1 resonance of the cellulose suggests a degree of substitution (D.S.) of only two. Thus, one may conclude that this type of comparison is not of use in quantitative assessments unless more time-consuming experiments are performed. Elemental analysis, differential refractometry and other studies on this system/sample clearly showed the degree of substitution to be very close to the theoretical value of 3.0. Comparison of the intensities of the aromatic, carbanilate carbon atoms with that of the free cellulose C_1 carbon atom yields values for the D.S. in the range of 2.5-3.6. The relative intensities of these individual carbanilate carbons should be in the ratio of 1:2:2:1 which is not the case. This explains the spread of values for the predicted D.S. The overall inequivalence of intensities for the peaks is presumably due to differences in relaxation rates for the individual carbon nuclei and to differences in their NOE factors.

In summary, it appears that highly resolved spectra may be accurately obtained in solution with resultant confident assignment of peak positions. Full carbanilation appears to be indicated, but exact quantitative calculation of the D.S. cannot be achieved unless spectral studies using longer pulse delays are carried out.

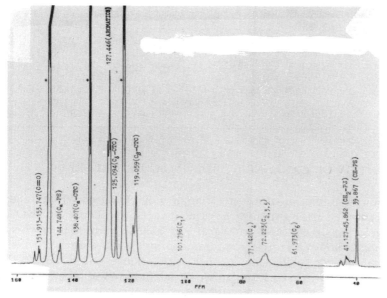

Figure 3. ^{13}C NMR WPCTC - Styrene Copolymer in Pyridine

TABLE 1 ^{13}C CHEMICAL SHIFT VALUES FOR CELLULOSE TRICARBANILATES

Carbon Atom	Increment	Calculated /ppm	Experimental /ppm	Relative intensity
C_α (CTC)	11.1	139.6	137.5	15
C_β (CTC)	-9.9	118.6	118.7	43
C_τ (CTC)	0.2	128.7	129.0	44
C_δ (CTC)	-5.6	122.9	124.0	17
C_1 (Cell)			102	6
C = 0 (CTC)			152-154	12

TABLE 2   INTEGRATED INTENSITIES OF SOLID STATE ^{13}C NMR SPECTRA FOR TRUE COPOLYMER

Copolymer Sample	Region Start/ppm	End/ppm	Integral
WP Copolymer	35.292	54.652	2.4886
	57.760	65.788	0.3993
	67.860	79.191	3.0427
	86.184	96.803	0.4719
	101.724	109.429	0.4788
	122.637	134.810	2.1058
	138.954	150.803	0.6042

13C NMR STUDY OF CARBANILATED (WP-G.CO-STYRENE)

Figure 3 shows the solution-state spectrum of a carbanilated WP-styrene copolymer achieved after grafting (radiation induced, total dose 24.5 k Gy apparent grafting 16.6% and true grafting 87.9%; styrene monomer concentration 20% W/V in tetrahydrofuran containing 4% water); Soxhlet extraction of poly(styrene) homopolymer with toluene and removal of ungrafted cellulose with Cadoxen (6,9); carbanilation was achieved at 105°C using phenyl isocyanate in pyridine for 12 hours.

The ^{13}C-NMR spectrum was obtained at room temperature after complete dissolution of the copolymer carbanilate in pyridine. The solvent peaks are identified on Figure 3 with an asterisk. The spectrum is clearly more complicated that that of WPCTC due to the presence of the grafted poly(styrene) branches. Assignments of peaks is relatively straightforward if reference is made to the previously mentioned CTC spectrum (Figure 2) and to published information relating to poly(styrene) (15-19). Some confusion concerns the exact position of the C PS peak which either overlaps with the C , CTC peak at 125.09 ppm or is grouped with the main aromatic resonance centred on 127.45 ppm.

Several features are noteworthy on closer scrutiny of Figure 3. One is that the low field carbonyl resonance appears as a triplet, suggesting yet again inequivalence of carbanilate group site attachment. Unfortunately, the splittings of other carbanilate carbon peaks are not as complete. This is obviously due to the lower temperature at which the spectrum was recorded. A most interesting feature is that the aliphatic region of the PS resonances is extremely well resolved. A strong methine resonance is seen at 39.867 ppm which is in extremely good agreement with the value of 39.79 ppm, calculated by Randall (16) using Grant and Paul substituent parameters (20). The multiple resonance seen between 41.127 - 45.862 ppm is that of the methylene carbon. Such fine splittings are, in principle, open to stereochemical, configurational interpretation. The approach taken in analysing such fine structure is usually to adopt the meso/racemic notation suggested by Frisch _et al_ for the addition of adjacent monomer units and to try to develop either a Bernoullian or Markovian statistical relationship to match the observed intensities (21). Such a task is beyond the scope of this work. However, by reference to studies of the stereochemical configuration of PS, it appears that splitting of the methylene resonance is occurring such that sequences of up to five (pentad) or six (hexad) carbon atoms may be resolved (16,17).

TABLE 3   QUANTITATIVE ANALYSIS OF A W.P. CELLULOSIC COPOLYMER BY ^{13}C SOLID-STATE N.M.R.

Sample	Peak	Intensity	C_a(PS)/$C_{1,4,6}$ (Cellulose)	% Calc. True Graft
WP Copolymer	C_a (PS)	0.6042	-	-
(112.2% true graft)	C_6 (Cellulose)	0.3993	1.51	97
	C_4 (Cellulose)	0.4179	1.28	82
	C_1 (Cellulose)	0.4788	1.26	81

A solution state ^{13}C NMR spectrum was obtained in dioxan for the carbanilated cellulose-styrene copolymer based on wood pulp (Figure 4). This spectrum was recorded at 97°C. These conditions yielded well resolved, narrow peaks. Again an attempt at quantification was made using the free C_1 resonance at 102 ppm. The relevant peak assignments are made on the spectrum together with the relative intensities of the C_1 cellulose carbon and the methin carbon of the PS branches. It should be noted that, in this "high temperature" spectrum, the C resonance of the PS occurs separately at about 126 ppm. The tentative ratio of C (methine)/C_1 (cellulose) for the WP copolymer is 3.3. Multiplication of this ratio by (100 x 104)/162 (molar masses of monomer units) allows determination of the graft level. This yields 210% for the copolymers based on WP. Thus, it is clear that, in the present spectrum the percentage of PS in the graft copolymer is overestimated. However, despite the need for refinement, we feel that the method shows potential.

CONCLUSIONS OF ^{13}C SOLUTION STATE NMR SPECTRAL STUDIES

Analysis of the cellulose graft copolymers using high resolution solution state ^{13}C NMR has shown promise and the main points are worthy of reiteration.

TABLE 4   INTEGRATED INTENSITIES OF SOLID STATE ^{13}C NMR SPECTRA FOR WP-G.CO. PS CARBANILATE TRUE COPOLYMERS

Sample	Region		Integral
	Start/ppm	End/ppm	
Carbanilated	36.199	54.522	1.7022
WP	59.961	67.990	0.2665
Copolymer	68.508	88.126	2.7360
	95.119	107.292	0.4543
	113.508	133.644	6.0658
	134.421	140.896	0.7134
	141.932	148.666	0.3077
	149.443	157.925	1.0012

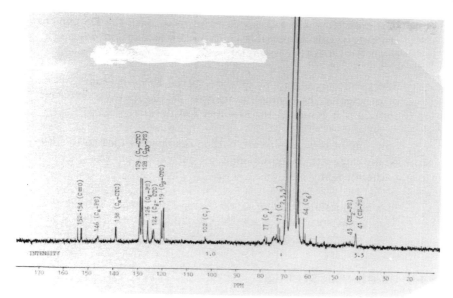

Figure 4. ^{13}C NMR WPCTC – Styrene Copolymer in Dioxan

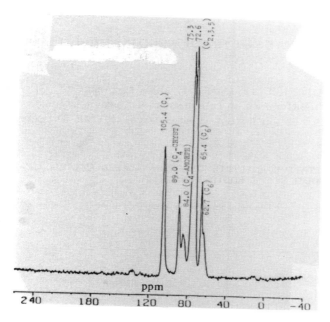

Figure 5. ^{13}C Solid State NMR: Wood Pulp Cellulose

(i)   It is clear that a quantitative analysis of both the DS and the grafting levels is possible, but longer relaxation delays are required. However, complete carbanilation is suggested.

(ii)   The tactic arrangement of the grafted PS is presumably one of atacticity, but complex analysis of the methylene carbon splittings is required to establish full details.

(iii)   Splitting of the carbanilate carbon resonances into up to three peaks suggests inequivalence of carbanilation site and full carbanilation.

(iv)   It is not possible to identify the sites of PS branch attachment. This is not surprising since the frequence of branching is low.

^{13}C-SOLID STATE NMR STUDIES

High resolution, cross polarisation (CP), magic angle spinning (MAS), ^{13}C solid state NMR spectroscopy was performed on native cellulose wood pulp (WP), WPCTC, WP-g.co-styrene and carbanilated WP-g.co-styrene.

CP MAS ^{13}C SOLID STATE SPECTRA OF WOOD PULP CELLULOSE AND ITS TRICARBANILATE (WPCTC)

Figure 5 shows the well resolved CP MAS spectrum of WP. Peak assignment is straightforward on comparison with previously published spectra of celluloses obtained from various sources (11-13). The spectrum

TABLE 5   QUANTITATIVE ANALYSIS OF CARBANILATED CELLULOSIC W.P. COPOLYMER BY ^{13}C SOLID-STATE NMR

Sample	Peak	$\dfrac{C=O}{C_{1,6}}$	$\dfrac{C_\alpha(CTC)}{C_{1,6}}$	$\dfrac{C_\alpha(PS)}{C_{1,6}}$	% True Graft -Calc.
WP-styrene copolymer	C=O (CTC)	-	-	-	-
carbanilated	C_α (CTC)	-	-	-	-
(true graft = 112.2%)	C_6 (Cell)	3.76	2.68	1.15	74
	C_1 (Cell)	2.20	1.58	0.68	43
	C_α (PS)	-	-	-	-

is lacking in the fine detail seen for other cellulose I samples. However, this observation has been noted before, by Vanderhart and Attalla, and has been interpreted as indicating higher amounts of disorder and small crystallite size when compared to highly crystalline cellulose materials (11). The peak assignments are given in Figure 5. They compare well with those already mentioned for the solution state spectra. It is interesting to note that the solid state spectrum provides a much better resolution for the cellulose peaks than do the solution state spectra. The date from Figure 5 will be used in the rest of this section as a basis for the assignment of all further cellulose peaks.

Figure 6 presents the CP MAS solid state spectrum of tricarbanilated WP (WP CTC). Again, a very well resolved spectrum is seen. Peak assignments are given in the spectrum. These compare very well with those previously seen for the solution state spectra. The cellulose-carbon resonances are all slightly shifted relative to the native WP. This is to be expected due to slight modification in the electronic environment on derivatisation. No splitting of the carbanilate carbons is seen in the solid state, but this is not surprising as such fine structure would not normally be detectable.

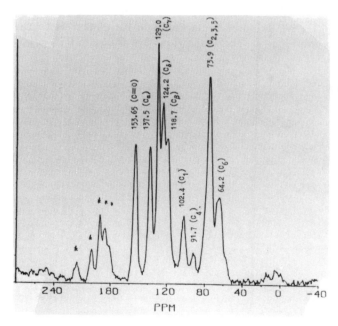

Figure 6. ^{13}C Solid State NMR : WPCTC

CP MAS ^{13}C SOLID STATE NMR SPECTRA OF THE WP CELLULOSE GRAFT COPOLYMER

Figure 7 gives the CP MAS ^{13}C spectrum of the "true" cellulose-g.co-styrene copolymer (ungrafted cellulose having been removed by extraction with Cadoxen) with a "true" graft level of 112.2%. The total side-band suppression spectrum (TOSS) is also given. The side-band suppression eliminates any stray peaks. The characteristic peaks due to the cellulose-carbon nuclei appear to be in good agreement to those seen already. The PS peaks appear at the positions expected from comparison with the solution state spectra. Thus, the aliphatic carbons are in the range 40 - 46 ppm; the C_β, C_τ and C_δ aromatic carbons are at 128 ppm and the C_α aromatic carbon is at 146 ppm.

Table 2 gives details of the integrated intensities for this copolymer. The C_1, C_4 and C_6 cellulose-carbon atoms appear separately for comparison with the free C_α aromatic carbon of the PS. The relevant data are presented in Table 3 and were used for calculation of grafting levels. The percentage grafting levels have been calculated by multiplying the C_α (PS)/$C_{1.4.6}$ (Cellulose) ratio by (100 x 104)/162.

Scrutiny of the results allows a number of observations to be made. In all cases, comparison of the intensities of C_α (PS) with C_6 (Cellulose) gives a better indication of the true graft level. Indeed, the result for the copolymer based on WP is quite remarkable in its closeness to the results obtained gravimetrically. The indications are that the PS portion

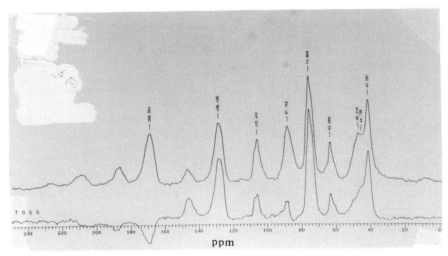

Figure 7. ^{13}C Solid State NMR "WP-g.co-S." Copolymer

of the copolymer is mobile since the cross polarisation spectrum receives greater contributions from the more rigid species. Since the C_6 cellulose carbon nucleus does not give as strong a contribution to the cross polarisation spectrum as the C_1 and C_4 nuclei, we have an indication that it may be more mobile. Such a feature seems reasonable as it is not restrained in the pyranose ring, but is pendant to it.

Figure 8 shows the cross polarisation magic angle spinning spectrum of the true WP-styrene copolymer and its spectrum when subjected to a non-quaternary suppression (NQS) experiment. The effect is that essentially only non-protonated carbons are seen in the spectrum. However, groups with rapid internal motion also show up. Thus, the experiment can be an aid to assignment.

The spectrum does not really provide any extra information other than confirming that the C_a PS signal at 146.5 ppm shows up strongly as expected, since it is unprotonated. Slight signals do occur for some of the cellulose carbon atoms. This indicates that grafting may have occurred with free radical sites formed by dehydrogenation.

CP, MAS, ^{13}C, SOLID STATE, NMR SPECTRA OF A CARBANILATED, W.P. CELLULOSE GRAFT COPOLYMER

The CP, MAS, ^{13}C, solid state, NMR spectra of the fully carbanilated copolymer of WP-styrene is given in Figure 9. The spectrum compares very favourably with those seen in the solution state. Assignments are made on the spectrum with reference to those in solution. The fine splittings of the solution spectra are not seen here. The peaks due to the cellulose-carbons tend to be a little broadened and slightly shifted relative to those of the CP, MAS spectra of the true copolymers. The peak due to the cellulose-C_4 carbon nucleus appears to have been lost, presumably being part of the broad peak centred at 72.9 ppm. The total sideband suppression spectrum was integrated and full details are given in Table 4. Relevant data and results of the calculations of the D.S. and percentage grafting levels are provided in Table 5. The D.S. was calculated by comparing the intensities of the free carbonyl and the C_a aromatic carbons of the CTC portion with those of the free C_6 and C_1 peaks of the pyranose ring. The percentage grafting was calculated by comparing the intensities of the C_a carbon of the PS at 146 ppm to those of the C_6 and C_1 carbons of the cellulose, followed by multiplication by $(104 \times 100)/162$.

Consideration of the data of Table 5 shows that a wide range of predictions for the D.S. is produced. This reflects the dispersion of cross polarisation rates for the different carbon atoms within the sample. However, it appears that the C_a (CTC)/C_6 (Cellulose) ratio gives the most

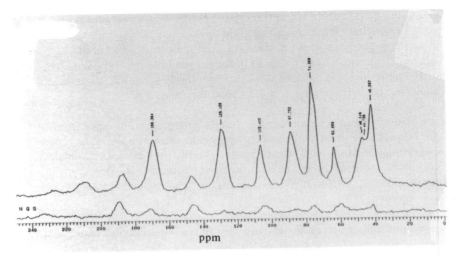

Figure 8. ^{13}C Solid State NMR (NQS) "Pure" WP-g.co-S Copolymer

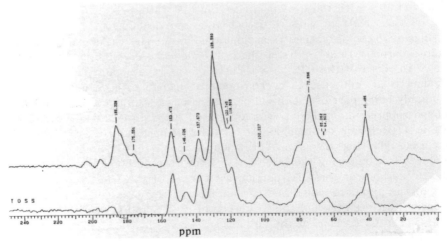

Figure 9. ^{13}C Solid State NMR : Carbanilated WP-g.co-S Copolymer

accurate estimation of the D.S. This predicts a value of 2.68 for the D.S. of the copolymer based on a little under the expected value of 3.0. Use of the data to calculate the true graft levels underestimates the amount of PS known to be present. Comparison of the C_a (PS) intensity to that of C_6 (Cellulose) gives the best results. This is also noted for the underivatised copolymers. The calculated value for the true graft level is somewhat further from the true graft level, obtained gravimetrically, than those calculated using the relative peak intensity of the underivatised copolymer.

NQS experiments were also performed for the carbanilated copolymer of WP-styrene. The spectrum is shown as Figure 9. No surprising features are seen. The unprotonated carbons of -C = O, C_a (PS) and C_a (CTC) appear at 153.5 ppm, 145.6 ppm and 137.7 ppm respectively, as expected. This helps to confirm their correct assignment. A small contribution to the NQS spectra is seen in the aromatic region at 118-120 ppm. This may reflect some degree of mobility of the pendant carbanilate groups. Other minor contributions are seen for the cellulose carbons which, again, may provide some evidence that grafting, involving dehydrogenation, occurs at these sites.

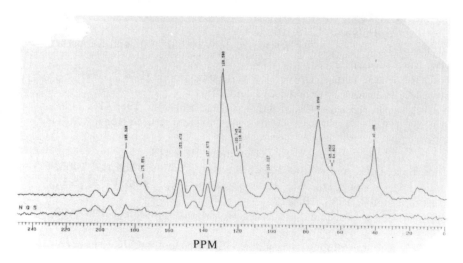

Figure 10. ^{13}C Solid State (NQS) NMR. CTCWP-g.co-Styrene Copolymer

CONCLUSIONS RELATING TO THE ^{13}C SOLID-STATE NMR STUDY

^{13}C solid-state NMR spectra of cellulose, cellulose-styrene graft copolymers and their fully carbanilated derivatives have been obtained. All the spectra show well resolved peaks and the results indicate that the method may be useful for the characterisation of such materials. The more important features seen are summarised below.

(i) Peak assignment is relatively straightforward and shows good agreement with the previously obtained solution state spectra.

(ii) Regions of varying mobility have been demonstrated by the measurement of CP spectra and proton relaxation times. This is to be expected for such heterogeneous systems.

(iii) Quantitative analysis of the spectra is difficult because of varying CP rates, due to regions of different mobility. However, comparison of the intensity of the C_α (PS) with that of the C_6 (cellulose) gives a reasonable indication of the grafting levels for copolymers based on cellulose I.

REFERENCES

1. Hebeish, A. and Guthrie, J.T., The Chemistry and Technology of Cellulosic Copolymers, Springer-Verlag, Heidelberg, **1981.**
2. Mishra, M.K., J.M.S.- Rev. Macromol. Chem. Phys. <u>C22</u> **1982/3,** 409 and 471.
3. Garnett, J.L., Jankiewicz, S.V., Long, M.A. and Sangster, D.F. J. Polymer Sci., **1986,** C24, 125.
4. J.T. Guthrie, Polymer **1975,** 16, 134.
5. Guthrie, J.T., Huglin, M.B., Richards, R.W., Shah, V.I. and Simpson, A.H., Euro. Polym. J. **1975,** 11, 527.
6. Hirst, R., Ph.D. Thesis, University of Leeds, Leeds (U.K.) **1988.**
7. Saito, M., Polym. J., **1983,** 15 213 and 249.
8. Wenzel, M., Burchard, W. and Schatzel, K., Polymer **1986,** 27, 195.
9. Guthrie, J.T., Ph.D. Thesis, University of Salford, Salford, (U.K.) **1971.**
10. Guthrie, J.T. and Percival, J.A., Polymer, **1977,** 18, 531.
11. Vanderhart, D.L., and Atalla, R.H., Macromolecules, **1984,** 17, 1465.
12. Cael, J.J., Kwoh, D.L.W., Bhattacharjee, S.S. and Patt, S.L., Macromolecules, **1985,** 18, 819.
13. Horii, F., Hirai, A. and Kitamaru, R., Macromolecules, **1987,** 20, 2117.
14. Sohar, P., Nuclear Magnetic Resonance Spectroscopy, C.R. Press, Vol. 2, Florida, 1983.

15. Johnson, L.F., Heatley, F. and Bovey, F.A., Macromolecules, **1970** 3, 175.
16. Randall, J.C., J. Polym. Sci., Polym. Phys. Ed., **1975,** 13, 889.
17. Randall, J.C., J. Polym. Sci., Polym. Phys. Ed., **1976,** 14, 2083.
18. Sato, H., Yanaka, Y. and Hatada. K., Makromol. Chem., Rapid Commun. **1982,** 3, 175.
19. Tonelli, A.E., Macromolecules, **1983,** 16, 604.
20. Grant, D.M. and Paul, E.G., J. Amer. Chem. Soc., **1964,** 86, 2984
21. Frisch, H.L., Mallows, C.L. and Bovey, F.A., J. Chem. Phys. **1966,** 45, 1565.

23

Characteristics of chemical celluloses from cold alkali extraction of *Pinus radiata* chemical pulps

Adrian F.A. Wallis and Ross H. Wearne - Division of Forest Products, CSIRO, Private Bag 10, Clayton, Victoria 3168, Australia.

ABSTRACT

Cold alkali extraction of bleached *Pinus radiata* bisulfite and kraft pulps gave chemical celluloses suitable for conversion to nitrocellulose. The bisulfite pulp was bleached by both a chlorine-extraction-hypochlorite (CEH) sequence and by a single stage peroxide treatment. Cold alkali extraction of the peroxide-bleached bisulfite pulp thus provides a method for preparing chemical cellulose without the use of chlorine-containing reagents. Extraction with increasing concentrations of alkali gave celluloses with increasing purities, but with increasing proportions of cellulose-II. The celluloses were characterised by their alkali solubilities, pentosan contents, cuene viscosities, molecular weight distributions, and crystallinities by x-ray powder diffractometry.

INTRODUCTION

As an alternative to imported cellulose for nitrocellulose production in Australia, we sought simple methods for converting Australian papermaking pulps to chemical cellulose. Previous workers have shown that chemical celluloses suitable for a range of uses could be made by cold alkali extraction of softwood paper-grade sulfite [1] and kraft [2] pulps, whereby the hemicelluloses were removed by dissolution. The method cannot be applied to hardwood pulps because a certain proportion of the glucuronoxylans are not removed by alkali [3]. We have demonstrated that cold alkali extraction of chlorine-bleached bisulfite [4] and kraft [5] pulps from *Pinus radiata* wood gave chemical celluloses with suitable properties for nitration. Furthermore, the nitrocelluloses prepared from these samples were found to have satisfactory characteristics [6].

A one-stage peroxide treatment at high consistency (steep bleaching) has been introduced to replace the chlorination-extraction-hypochlorite (CEH) sequence for the commercial bleaching of *P. radiata* bisulfite pulp in Australia. A charge of 4% hydrogen peroxide (based on pulp) increased the brightness of the pulp from 55% to 85% [7]. Unlike the brightening effect of alkaline peroxide on mechanical pulps in which the lignin is retained in the pulp, with chemical pulps the treatment has the effect of removing the residual lignin in the pulp [8]. It was thus of interest to ascertain whether chemical cellulose could be obtained from a cold alkali extraction of peroxide-bleached pulp, to provide a route to chemical cellulose which does not involve the use of chlorine-containing reagents.

Commercial chemical celluloses are characterized by their resistance to dissolution in 10% sodium hydroxide (%R_{10} > 90%), and low contents of hemicelluloses as reflected in low pentosan values, extractives (dichloromethane solubles), minerals, and residual lignin. Physical properties should be appropriate for different end uses; high brightness is required for samples to be converted to fibres, films or plastics, and viscosities, or molecular weight distributions, are important parameters for all end uses. In this paper, the preparation and properties of chemical celluloses prepared by cold alkali extraction of bleached *P. radiata* pulps is discussed.

EXPERIMENTAL

The nitration-grade cellulose was a commercial sample derived from sulfite pulping of a softwood. Chemical celluloses were prepared from chlorine-bleached *P. radiata* bisulfite and kraft pulps by cold alkali extraction [4,5].

For peroxide bleaching, a *P. radiata* bisulfite pulp (Kappa no. 31) was pretreated with 0.2% EDTA (pulp basis) at 2% consistency for 30 min at pH 7.5. The pulp was reacted with a solution containing 20 g/L NaOH, 10 g/L Na_2SiO_3, 1 g/L DTPA and 13.34 g/L H_2O_2 at 25% consistency. After 48 h at 45°C, the pulp was washed with purified water and air dried.

Cold alkali extraction of the pulps was effected by treatment of the pulps with varying concentrations of NaOH at 12% consistency for 1 h at 20°C.

Klason lignin contents of the pulps were determined by Australian Standard AS1301.11s-78. Kappa numbers of the pulps were acquired according to the Australian Standard AS1301.201m-86. Solubilities of cellulose samples in 18% and 10% NaOH solutions were obtained by ASTM standard method D1696-61. Pentosan contents of the pulps were determined by TAPPI classical method T223 cm-84. Dichloromethane solubles were by TAPPI official method T204 os-76. Cuene viscosities on 0.5% solutions of cellulose in 0.5M cupriethylenediamine hydroxide were obtained by TAPPI official method T230 om-82. Brightnesses of pulps were determined by a method based on ISO 3688-1977 [4].

The molecular weight distributions of the pulps were determined from their tricarbanilates by HPSEC in tetrahydrofuran with crosslinked polystyrene columns [9]. The tricarbanilates were prepared by reaction of the celluloses with phenylisocyanate in DMSO.

The contents of cellulose-I and cellulose-II polymorphs in the pulp samples were determined by x-ray powder diffractometry with the aid of a microcomputer according to Evans *et al.* [4]. A Siemens Kristalloflex 4 diffractometer with a type F horizontal goniometer and scintillation detector was used in the symmetrical transmission mode. Nickel-filtered copper $K\alpha$-radiation was employed at 30 kV and 20 mA.

RESULTS AND DISCUSSION

Laboratory steep bleaching of *P. radiata* bisulfite pulp (Kappa no. 31; brightness 57%) with 4% hydrogen peroxide gave a pulp with brightness 85%. The total lignin content of the pulp was reduced from 5.7% to 2.6% (compared to 1.7% for the CEH-bleached pulp). Other analytical characteristics of the peroxide-bleached pulp were similar to those of the CEH-bleached pulp (Table 1).

Table 1. Characteristics of *P. radiata* pulps and chemical celluloses

Sample	w/w% NaOH extraction	Alkali solubility			Pentosan %	Cuene viscosity mPa.s
		%R_{10}	%S_{18}	%S_{10}-S_{18}		
nitration-grade cellulose	--	90.3	6.2	3.4	2.3	13.8
CEH-bleached bisulfite pulp	-	86.3	11.8	1.9	6.2	17.1
	6.0	91.2	7.4	1.5	2.8	15.4
	8.0	94.3	5.3	0.5	1.9	15.4
	10.0	95.9	3.6	0.6	1.3	17.7
peroxide-bleached bisulfite pulp	-	87.8	10.9	1.3	6.0	22.1
	6.0	92.6	6.6	0.9	2.1	17.7
	8.0	94.2	4.5	1.3	1.8	16.9
	10.0	96.4	3.3	0.3	1.6	12.6
CEH-bleached kraft pulp	--	88.1	12.3	-0.4	3.7	16.5
	8.0	96.2	4.1	-0.3	1.9	14.6
	10.0	97.4	3.6	-1.0	2.2	15.8

The *P. radiata* pulps were extracted with 6-8 w/w% sodium hydroxide at 20°C for 1 h, conditions typical of cold alkali extractions. Extraction with increasing concentrations of alkali gave chemical celluloses with increasing purities, as reflected in their alkali solubilities (increasing %R_{10} and decreasing %S_{18} and %S_{10}-S_{18} values) and in their decreasing pentosan contents (Table 1). For alkali extraction of the CEH-bleached pulps, there was little change in cuene viscosities, although for the peroxide-bleached pulp, the viscosities were progressively lowered after extraction with increasing concentrations of alkali. A possible explanation for this is the scission of bonds between the cellulose and the residual lignin-carbohydrate complex in the pulp during the cold alkali extraction.

For the CEH-bleached bisulfite pulp, extraction with 8% alkali gave a cellulose sample with analytical characteristics which matched that of the reference nitration-grade cellulose. Extraction of the peroxide-bleached pulp with only 6% alkali gave a cellulose sample with similar analytical characteristics to the commercial cellulose. After cold alkali extraction, the *P. radiata* bisulfite and kraft pulps had lower dichloromethane solubles than the reference pulp (< 0.12%). The brightnesses of the pulps were not lowered by the alkali extractions, and after washing the pulps with dilute sulfurous acid, metal ion contents were lower than those of the reference cellulose. Cold alkali extraction of the peroxide-bleached bisulfite pulps provides a method for chemical cellulose preparation without the use of chlorine-containing reagents.

A *P. radiata* kraft pulp partially bleached with a CEH sequence to a brightness of 76% was also subjected to cold alkali extraction. With 8% alkali, a cellulose sample with high resistance to dissolution in alkali and low in pentosan was obtained, although its low brightness would be a disadvantage for many applications.

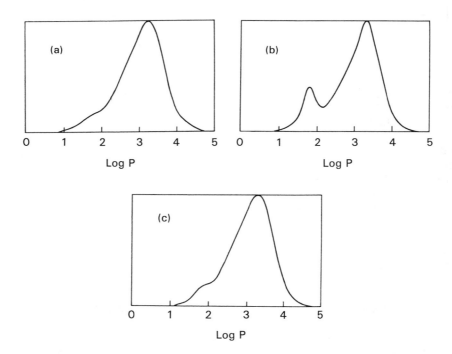

Figure 1. Molecular weight distributions of (a) nitration-grade cellulose; (b) CEH-bleached bisulfite pulp; (c) cold alkali extracted bisulfite pulp.

Although viscosities of the celluloses are good indicators of their molecular weights, they give only single average values. More detail is provided by the molecular weight distributions, which for cellulose can be very important. The MWDs were determined by high performance size exclusion chromatography of their tricarbanilates [9]. Figure 1 shows the MWDs of the nitration-grade cellulose, the CEH-bleached bisulfite pulp and the pulp after cold alkali extraction. The bleached pulp has a bimodal MWD, and the peak at lower molecular weight is attributed to the hemicelluloses. In the reference cellulose and the alkali-extracted bisulfite cellulose, the hemicellulose peaks are largely removed, and both samples have similar distributions.

During cold alkali extractions, partial mercerisation often occurs, involving a shift in the crystalline structure of cellulose from cellulose-I to cellulose-II. Because mercerisation can cause decreased reactivities of the cellulose, attempts have been made to reduce the shift during cold alkali extraction [10]. The proportions of crystalline and amorphous components in the cellulose samples were determined by x-ray powder diffractometry using multiple linear regression analysis of the samples and appropriate standards [4]. The diffractograms of the CEH-bleached bisulfite pulp and the pulp after extraction with 11% alkali are given in Figure 2. Analysis of a series of CEH-bleached bisulfite pulps shows that after extraction with increasing concentrations of alkali, increasing proportions of cellulose-II are found in the samples (Figure 3).

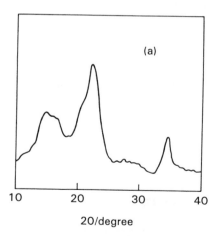
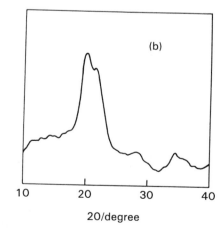

Figure 2. X-ray diffractograms of (a) CEH-bleached bisulfite pulp; (b) cold alkali extracted bisulfite pulp.

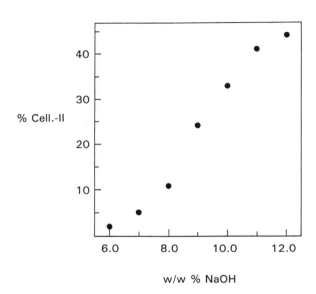

Figure 3. Content of cellulose-II in cellulose samples extracted with varying concentrations of sodium hydroxide.

ACKNOWLEDGEMENT

The authors acknowledge the valuable contributions of Dr R. Evans in the MWD and x-ray diffractometry experiments.

REFERENCES

1.   Wells, F.L.; Schattner, W.C.; Ekwell, L.E. *Tappi* **1971**, *54*(4), 525.
2.   Stockman, L. *Sven. Papperstidn.* **1953**, *56*(14), 523.
3.   Meller, A. *Tappi* **1950**, *33*(5), 248; *ibid.* **1952**, *35*(4), 178.
4.   Evans, R.; Wearne, R.H.; Wallis A.F.A. *Appita J.* **1990**, *43*(2), 130.
5.   Wallis, A.F.A.; Wearne, R.H. *Appita J.* **1990**, *43*(5), 355.
6.   Sioumis, A.A.; Wallis, A.F.A.; Puri, V.P. *Polym. Intern.* **1991**, *25*(4), 203.
7.   Trafford, J. *Appita J.* **1990**, *43*(5), 358.
8.   Lachenal, D.; de Choudens, C.; Monzie, P. *Tappi* **1980**, *63*(4), 119.
9.   Evans, R.; Wearne, R.H.; Wallis, A.F.A. *J. Appl. Polym. Sci.* **1989**, 37(12), 3291.
10.   Wallis, A.F.A.; Wearne, R.H. *Tappi J.* **1990**, *73*(6), 226.

24

The heterogeneous, dilute-acid hydrolysis of cellulose

Chyong-Huey Lin*, Anthony H. Conner** and Charles G. Hill, Jr.* -
*Department of Chemical Engineering, University of Wisconsin-Madison, Madison, Wisconsin 53706, USA. **USDA-Forest Service, Forest Products Laboratory, Madison, Wisconsin 53705-2398, USA.

ABSTRACT

Saccharification of cellulose is of interest to researchers concerned with the development of alternative sources of liquid fuels and chemicals from cellulose. The results reported in this paper focus on the kinetics of the dilute-acid hydrolysis process. Thirteen pre-hydrolyzed samples of cellulose, including native, mercerized, and regenerated materials were hydrolyzed in 1%, 1.5%, and 2% sulfuric acid at 150, 160, 170, and 180°C. Pseudo first-order rate constants and weight average degrees of polymerization were determined for each sample. For all cellulose samples, data from several experiments were used to determine the dependence of the rate of hydrolysis on sulfuric acid concentration. The results obtained in this study indicate that Sharples end-attack model is consistent with kinetic data for the hydrolysis of cellulose II samples, but is not applicable to the hydrolysis of cellulose I samples.

INTRODUCTION

Cellulose is not soluble in dilute-acid; therefore, the dilute-acid hydrolysis of cellulose is a heterogeneous reaction. Many researchers have studied this reaction under a variety of conditions[1-12]. For both cellulose I and cellulose II, the dilute-acid hydrolysis can be modeled as a pseudo-homogeneous, first-order reaction. In addition, celluloses with the same crystalline morphology, but obtained from different sources, often hydrolyze at widely different rates. In order to be able to model the production of glucose from different sources of cellulose without the necessity for independently studying the hydrolysis of each substrate, it would be desirable to have a mathematical model of the dilute-acid hydrolysis which could be used to explain the observed first-order kinetics and provide a framework for correlating kinetic data for a variety of cellulosic substrates.

Sharples[7,8] proposed a model for the dilute-acid hydrolysis of cellulose which explains both the first-order kinetics and the fact that different rates of hydrolysis are observed for celluloses from different sources. In previous papers[1,2], we reported kinetic data (rate constants and activation energies) for the dilute-acid hydrolysis of a variety of cellulosics and examined the applicability of Sharples model. We reported that Sharples end-attack model is consistent with kinetic data for cellulose II samples, but is not appropriate for characterizing the reactions of cellulose I samples. The present paper extends our previous study of the dilute-acid hydrolysis of cellulose to a wider range of reaction conditions. In particular, the present paper focuses on the effects of reaction temperature and acid concentration.

EXPERIMENTAL

Prehydrolyzed cellulose samples were subjected to batch hydrolysis in 1%, 1.5%, and 2% sulfuric acid for varying lengths of time at 150, 160, 170, and 180°C. For each of the hydrolyzed samples, both the total weight loss and the cellulose weight loss (based on sugar analysis[13]) were determined. The cellulosic materials used for the hydrolysis experiments and the experimental methods for mercerization, prehydrolysis, batch hydrolysis and size-exclusion-chromatography are the same as those employed previously[2]. In addition, four other cellulose samples (2 cotton linters, microcrystalline cellulose, and spruce pulp) were also hydrolyzed in 1% sulfuric acid. Table I indicates the various samples employed and their characteristics.

Table I – Characteristics of prehydrolyzed cellulose samples.[a]

Cellulose Sample	DP_w	2/b
CELLULOSE I		
Cotton	171.3±12	170.4
Ramie	188.3± 7	187.5
Linen	164.1± 8	163.2
α–cellulose	98.4± 2	97.0
Cotton linter 1	124.7± 2	123.6
Cotton linter 2	164.3± 9	163.2
Microcrystalline		
cellulose	118.0± 1	116.8
Spruce pulp	81.0± 4	79.4
CELLULOSE II		
Mercerized cotton	75.3± 5	73.6
Mercerized ramie	99.8± 5	98.4
Mercerized linen	101.7± 3	100.3
Mercerized		
α–cellulose	44.8± 1	42.2
Rayon	16.7± 1	12.0

[a]DP_w = weight average degree of polymerization; b = distribution constant (see equation (3)).

RESULTS AND DISCUSSION

Kinetic Studies of the Acid Hydrolysis of Cellulose. First-order rate constants for each of the prehydrolyzed-cellulose samples were determined from linear regression analyses of the dilute-acid hydrolysis data. The dependence of the rate of hydrolysis on acid concentration and temperature can be expressed as

$$k = k_0[H^+]^n exp(-E/RT) \tag{1}$$

where

k is the pseudo first–order rate constant (min^{-1}).
k_0 is the Arrhenius preexponential factor (min^{-1}).
n is the order of the reaction with respect to $[H^+]$.
$[H^+]$ is the hydrogen ion concentration (%).
E is the activation energy for the reaction (cal/mol).
R is the gas constant (1.987 cal/mol K).
T is the absolute temperature (K).

The acidity of the hydrolysis reaction, $[H^+]$, is dependent on many factors – the concentration of acid solution employed, the neutralizing capacity of the substrate (ash content), and the extent of the secondary ionization of sulfuric acid. Since less than 1% of the bisulfate ion dissociates under the conditions usually employed for hydrolysis of cellulose[14], one may treat this system as if only the first ionization of H_2SO_4 occurs. The neutralizing capacity differs from substrate to substrate. The ash contents of the prehydrolyzed-cellulose samples used in this study (determined by ASTM Method No. D 1102[15]) are negligible (ranging from 0 to 0.03%).

The dilute-acid hydrolysis data were analyzed using the regression procedure in the SAS statistical package[16]. We treated the data assuming that the values of both n and the activation energy were the same for all samples. $[H^+]$ was taken as the percent sulfuric acid concentration. The value of n obtained in this manner is 1.172 ± 0.043. The corresponding value of the activation energy is 40.6 ± 0.5 Kcal/mol. Saeman[3] reported that the value of n for cellulose from Douglas-fir was 1.34 when $[H^+]$ was taken as the percent acid concentration. The corresponding value reported by Bhandari et al.[11] for cellulose contained in corn stover was 2.74. The value of n reported by Conner et al.[17] for Douglas-fir lignocellulose was 1.218 when $[H^+]$ was taken as the molal concentration of acid. The Arrhenius preexponential factor k_0 was also determined for each sample as part of the regression analysis.

By inserting the values of n and activation energy obtained from statistical treatment of the hydrolysis data, equation (1) can be rewritten as

$$k = k_0[H^+]^{1.172} exp(-40600/RT) \tag{2}$$

Using Equation (2) and the value of k_0 determined for a given sample of cellulose, one can calculate the reaction rate constant, k, for that sample of cellulose as a function of temperature and acid concentration. Figure 1 is a plot of the calculated rate constants versus the experimental rate constants. Examination of this plot indicates that the mathematical model developed from the statistical analysis can be used to predict the rate constant for the dilute sulfuric acid hydrolysis of cellulose.

Size-Exclusion-Chromatography Analysis. The SEC analysis used in this study is basically the same as that employed in the previous investigation. The universal calibration technique employing polystyrene standards was used[18]. The Mark-Houwink coefficients used in this work were K=0.0112 and α=0.72 for polystyrene in tetrahydrofuran (THF)[19] and K=0.0053 and α=0.84 for cellulose tricarbanilate (CTC) in THF[20].

The basic hypothesis on which the end-attack model of Sharples[7,8] is based is that differences in the rates of hydrolysis of cellulosic materials result from differences in the degrees of polymerization of these materials. In terms of this model, the relationship which should exist between the rate constant k and weight average degree of polymerization DP_w is

$$\frac{1}{k} = (\frac{\rho}{4B})(\frac{2}{b}) = (\frac{\rho}{4B})\left[DP_w - \frac{2S^2}{(DP_w + 2S)}\right] \tag{3}$$

where

ρ is the density of cellulose.
B is an intrinsic rate constant.
b is the distribution constant.
S is the solubility limit.

The average degree of polymerization (DP_w) from the SEC analysis and values of 2/b calculated from eq. (3) for cellulose samples hydrolyzed in dilute sulfuric acid are listed in Table I. The solubility limit S was taken to be 9 in accordance with Sharples suggestion[7].

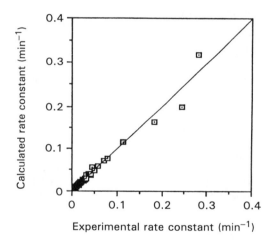

Figure 1. Comparison of experimental and calculated rate constants.

Plots of 1/k versus 2/b are shown in Figures 2 and 3 for cellulose I and cellulose II samples hydrolyzed in 1% H_2SO_4, respectively. Similar plots were obtained for cellulose samples hydrolyzed in 1.5% and 2% H_2SO_4. Only in the case of the cellulose II samples do the plots go through the origin as required by Sharples model. Since cellulose I and cellulose II polymorphs have different chain configurations and hydrogen bonding patterns, it is not unexpected that the data from the cellulose I samples and cellulose II samples show different linear relationships. However, for Sharples model to apply, the straight lines have to pass through the origin. None of the plots for the cellulose I samples seems to go through the origin. These data indicate that Sharples model is consistent with data for the dilute-acid hydrolysis of the cellulose II, but not with data for the hydrolysis of the cellulose I samples. This observation is consistent with data we reported in previous papers[1,2].

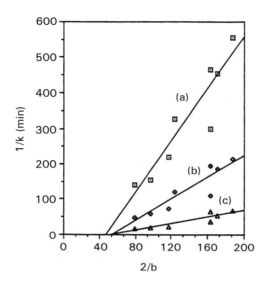

Figure 2. Plots of 1/k versus 2/b for cellulose I samples hydrolyzed in 1% H_2SO_4 at (a) 160, (b) 170, and (c) 180 ° C.

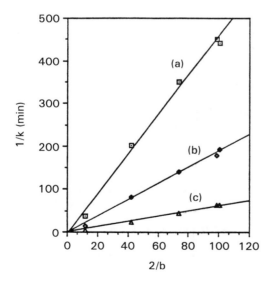

Figure 3. Plots of 1/k versus 2/b for cellulose II samples hydrolyzed in 1% H_2SO_4 at (a) 160, (b) 170, and (c) 180 ° C.

CONCLUSIONS

The conclusion drawn in previous studies that the hydrolysis behavior of cellulose II samples is consistent with Sharples end-attack model is confirmed. Differences in the rates of hydrolysis of cellulose II samples result from differences in the degrees of polymerization of these samples. Sharples model is not appropriate for use with cellulose I samples. In these samples, the degree of polymerization is not the only factor which influences the rates of hydrolysis of celluloses from different sources. Statistical treatment of the kinetic data provides a general mathematic expression for the pseudo first-order rate constants as a function of acid concentration and temperature. This mathematical model can be used to predict the rate constants of the samples used in this study under various reaction conditions.

ACKNOWLEDGMENTS

Financial support for this research was provided by the U.S.D.A. Forest Products Laboratory and the U.S.D.A. Forest Service, Competitive Grant No. 85-08120. The authors would like to thank Martin Wesoloski and Virgil Schwandt for performing the ash content and sugar analyses. The authors also want to thank Steve Verrill for his help in statistical analysis of the data. Helpful discussions with Dr. R.H. Atalla are also appreciated.

REFERENCES

1. B.F. Wood, A.H. Conner, and C.G. Hill, Jr., *J. Appl. Polym. Sci.*, 37, 1373 (1989).
2. C.-H. Lin, A.H. Conner, and C.G. Hill, Jr., *J. Appl. Polym. Sci.*, 42, 417 (1991).
3. J.F. Saeman, *Ind. Eng. Chem.*, 37(1), 43(1945).
4. H.P. Philipp, M.L. Nelson, and H.M. Ziifle, Text. Res. J., 17(11), 585 (1947).
5. D.H. Foster and A.B. Wardrop, Austr. J. Sci. Res., A4, 412 (1951).
6. M.A. Millett, W.E. Moore, and J.F. Saeman, *Ind. Eng. Chem.*, 46(7), 1493 (1954).
7. A. Sharples, Trans.Farada Soc., 53 1003 (1957).
8. A. Sharples, Trans.Faraday Soc., 54, 913 (1958).
9. M.L. Nelson, *J. Polym. Sci.*, 18, 351 (1960).
10. E.H. Daruwalla and M.G. Narsian, Tappi, 49(3), 106 (1966).
11. N. Bhandari, D.G. Macdonold, and N.N. Bakhshi, Biotechnol. Bioeng., 26, 320 (1984).
12. M. Marx-Figini, *J. Appl. Polym. Sci.*, Appl. Polym. Symp., 37, 157 (1983).
13. R.C. Pettersen, V.H. Schwandt, and M.J. Effland, *J. Chromatogr. Sci.*, 22, 478 (1984).
14. M.H. Lietzke, R.W. Stoughton and T.F. Young, *J. Phys. Chem.*, 65, 2245 (1961).
15. American Society for Testing and Materials, Standard Test Method for Ash in Wood, ASTM D 1102-84.
16. SAS Institute Inc., Box 8000, Cary, NC 27511-8000, USA.
17. A.H. Conner, B.F. Wood, C.G. Hill, Jr., and J.F. Harris, *J. Wood Chem. Technol.*, 5(4), 461 (1985).
18. Z. Grubisic, P. Rempp, and H. Benoit, *J. Polym. Sci.*, Polm. Lett. Ed., 5, 753 (1967).
19. M. Kolinsky and J. Janca, *J. Polym. Sci.*, Polym. Chem. Ed., 12, 1181 (1974).
20. J. Danhelka and I. Kossler, *J. Polym. Sci.*, Polym. Chem. Ed., 14, 287 (1976).

25

The application of thermochemical approach in investigation of cellulose interactions

E.A. Antifeyev, V.E. Gusev, G.M. Poltoratsky, D.A. Sukhov and N.N. Volkova* - Leningrad Technological Institute of Pulp and Paper Industry, Sankt-Peterburg, Ivana Chernikh, 4, 198092, Russia.

ABSTRACT

The thermokinetic approach was used to investigate the interactions of different types of cellulose with FeTNa.The integral ΔH of interaction includes the heats of moisture,swelling,decomposition of structure,solvation and complex formation.On the basis of the form of $\Delta H = f(time)$ curves the process was divided into several stages.The correlation between specific surface area and values of ΔH on the first surface diffusion stage was found out.The values of ΔH of moisture of different phase-modification cellulose showed the dependence of ΔH value from total content of disordered parts of C1 and C2.By the comparision of IR-spectra and thermochemical data for some diols in dilute water and CCl_4 solutions correlation between heats of interaction and energies of intramolecular hydrogen bonds was observed.The stage of complex formation was studied by sorption of some ions on the surface of powder cellulose followed by dissolving in FeTNa.On the basis of this complex investigation the mechanism of interaction of cellulose with FeTNa was proposed.

INTRODUCTION

Chemical as well as physical properties of cellulose can only be understood

*) Present address:Thermochemistry,Chemical Center,PO Box 124,S-22100 Lund, Sweden

by the combined knowledge of chemical nature of cellulose molecules and their structural and morphological arrangment/ 1 /.One of the approaches is to investigate the solution properties of cellulose and compare them with data from other methods.In this case it is very important to have a good solvent for different types of cellulose and it was this reason why we chose the iron tartrate complex in sodium hydroxide(FeTNa)/ 2 /.The main idea of this investigation was to try to find out the mechanism of interaction of cellulose with this solvent by the use of a thermokinetical approach.

RESULTS AND DISCUSSION

The enthalpies of interaction (ΔH) were determined by use of the adiabatic calorimeter / 3 /, the volume of the solvent in the calorimetric cell was 30ml, the mass of cellulose samples-100-200mg.Calibrations were performed by dissolution of KCl in water.All values are accurate within 2-4%.The thermokinetic values of ΔH of two types of cellulose(cotton,sulphate cord) with solution of FeTNa(c=0.3mol/l)are shown in figure 1.

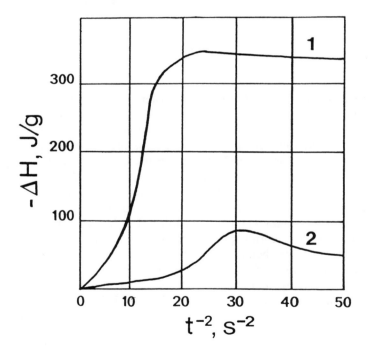

Figure 1 The ΔH of interaction as a function of time(cellulose+FeTNa)
    1-sulphate cord
    2-cotton

According to the model / 4 / on the basis of $\Delta H = f(t-2)$ curves the process can be divided into three stages:an external diffusion stage,an internal diffusion stage and the stage of chemical reaction.In the first stage the ΔH values are correlated with the specific surface areas(sulphate cord-12.8m^2/g, cotton-4.8m^2/g)of the cellulose. Besides that the heat effect on this stage includes the heats of breaking hydrogen bonds in the fibers structure and formation of new hydrogen bonds with solvent molecules.By the comparison of IR-spectra and thermochemical data for propan-diols in dilute water and CCl_4 solutions the correlation between heats of interaction and energies of intramolecular hydrogen bonds was observed, the data was in a good agreement with previous results / 5 /.

It is clear that the structural nature of cellulose plays a main role in the thermodynamical behavior of the system, and this is very important to understand the process of the internal diffusion stage(corresponding to the moistering and swelling processes).

It is known that cellulose after the treatment by NaOH solutions has a complicated phase structure, which is characterized by the existence of two modifications parts(C1-native cellulose and C2-mercerized cellulose).Our spectral investigations(Sukhov D.A.) also revealed some structural differences between disordered parts of C1 and C2.On the basis of this results the structure of cellulose can be presented as a four component system(figure 2):

C=C1or+C2or+C1dor+C2dor

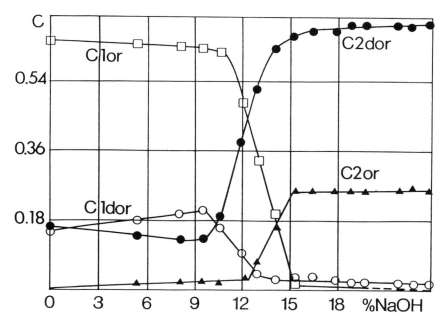

Figure 2 The changes of concentrations of cellulose components in the process of cotton mercerization(**or**-ordered parts of cellulose,**dor**-disordered parts of cellulose)

In order to make clear which structural units exert the main influence on thermodynamical behavior of cellulose in liquids we measured the ΔH of moistering in water of some celluloses(natural and hydrolyzed cotton,wood sulphite cellulose).The hydrolysis of natural cotton was made using $0.5mol/lH_2SO_4$ solution under $87°C$ during 3 hours.Then all cellulose samples were mercerized by NaOH solutions(with concentration 2-20 weight per cent) to change step by step their structure-modification composition.After mercerization all samples was washed with water and dried up to constant weight under $105°C$ and vacuum(10-3mm Hg).The results are given in figure 3.

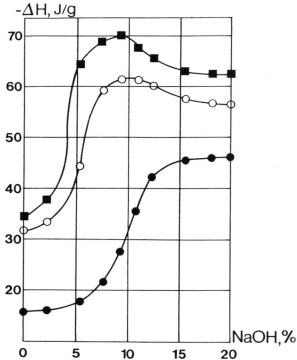

Figure 3 The ΔH for the process of moistering in water
● -natural cotton
○ -hydrolized cotton
■ -wood sulphite cellulose

One can see that the main effect on ΔH values is given by the change of composition of disordered parts C1 and C2.Some authors/6,7/ confirm the existence of a straight-line dependence between ΔH of moistering and total content of crystal parts of C1 and C2,on the basis of our obtained data we can't give the same conclusion.

The stage of chemical reaction(complex formation)was studied by the sorption of Fe-ions in different forms on the surface of powder cellulose followed by dissolution in FeTNa.To keep the cellulose structure unchanged the sorption was carried out at pH less than 4.The samples consisted 0.5-1.2mmol Fe/g cellulose in the form of thiocyanate,sulphosalicylate and tartrate.The results are presented in figure 4.

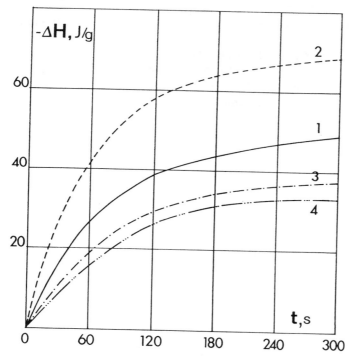

Figure 4 The ΔH of interaction of cellulose with FeTNa after
treatment with some Fe-complexes
1-initial powder cellulose
2-tartrate
3- sulphosalicylate
4-thiocyanate

The analysis of results showed that the previous sorption will change the values of ΔH of dissolving in FeTNa.In comparison with ΔH for initial powder cellulose only Fe-tartrate highly increased the heat of dissolving.These data confirm the assumption about the ligand exchange mechanism of dissolving in FeTNa on the last stage.

On the basis of this complex research we conclude that the thermokinetic method is really useful in cellulose studies and can gives additional information in comparison with a pure thermodynamic approach.

REFERENCES

1.Karachevtsev V.G., Kozlov N.A. Vysikomol.soyed.,1974,A16:8,1892-1897
2.Hudson S.M., Cuculo J.A. J.Macromol.Sci.-Rev.Macromol.Chem., 1980, C18(1),22-25
3.Ivanov A.V.Cand.dis.,Technol.Inst.of Pulp and Paper Ind.,Leningrad, 1984
4.Kalinin N.N. Zh.Prikl.Khim.,1984,57(11),2538-2542
5.Busfield W.K., Ennis M.P., McEwen I.J. Spectrochim.Acta,1973,1259-1264
6.Zenkov I.D. Khim.volokna,1989,2,58-59
7.Tsvetkov V.G., Ioelovich M.Ya.,Kaimin I.F.,ReizinshR.E. Khim.dreves., 1980, 5, 12-15

26

The synthesis, properties and polymerization of levoglucosan oligomers cured under the action of UV and radioactive irradiation

R.Ja. Pernikis and I.N. Zaks - Institute of Wood Chemistry, Latvian Academy of Sciences, 226006, Riga, Latvia.

ABSTRACT

Reactive oligomers based on levoglucosan – a product obtained during cellulose-containing materials processing have been synthesized and characterized. The process which takes place during the interaction between levoglucosan, bis-chloroformates and 2-hydroxyethylmethacrylate as well as between levoglucosan and 2-hydroxyethylmethacrylate chloroformate have been studied by size exclusion chromatography. The properties of the polymers have been determined. It has been established that the synthesized products are of interest as materials for optical purposes.

INTRODUCTION

The processing of plant raw materials is characterized by obtaining certain valuable compounds for the chemical industry. These products can be used as monomers in the process of polymer materials manufacturing.

A family of acrylates, carbonate acrylates, epoxides, urethane epoxides on the basis of levoglucosan was synthesized. Carbonate acrylates are of particular interest. This paper deals with carbonate acrylate oligomer synthesis, the influence of the synthesis conditions on their content in the final product and the properties of polymers. Liquid chromatography is most suitable for this purpose [1].

EXPERIMENTAL

The procedure of the synthesis is described in detail in [2]. A Waters ALC-200 liquid chromatograph and a LC column filled with Ultrastyrogel 100, 500 and 1000 Å were used. The flow rate was 1 ml/min. The samples were prepared by dissolving carbonate methacrylates in tetrahydrofuran. Some of the oligomers were identified simply by elution time of specially prepared substances.

RESULTS AND DISCUSSION

Carbonate acrylate oligomers from the formulae I and

$$R \ [OCO \ R'OCO(CH_2)_2OC-C=CH_2]_3 \qquad (I)$$
$$O O O \ X$$

where

$$R - $$

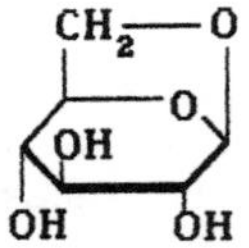

$$X - H, \ CH_3;$$
$$R' - (CH_2)_2 \ -(LgE); \quad (CH_2)_2O(CH_2)_2 - (LgD);$$

$$(CH_2)_2O(CH_2)_2O(CH_2)_2O(CH_2)_2 - (LgT)$$

and

$$R \ [OCO(CH_2)_2OC-C=CH_2]_3 \quad (LgOCM) \qquad (II)$$
$$O \ CH_3$$

II were synthesized and characterized by chemical analysis and liquid chromatography. LgE, LgD, LgT were synthesized by interaction between levoglucosan (Lg), bis-chloroformates of glycols (BHF) and 2-hydroxyethylmethacrylate (HEMA) as is shown in the scheme I:

$$R(OH)_3 + 3 \ Cl\underset{O}{\underset{\|}{C}}O(CH_2)_2O\underset{O}{\underset{\|}{C}}Cl + 3 \ HO(CH_2)_2O\underset{O}{\underset{\|}{C}}-\underset{CH_3}{\underset{|}{C}}=CH_2 \quad ----> $$

$$R \ [O\underset{O}{\underset{\|}{C}}O(CH_2)_2O\underset{O}{\underset{\|}{C}}O(CH_2)_2O\underset{O}{\underset{\|}{C}}-\underset{CH_3}{\underset{|}{C}}=CH_2]_3$$

Scheme I

Yet, the process of condensation telomerization is complicated due to the participation of three components of different functionality. It proceeds in several stages and results in the formation of a large number of products of different molecular weight and oligomer composition (Fig.1, curve I). The latter reduces the optical properties of polymers and their reproducibility. [4].

As has been shown [5], in the presence of pyridine the interaction is carried out including the stage of ion adduct (III) appearance

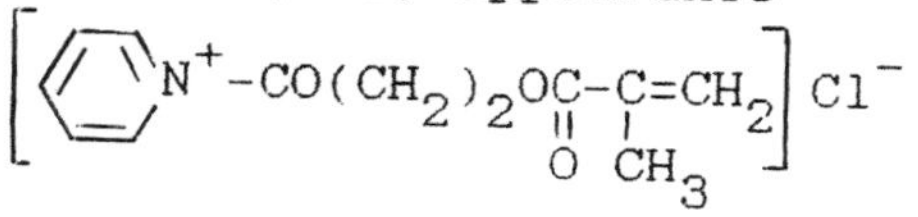

$$\left[\langle \hspace{-4pt} \bigcirc \hspace{-4pt} \rangle N^+ -CO(CH_2)_2O\underset{O}{\underset{\|}{C}}-\underset{CH_3}{\underset{|}{C}}=CH_2 \right] Cl^- \qquad (III)$$

which is consumed in the end reaction with HEMA, as well as in the supplementary reaction with H_2O present in the initial compounds.

During the interaction of 1 mole of BHF with 1 mole of HEMA, the following compound (IV) is formed:

$$Cl\underset{O}{\underset{\|}{C}}O(CH_2)_2O(CH_2)_2O\underset{O}{\underset{\|}{C}}O(CH_2)_2O\underset{O}{\underset{\|}{C}}-\underset{CH_3}{\underset{|}{C}}=CH_2 \qquad (IV)$$

which appears in the chromatogram as Peak 1. Its content in the synthesis products makes up 1-9%. The content of the product of interaction of BHF with 2

$$CH_2=\underset{H_3C}{\underset{|}{C}}-\underset{O}{\underset{\|}{C}}O(CH_2)_2O\underset{O}{\underset{\|}{C}}O(CH_2)_2O(CH_2)_2O\underset{O}{\underset{\|}{C}}O(CH_2)_2O\underset{O}{\underset{\|}{C}}-\underset{CH_3}{\underset{|}{C}}=CH_2 \qquad (V)$$

moles of HEMA (V) is the highest in the oligomer (60%). This product appears in the chromatogram as Peak 2. Diethylene glycol (VI) - the end product of

$$Cl\underset{O}{\underset{\|}{C}}O(CH_2)_2O(CH_2)_2O\underset{O}{\underset{\|}{C}}Cl \overset{2H_2O}{-------->} HO(CH_2)_2O(CH_2)_2OH \qquad (VI)$$

BHF hydrolysis reacts with 2 mole of BHF.

$$HO(CH_2)_2O(CH_2)_2OH\ +\ 2ClCO(CH_2)_2O(CH_2)_2OCCl\ \xrightarrow[-2HCl]{-------->}$$

$$ClCO(CH_2)_2O(CH_2)_2OCO(CH_2)_2O(CH_2)_2OCO(CH_2)_2O(CH_2)_2OCCl$$

Next in the reaction with HEMA, the product (VII) is

$$CH_2=C-C(CH_2)_2OCO(CH_2)_2O(CH_2)_2OCO(CH_2)_2O(CH_2)_2-$$
$$\quad\ CH_3$$
$$-OCO(CH_2)_2O(CH_2)_2OCO(CH_2)_2C-C=CH_2 \qquad\qquad (VII)$$
$$\qquad\qquad\qquad\qquad\qquad\qquad\quad O\ CH_3$$

obtained appearing in the chromatogram as Peak 3.

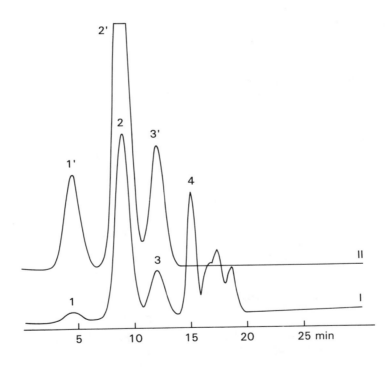

Figure 1. Chromatograms of levoglucosan carbonate methacrylates: I-product, obtained by reaction between Lg, BHF and HEMA, II-product, obtained by reaction between Lg and HEMA/CF.

Further the peaks corresponding to the high molecular compounds can be observed in the chromatogram. Only Peak 4 corresponds to the main product, its content not exceeding 20%.

The reaction between levoglucosan and 2-hydroxyethyl methacrylate chloroformate (BHF/CF) carried out as shown in scheme II:

$$R(OH)_3 + 3ClCO(CH_2)_2OC-C=CH_2 \xrightarrow[-HCl]{} R[OCO(CH_2)_2OC-C=CH_2]_3$$

(with O below the first carbonyl, O and CH$_3$ below the methacrylate; product with O below OCO, O and CH$_3$ below methacrylate)

Scheme II

gave a product more homogeneous in composition as compared with the products obtained by the previous method (Fig.1, curve II).

As in the case of the interaction of Lg with BHF and HEMA, an ionic adduct (III) is formed which is hydrolyzed to form HEMA

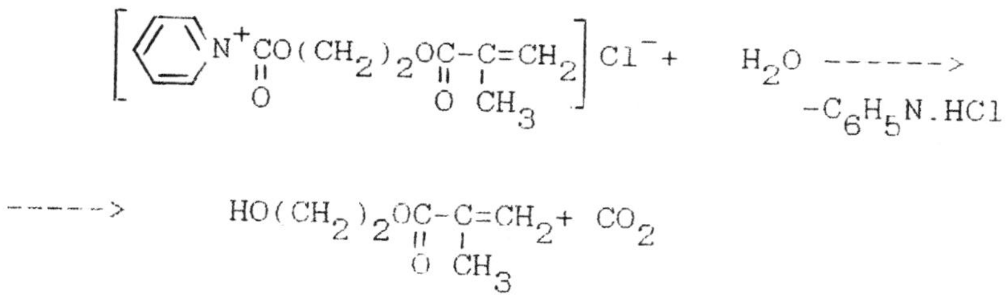

$$\left[\langle\bigcirc\rangle N^+CO(CH_2)_2OC-C=CH_2\right]Cl^- + H_2O \xrightarrow{} $$

$$-C_6H_5N.HCl$$

$$\xrightarrow{} HO(CH_2)_2OC-C=CH_2 + CO_2$$

(with O and CH$_3$ below)

The way HEMA reacts with ion adduct is easier than the way Lg does with HEMA/CF that results in the formation of the product (VIII), which appears in

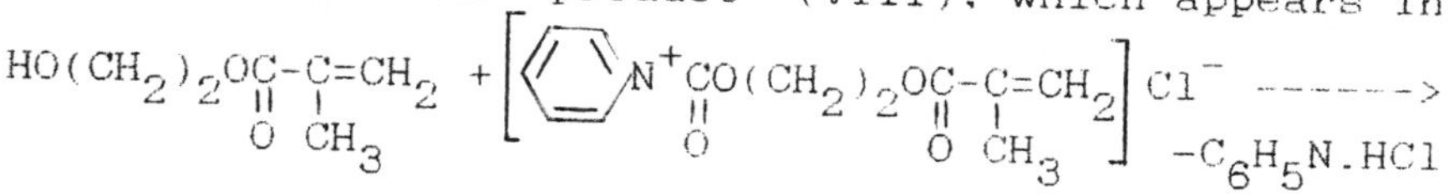

$$HO(CH_2)_2OC-C=CH_2 + \left[\langle\bigcirc\rangle N^+CO(CH_2)_2OC-C=CH_2\right]Cl^- \xrightarrow{} $$

$$-C_6H_5N.HCl$$

$$CH_2=C-CO(CH_2)_2OCO(CH_2)_2OC-C=CH_2 \qquad (VIII)$$

(with H$_3$C, O below the first; O below OCO; O and CH$_3$ below)

chromatograms as Peak 1´.Peak 2´and Peak 3´correspond to disubstituted and trisubstituted derivatives of Lg. Depending on the reaction conditions, the content of these products is within 25-47% for Lg with two substituted hydroxyl groups and 38-62% for Lg with three substituted hydroxyl groups. Thus the carbonate methacrylate oligomer with the main substance content of 62% has been synthesized, which can be explained by the presence of steric factors. The properties of polymers can be changed over a wide range by varying the oligomer block length and structure. The properties of oligomers and those of polymers produced are shown in Table 1.

Table 1

Properties of carbonate(meth)acrylates

Characteristics	LgE	LgD	LgT	LgOCM
Refractive Index	1.472	1.478	1.473	1.477
Photocuring Time, s (inert atm.,2% init.)	20	25	30	10
Volume Shrinkage, %	6.2	4.5	4.2	7.0
Tensile Strength, MPA	76	70	18	85
Elongat. at Break, %	1.5	2.5	10.0	0.5
Glass Trans. Temp., $^\circ$C	100	60	20	125
Temp. of Decomp., $^\circ$C	260	215	195	270
Birefringence, nm/mm	10	10	10	6.2

The results tend to indicate that LgOCM polymer is characterized by high mechanical properties, low volume shrinkage and adequate optical properties. This can be attributed to the effects of carbonate methacrylate structure and made it possible to improve greatly the properties of polymers.

Novel multifunctional carbonate methacrylate oligomers on the basis of raw materials of plant origin were synthesized and characterized. Particular properties of their polymers (low birefringence, volume shrinkage, tensile strength) provide their application in manufacturing details for optical purposes.

REFERENCES

1. Jen-Tay Gu and Shu-Lang Huang *J. Appl. Polym. Sci.* **1990**, *40*, 555.
2. Zaks, I.; Penczek, P.; Rot, A.; Pernikis, R. *Die Angewandte Makromol.Chem.* in press.
3. Grishchenko, V.K.; Masluk, A.P.; Gudzera, S.S. *Liquid photopolymerizable compositions*, Naukova dumka, Kiev, **1985**, 208 (in Russian).
4. Kloosterboer, I.G.; Lijten, G.F. and Boots, H.M. *Makromol.Chem. , Macromol. Symp.* **1989**, *24*, 223.
5. Sivergin, Ju.M.; Pernikis,R.Ja.; Kireeva, S.M. *Polycarbonate methacrylates*, Zinatne, Riga, **1988**, 213 (in Russian).

27

Chemical synthesis of a comb-shaped polysaccharide

H. Ichikawa and K. Kobayashi* - Research Center, Daicel Chemical Industries Ltd., Shinzaike, Aboshi-ku, Himeji, 671-12, Japan. *Faculty of Agriculture, Nagoya University, Furo-cho, Chikusa-ku, Nagoya, 464-01, Japan.

ABSTRACT

A new comb-shaped, branched polysaccharide has been obtained via ring-opening polymerization of a 1,6-anhydro disaccharide derivative. 1,6-Anhydro-3-O-(2,3,4,6-tetra-O-benzyl-β-D-galactopyranosyl)-2,4-dideoxy-β-D-threo-hexopyranose (**Gal-Thr**) was synthesized from cellulose as starting material. PF5-initiated ring-opening polymerization of **Gal-Thr** proceeded rapidly to give a stereoregular polysaccharide derivative with relatively high molecular weight, although an analogous disaccharide derivative showed no polymerizability. The high reactivity of **Gal-Thr** was probably due to the 2,4-dideoxygenated structure. Debenzylation of the polymer gave a comb-shaped polysaccharide which has pendant galactose branches regularly substituted on each sugar units in the main chain. The result of the force field calculation indicated that the polysaccharide had a linear conformation.

INTRODUCTION

Some polysaccharides found in nature have branched structures, and these branches affect physical and biological functions in these polysaccharides. The branched structure is complicated in certain cases, and hence structurally well-defined model compounds are required for elucidation of structure-property relationship on branched polysaccharides. As a model compound of branched polysaccharides, we have synthesized a comb-shaped polysaccharide which has a linear main chain and pendant monosaccharide units regularly substituted on each sugar unit in the main chain. Synthesis of comb-shaped polysaccharide has been attempted by two different routes: first route is ring-opening polymerization of anhydro disaccharide derivatives followed by deprotection (1,2), and second is synthesis of regiospecifically protected linear polysaccharides followed by stereoselective glycosidation and deprotection

(3,4). The former is favorable for the synthesis of the comb-shaped
polysaccharides that are substituted completely and stereoselectively with
monosaccharide branches. Disaccharide derivatives so far prepared, however,
showed low polymerizability (1,2), and polysaccharides with high molecular
weight have not been obtained.

We have paid attention to high polymerizability of a 2,4-dideoxygenated
anhydrosugar derivative (5), and applied this to the synthesis of new
polysaccharides. We have synthesized a disaccharide derivative, which contained
the reactive deoxysugar as polymerizable unit, from cellulose as starting
material, and polymerized and deprotected to obtain a new comb-shaped
polysaccharide with relatively high molecular weight. We have estimated the
polysaccharide conformation on the basis of NMR spectroscopy and force field
calculation.

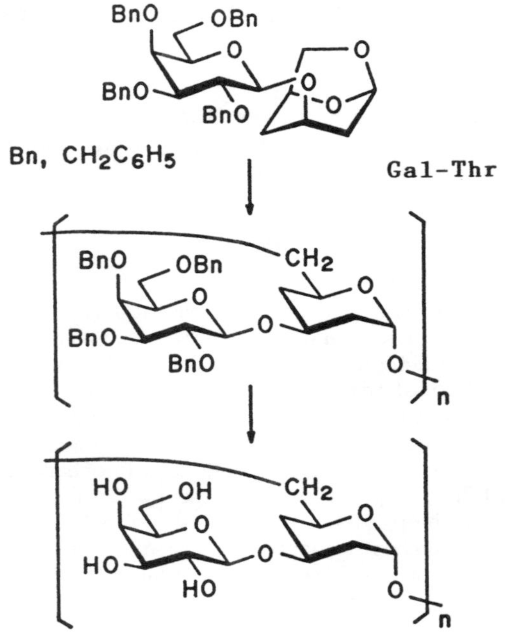

Scheme I
Synthesis of 2,4-Dideoxy-3-O-(β-D-galactopyranosyl)-(1→6)-α-D-threo-hexopyranan

EXPERIMENTAL
**1,6-Anhydro-3-O-(2,3,4,6-tetra-O-benzyl-β-D-galactopyranosyl)-2,4-dideoxy-
β-D-threo-hexopyranose (Gal-Thr)**
Microcrystalline cellulose was pyrolyzed under reduced pressure to give
1,6-anhydro-β-D-glucopyranose (levoglucosan) in a 37% yield.
1,6-Anhydro-2,4-di-O-p-toluenesulfonyl-β-D-glucopyranose was prepared from

levoglucosan by treatment with two equivalents of p-toluenesulfonyl chloride
in pyridine. The tosylate was treated with lithium triethylborohydride in dried
THF, and then the reaction mixture was purified by silica gel chromatography
to isolate 1,6-anhydro-2,4-dideoxy-β-D-threo-hexopyranose. Glycosidation of
2,3,4,6-tetra-O-acetyl-α-D-galactopyranosyl bromide to the anhydrosugar in the
presence of silver silicate gave exclusively β-linked glycoside. The glycoside
was deacetylated followed by benzylation to give **Gal-Thr.** The details of the
procedure are reported in the paper described previously (6).

**1,6-Anhydro-2,4-di-O-benzyl-3-O-(2,3,4,6-tetra-O-benzyl-β-D-galactopyranosyl)-
β-D-glucopyranose (Gal-Glc)**

1,6-Anhydro-2,4-di-O-benzyl-β-D-glucopyranose was synthesized from levoglucosan
according to the procedure described previously (7). Following glycosidation,
deacetylation, and benzylation were similar to those described above.

POLYMERIZATION AND DEBENZYLATION

Polymerization and debenzylation were carried out according to the method
described previously (8).

FORCE FIELD CALCULATION

The optimized conformation of the comb-shaped polysaccharide was estimated
through force field calculations on a computer in Research Center, Daicel
Chemical Industries, Ltd. We adopted a systematic search method, using SYBYL
5.32 Maxmin 2 as software.

GENERAL

1H- and 13C-NMR spectra were recorded with a JEOL JNM-FX-200 Fourier-transform
NMR spectrometer operating at 200 and 50 MHz, respectively. Number average
molecular weights of the benzylated polysaccharides were estimated by gel
permeation chromatography with Hitachi 634A, using chloroform with polystyrene
as the standard.

RESULTS AND DISCUSSION
POLYMERIZATION

Polymerization of **Gal-Thr** was carried out using high vaccum technique (8)
(Table 1). Phosphorus pentafluoride and dichloromethane were employed as
initiator and solvent, respectively. When the initiator concentration was higher
than 10 mol% to **Gal-Thr**, polymerization occurred rapidly to give polymer as
a methanol-insoluble fraction. Polymerization at -78°C afforded the polymer

Table 1.   Polymerization of **Gal-Thr** and **Gal-Glc**[a]

Monomer	Concn. mol/1	PF5 mol%	Temp. °C	Time min	Yield %	M_n[b] x10^{-3}
Gal-Thr	0.7	20	-78	6	67	27
Gal-Thr	0.7	20	-78	60	71	18
Gal-Thr	0.7	20	-60	60	42	4.3
Gal-Thr	1.0	10	-60	60	57	5.9
Gal-Thr	0.7	10	-60	60	33	5.1
Gal-Thr	0.5	10	-60	1440	56	4.3
Gal-Glc	0.5	10	0	1440	0	-
Gal-Glc	0.5	10	-60	6000	0	-

[a] Monomer, 0.5mmol; Solvent, dichloromethane.
[b] Determined by GPC.

of Mn = 27000 (Dp = 40) in a 67% yield. For the first time a comb-shaped
polysaccharide derivative was obtained with relatively high molecular weight.
We synthesized also an analogous disaccharide derivative, 1,6-anhydro-2,4-
di-O-benzyl-3-O-(2,3,4,6-tetra-O-benzyl-β-D-galactopyranosyl)-β-D-glucopyranose
(Gal-Glc), and attempted homopolymerization of Gal-Glc. Gal-Glc however showed
no polymerizability under similar conditions. Polymerization of an equimolar
mixture of Gal-Thr with 1,6-anhydro-2,3,4-tri-O-benzyl-β-D-glucopyranose (TBLG)
gave a copolymer which consisted of Gal-Thr unit as major fraction (-78°C, 4min;
yield, 10%; mole fraction of Gal-Thr in the copolymer, 0.68). Copolymerization
of Gal-Glc with TBLG gave no copolymer, but homopolymer of TBLG was produced.

The high reactivity of Gal-Thr relative to Gal-Glc can be explained in terms
of 2,4-dideoxygenated structure. Because of absence of the bulky, electron-
attracting benzyloxy groups, the reaction center in the polymerization is less
sterically hindered and ring oxygen of the monomer is more basic, which are
advantageous for cationic ring-opening polymerization.

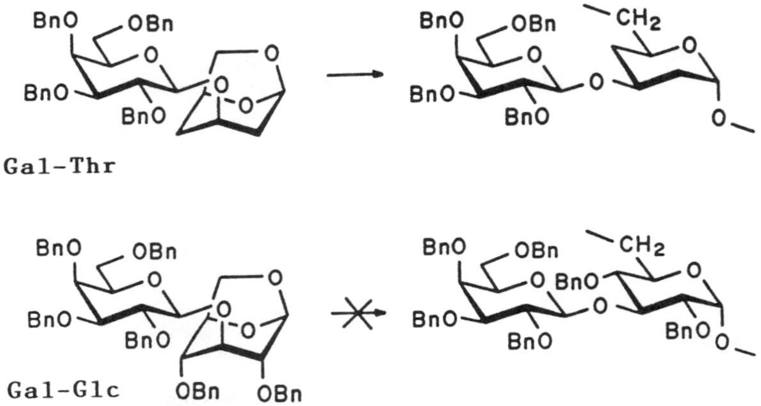

Figure 1. Polymerization of Gal-Thr and Gal-Glc

DEBENZYLATION

The polymer was debenzylated with sodium in liquid ammonia at -33°C, and
thus a white powdery product was obtained in a 90% yield. It was partially
soluble in water and dimethyl sulfoxide, and insoluble in other common organic
solvents. It was unexpected for us that the deprotected polysaccharide showed
low solubility in water, since it is well-known that branched polysaccharides
normally have higher solubility in water than linear ones.

Figure 2 shows 13C-NMR spectrum of water-soluble fraction of the deprotected
product. All signals can be assigned to the carbons of the comb-shaped
polysaccharide, 2,4-dideoxy-3-O-(β-D-galactopyranosyl)-(1→6)-α-D-threo-
hexopyranan. Main chain: C-1, 97.63; C-2, 35.27; C-3, 71.72; C-4, 34.45; C-5,
67.12; C-6, 68.62; Branch: C'-1, 101.23; C'-2, 70.75; C'-3, 72.85; C'-4, 68.62;
C'-5, 75.14; C'-6, 60.94. Following coupling constants confirm that the main
chain had α-stereoregularity and the branch β-stereoregularity. J(1H-13C):
main chain C1, 171.4Hz; branch C'-1, 155.8Hz. J(1H-1H): main chain H-1(5.18ppm),
less than 1Hz; branch H'-1(4.57ppm), 7.6Hz.

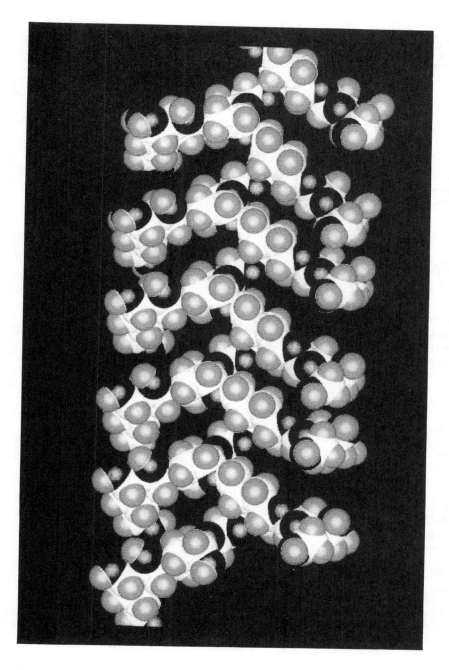

Figure 3.   Optimized Conformation of the Comb–Shaped Polysaccharide.

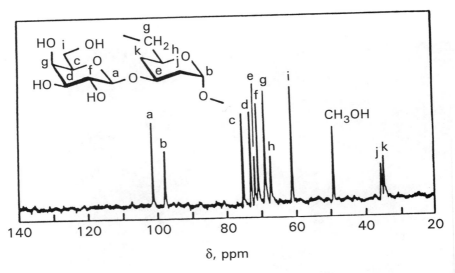

Figure 2. ^{13}C-NMR spectrum of 2,4-dideoxy-3-O-(β-D-galactopyranosyl)-(1→6)-α-D-threo-hexopyranan (in D_2O, 5%; reference, methanol, 49.0ppm)

CONFORMATION ANALYSIS

We observed that the two coupling constants are not equivalent between H-5 and two hydrogens (H-6a, H-6b) attached to C-6 in the main chain (8.4 and 5.6Hz), and hence speculated that the main chain of the comb-shaped polysaccharide had a characteristic conformation, which may be similar to that of dextran (9).

In order to gain further information, we have done the force field calculation about the main chain- and side chain-conformation of the polysaccharide. Figure 3 shows the space-filling model of the polysaccharide with the optimized conformation. The polysaccharide may have a linear conformation in two-fold screw relationship (9), as we expected. The pendant galactose branches are located alternately in the two opposite direction. We suppose that the pendants are favorably located to form hydrogen-bonding each other. The low solubility of the polysaccharide in water may be responsible for the ordered structure.

REFERENCES

1) B.Veruovic and C.Schuerch, Carbohydr.Res., **14**, 199(1970).
2) V.Masura and C.Schuerch, Carbohydr.Res., **15**, 65(1970).
3) H.Ito and C.Schuerch, J.Am.Chem.Soc., **101**, 5797(1979).
4) T.Uryu, M.Yamanaka, M.Hemmi, K.Hatanaka, and K.Matsuzaki, Carbohydr.Res., **157**, 157(1986).
5) K.Kobayashi, H.Sumitomo, H.Ichikawa, and H.Sugiura, Polym.J., **18**, 927(1986).
6) H.Ichikawa, K.Kobayashi, and H.Sumitomo, Macromolecules, **23**, 1884(1990).
7) C.Zemplén, Z.Csürös, and S.Angyal, Ber.Dtsch.Chem.Ges., **70**, 1848(1937).
8) C.Schuerch and T.Uryu, Macromol.Synth., **4**, 151(1972).
9) C.Guizard, H.Chanzy, and A.Sarko, Macromolecules, **17**, 100(1984).

28

Functionalization of cellulose by photografting: reactivity of glycidyl methacrylate-grafted cellulose

H. Kubota and Y. Ogiwara - Department of Chemistry, Gunma University, Kiryu, Gunma 376, Japan.

ABSTRACT

Photografting was applied to functionalize cellulose, that is, epoxy groups were introduced into the cellulose substrate by photografting of glycidyl methacrylate (GMA). The GMA-grafted cellulose was subjected to the following examinations ((1) to (3)) in comparison with epoxy-activated cellulose prepared by reaction with epichlorohydrin in order to establish whether photografting is a useful means of functionalization of cellulose:

(1) Reactivity of the GMA-grafted cellulose towards diamines such as $H_2N(CH_2)_nNH_2$ and $H_2N(CH_2CH_2NH)_n$.

(2) Ability of the aminated celluloses prepared by examination (1) to absorb cupric ion.
(3) Catalytic activity of the aminated cellulose/cupric ion complexes prepared by examination (2) towards decomposition of hydrogen peroxide.

INTRODUCTION

Grafting is known to be a useful means for the functionalization of polymeric substrates. The technique is classified into two; one where monomer having a desired function is directly grafted on the substrate, and the other where monomer with a reactive group is first grafted and the resultant reactive group is used as a reaction site for further functionalization. It is conceivable that the latter method is useful from the viewpoint of introducing wide varieties of function into polymeric materials.

In this study, photografting was applied to the functionalization of cellulose since the authors have studied photografting [1,2] of vinyl monomers on cellulose and polyolefins. The epoxy group was chosen as the reactive group and photografting of glycidyl methacrylate (GMA) was performed in aqueous medium. In order to understand the mechanism of photografting for the functionalization of cellulose, the GMA-grafted cellulose was examined in terms of reaction with diamines, absorption of cupric ion with the aminated cellulose, and decomposition of hydrogen peroxide with the aminated cellulose/cupric ion complexes in comparison with epoxy-activated cellulose prepared by reaction with epichlorohydrin.

EXPERIMENTAL

Dissolving pulp from softwoods (NDP) was used as the cellulose sample. Photografting was performed by irradiating 0.50g NDP, 40ml 0.01wt% hydrogen peroxide aqueous solution and 2ml GMA contained in Pyrex glass tube under nitrogen gas at 60°C for 5 to 7 min to yield GMA-grafted cellulose (G-Cell) with 30 to 50% grafting which contain epoxy contents of 1 to 2 mmol/g. cellulose.

The irradiation was carried out in a Riko rotary photochemical reactor RH400-10W with a high-pressure mercury lamp (400W).

Epoxy-activated cellulose (E-Cell) was prepared as follows: 0.50g NDP was treated with 3.2M KOH aqueous solution at 55°C for 60 min. The pretreated sample was added to 16ml DMSO/water (1:1) containing 3.2M epichlorohydrin and 0.024M tetraethylammonium hydroxide and the reaction was carried out at 55°C for 120 min to form E- Cell with an epoxy content of 0.6 mmol/g. cellulose.

RESULTS AND DISCUSSION

Figure 1 shows the reaction of epoxidized cellulose with diamines. Two kinds of diamine are
used: one is an amine which contains amino groups
at both ends of the molecule and the other is a
polyamine. The amount of amine residue introduced into the substrate was determined by nitrogen analysis. Aminated epoxy group was higher
for E-Cell than G-Cell. On the other hand, hydrolysis reaction of the epoxy groups also proceeds
as a side reaction. The proportion of unreacted
epoxy groups was higher for G-Cell than E-Cell.
This tendency was common for both diamines. The
result of G-Cell suggests the existence of epoxy
groups on the grafted chains which can not contribute to the amination. It may be due to a loop
structure [3] of grafted chains.

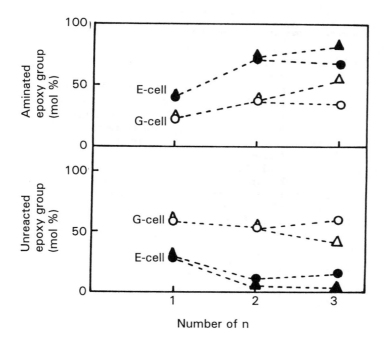

Fig. 1. Reaction of epoxidized cellulose with di-
amines. ($\blacktriangle$,$\triangle$) $H_2N(CH_2CH_2)_nNH_2$, ($\bullet$,$\circ$)
$H_2N(CH_2CH_2NH)_nH$.
Epoxy contents of G-Cell and E-Cell are 1.1 and 0.4
mmol/g.cell. Concentrations of diamine are 1.1 and
0.4 M for G-Cell and E-Cell, respectively.
Reaction was carried out at 70°C for 8 hr in DMF.

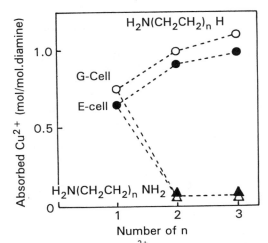

Fig. 2. Absorption of Cu^{2+} with aminated celluloses. Concentrations of cupric ion and ligand are 1.0×10^{-3} M and 8.0×10^{-4} M, respectively. Absorption reaction was carried out at 30°C for 24 hr at pH=5.

Figure 2 shows absorption of cupric ion with the aminated celluloses. The absorbed amount was determined by chelate titration. Only the polyamine-introduced sample showed ability to absorb cupric ion and the amount increased with increasing value of 'n' in the molecule. The mole numbers of ethylenediamine (En), diethylenetriamine, and triethylenetetramine (Trien) necessary for complexation with one mole of cupric ion are estimated as 0.5, 0.75, and 1.0, respectively. The value of En was considerably higher than the estimated one. This suggests that the complex of En contains incompletely coordinated species (structure II) besides the complex with structure I. The absorbed amount was nearly equal for G-Cell and E-Cell.

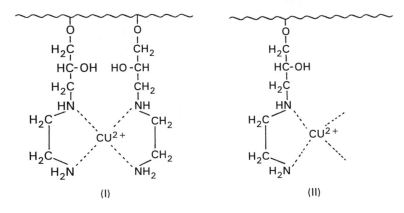

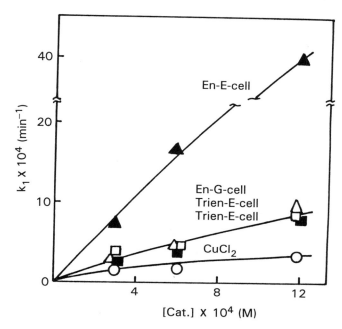

Fig. 3. Relationship between rate constant k_1 and catalyst concentration. Concentration of hydrogen peroxide is 2.5×10^{-3} M. Cu^{2+}/Ligand=1.6, 30°C.

Figure 3 presents the results of catalytic activity of complex prepared by the aminated cellulose and cupric ions. The pseudo-first order rate constant k_1 is used as a measure of catalytic activity. All complexes showed catalytic activity towards decomposition of hydrogen peroxide. It is known that the catalytic activity of metal complexes towards decomposition of hydrogen peroxide originated in the complex species with incompletely coordinated structure. En-introduced E-Cell (En-E-Cell) showed the highest activity. The activity of G-Cell was lower than E-Cell. It is supposed that the result is ascribed to formation of less of the complex species with structure II in G-Cell than E-Cell.

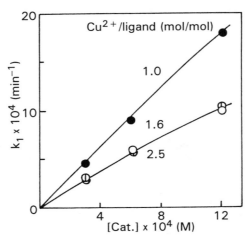

Fig. 4. Effect of Cu^{2+}/Ligand on catalytic activity
of En-G-Cell.

Factors influencing the catalytic activity of G-
Cell were examined, and the effect of the ratio
of absorbed amount of cupric ion to En ligand
content is shown in Figure 4. The activity increased with decreasing the ratio. At the lower
ratio, the complex species with structure II seems
to be formed easily. The effect of the ratio of
En ligand to epoxy content was examined and observed that the activity increased with decreasing
the ratio. This means the existence of even distribution of ligand molecules on grafted chains.
When the distance increases, the ligand molecules
are not able to participate in complex formation.
This leads to the formation of complex species
with structure II. Thus, it was possible to increase catalytic activity of G-Cell by controlling the preparation conditions of complex. It
is concluded that photografting is useful for introducing epoxy groups into the cellulose substrate, but it remains to be proved that epoxy
groups which do not contribute to amination exist on grafted chains.

REFERENCES

1. Kubota,H.; Ogiwara,Y.; Hinohara,S. J. Appl.
    Polym. Sci., **1987**, 33, 3045. 34, 1277.
2. Kubota,H.; Hata,Y. J. Appl. Polym. Sci.,
    1990, 41, 689.
3. Okamoto,J.; Sugo,T.; Katakai,A.; Omichi,H.
    J. Appl. Polym. Sci., **1985,** 30, 2967.

29

Adhesion by calculation of secondary forces of UF oligomers with cellulose I and amorphous cellulose: theoretical and applied composite results

A. Pizzi - Department of Chemistry, University of the Witwatersrand, Johannesburg, South Africa.

ABSTRACT

Specific adhesion of urea—formaldehyde (UF) oligomers of progressively increasing molecular weight on to wood cellulose were modelled by calculation of the secondary forces interactions between each UF species and simple models of amorphous cellulose and of the surface of crystalline Cellulose I. The results obtained indicate that adhesion to the two forms of cellulose can be used to model a great part of wood adhesion. The theoretical results obtained may indicate the route to better UF resins as regards wood adhesion. Applied results on particleboard confirmed the indications obtained by theoretical computational methods.

INTRODUCTION

Recently the adhesion of urea—formaldehyde[1,2] and phenol—formaldehyde resins[3] to wood cellulose has been modelled as a surface physico—chemical phenomenon due to the sum of secondary forces interactions between the PF and UF resin and the cellulose.

Urea formaldehyde (UF) resins are today the most extensively used binders for lignocellulosic materials, especially for interior wood adhesives. Studies of the three—dimensional space conformation of UF resins have already been successfully carried out[4-6]. However, it has also already been determined that the configuration of polymeric resin oligomers when taken in isolation does not correspond to their minimum energy configuration when interacting with the surface of a substrate[1-3 7]. This indicates that while previous studies of the internal UF resin configurations taken in isolation are of considerable importance to the chemistry of UF resins in itself.

Their configurations and energies of interaction, hence of adhesion, with a particular substrate are of some applied and theoretical interest. It is then worthwhile to follow the development and variations of the energy of adhesion of UF resins with cellulose through the various initial stages of UF condensation. This entailed calculation of the theoretical energies of adhesion of urea through to the various methylol ureas, UF dimers and trimers, and their monomethylol derivatives. The study was carried out for both crystalline Cellulose I and amorphous cellulose to give enough insight into what is likely to occur at the resin—substrate interface as the condensation of the UF resin proceeds.

Although a molecular mechanics study of the adhesion of UF condensates to cellulose involves very extensive calculations and must, by its definition, be subjected to both physical and mathematical limitations, it was nevertheless attempted with interesting results.

As cellulose constitutes as much as 40% of wood (and total carbohydrates, crystalline and amorphous, as much as 75%), this study also infers partial applicability to a wood substrate.

EXPERIMENTAL

Urea, monomethylolurea, the two possible dimethylolureas, trimethylolurea, methylene bis urea (UF dimer), the two possible monomethylol—methylene bis ureas (monomethylol UF dimers), and the two possible trimethylene tris ureas (UF trimers) were used in this study. The study was stopped at this stage as all the most important trends had been determined by this sequence. All these UF oligomers are shown in Fig.1. The structure of five chains, four glucose residues each, of a schematic elementary Cellulose I crystallite, already reported[8], and with a single strand of amorphous cellulose composed of four anhydroglucose residues[9], were used as the substrate. The computer program and techniques used for the calculation of van der Waals, H—bonding, electrostatic, and torsional interactions between each pair of non—covalently bonded atoms within the UF species itself, and between the UF species and the cellulose substrate, have also been reported[9] [10]. The number of degrees of freedom for such calculations is considerable and an identical technique to that previously reported was also used with some opportune variations. Four positions on the surface of the cellulose crystallite were taken. These are shown in Fig.1. Positions A, B, C, and D correspond to geometrical central points of each of the four surfaces of the elementary cellulose crystal. The approach to the A, B, C, and D positions was taken as the plane connecting the longitudinal axis of the crystal with the axis on the crystal surface which is parallel and equidistant from, and in the same plane of, the longitudinal axis of the two cellulose chains on the particular crystal surface taken into consideration.

The central position of all the UF oligomers was taken as the carbon atom of the methylene bridge connecting the two urea residues of methylene bis urea (UF dimer). All the other UF oligomers were calculated on this basis. This means, for example, that for monomethylolurea the central point was taken as the carbon of the methylol group. The starting position of all the

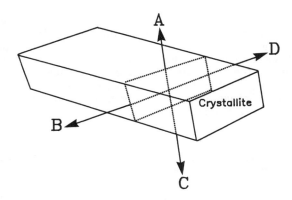

Figure 1.   Schematic positioning sites of the UF species on a Cellulose
             I crystallite

UF species' rotational angles was taken as that in which the UF oligomer
molecule is planar with the O of all the C=O groups on the same side of
the molecule and the H of the N—H groups with the hydrogen pointing in
the opposite direction; hence at 180° from the C=O groups. The plane in
which the planar UF oligomer lies was taken as parallel to the plane of the
cellulose crystallite surface in the cases of positions A and C; and
perpendicular to the plane of the cellulose crystallite surface in the cases of
positions B and D. The study concentrated throughout on positions A and
C as more representative of the UF—cellulose crystallite interaction, as
these surfaces are more extensive than those at positions B and D. The
study on positions B and D was limited to three oligomers to understand
what their influence could be in a statistically orientated wood surface.
Their influence is less in the case of pure cellulose.

Each of the UF oligomers was aligned, in turn, on positions A, B, C, and D
at a distance of 12 Å and then allowed to approach the cellulose surface in
increments of 0.5 Å. The $\phi°$, $\psi°$, 1°, 2° and 3°, 4° internal rotational
angles of the UF molecules were rotated by 360° in 20° increments and
then refined in 5° increments. Rotation of the angles $\phi°$ and $\psi°$ was
conducted so as to rotate both groups of atoms for each angle connected
to the central methylene linkage, and not only one group for each angle as
was done previously for the PF dimers[3]. This allowed rotation of the
molecule in two orthogonal planes to be eliminated, decreasing the amount
of computation while monitoring the same number of degrees of freedom
of the molecule. The UF oligomer conformation and the value of the
minimum total energy of interaction between the UF molecule and cellulose
were determined for each of the incremental 0.5 Å distances.

The procedure was repeated for shifts of the UF species along the
longitudinal axis of the cellulose crystallite of + 2.5 Å and −2.5 Å (half a
glucose ring span).

Contrary to the previously reported PF case[7], the 20 glucose residues of the elementary model of the cellulose crystallite were all in the conformation characteristic of the body of a full—size cellulose crystallite, rather than some of them being in the conformation of the terminal glucose residues in a full—scale crystallite. The distance of the UF species from the surface of the cellulose crystallite was taken, for ease of reporting, as the distance between the plane containing the two oxygens of the two central β—glucosidic linkages of two vicinal cellulose chains and the carbon atoms of the methylene linkage of the UF oligomer or the carbon atom of the methylol group in the case of methylolureas.

RESULTS

The results can be reported in several ways, namely as averages of all the minimum energies of interactions at the 12 sites used on the crystalline and amorphous cellulose, or exclusively as the value of the minimum energy of interaction calculated at one of the sites only, or even as averages of the minimum energies per number of atoms of the particular UF oligomer under consideration. While the extended results have been reported elsewhere[1][2], the results obtained in their different forms are reported in Schemes 1,2 and 3 on crystalline Cellulose I and in Schemes 4 and 5 for a single strand in the most energetically favourable conformation[9] of amorphous cellulose. The energies reported are in kcal/mole.

Energies of interaction of UF oligomers on Crystalline Cellulose I surface[1].

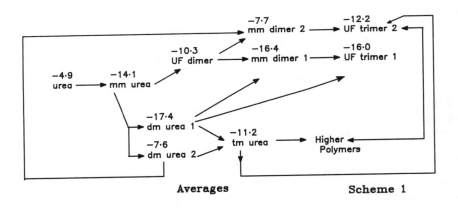

Averages   Scheme 1

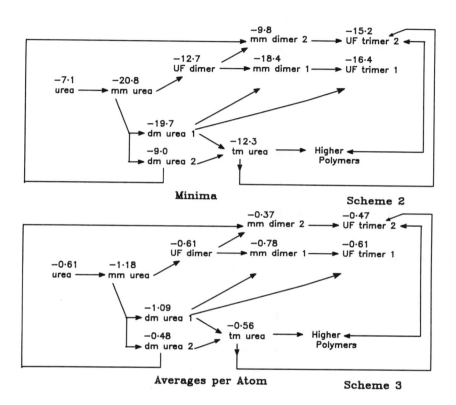

Energies of interaction of UF oligomers on a single, most favourable conformation[2], of amorphous cellulose strand.

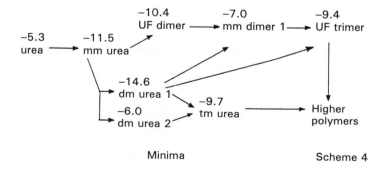

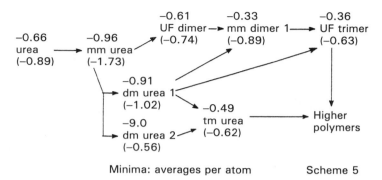

Minima: averages per atom   Scheme 5

* ( ) value for crystalline cellulose I

From these results it would, in theory, be possible to determine the specific adhesion of a UF resin to the sum of amorphous and crystalline cellulose. The values used are shown in Table 1.

Table 1: Average theoretical values of the energy of adhesion per atom of UF oligomers for urea and urea—formaldehyde condensates[10] [11].

Chemical species	Average energy of adhesion in kcal/mol per atom of UF species to:	
	Amorphous Holocellulose	Crystalline Cellulose I
Urea	0.66	0.61
Monomethylolurea	0.96	1.18
N,N'—Dimethylolurea, linear	0.91	1.09
N-Dimethylolurea, branched	0.38	0.48
Trimethylolurea	0.49	0.56
Methylenebisurea (UF dimer)	0.61	0.61
Dimethylenetrisurea(UF trimer)	0.36	0.61

(for linear and branched dimethylolurea rows: –0.94)

The method is based on matching the relative proportions of UF oligomers present at any condensation stage in a mixture, on a concentration or even on a percentual basis, with the average energy of adhesion of such oligomers to the amorphous and crystalline wood carbohydrates (of Table 1). The adhesion energy values in Table 1 have been reported "by atom" as the value of the energy of adhesion for each whole oligomer increases with increasing molecular weight and degree of polymerization of the oligomer. Thus, a UF dimer has an energy of adhesion of 19 atoms × 0.61 kcal/mol) = 11.59 kcal/mol, while a UF trimer, also on crystalline cellulose, has a total energy of adhesion of (29 atoms × 0.61 kcal/mol) = 17.69 kcal/mol. Thus, the energy of adhesion expressed per atom appears to be a better index of adhesion "efficiency" for a particular oligomer than the energy of adhesion of the whole molecule which is likely to be only an indication of increasing/decreasing molecular weight. On curing, however, the disappearance of the higher adhesion efficiency species such as monomethylol– and dimethylol–ureas is concomitantly accompanied also by the disappearance of the low molecular weight species of poor adhesion efficiency. This, together with the increase in total adhesion energy apported by the increasing molecular weight of the polymers may cause only a slight or even no drop in the total energy of adhesion of the system.

For a graph of the relative proportions of UF oligomers as a function of reaction time, as shown in Fig.2, the following formula was used:

$$\text{Adhesion} = \frac{(\% \text{ urea})}{100} \times (nE_{amorph} + [1-n]E_{cryst}) + \frac{(\% \text{ mm urea})}{100} \times$$

$$\times (nE_{amorph} + [1-n]E_{cryst} + \frac{(\% \text{ dm urea})}{100} \times (n\{0.75\ E_{amorph}(dmu\ 1) +$$

$$+ 0.25\ E_{amorph}(dmu\ 2)\} + [1-n]\{0.75\ E_{cryst}(dmu\ 1) + 0.25\ E_{cryst}(dmu\ 2)\}) +$$

$$+ \frac{(\% \text{ tm urea})}{100} \times (nE_{amorph} + [1-n]E_{cryst}) + \frac{(\% \text{ UF polymers})}{100} \times$$

$$\times (nE_{amorph}(polymers) + [1-n]E_{cryst}(polymers)) \qquad \text{(equation 1)}$$

where E_{amorph} and E_{cryst} are the average energies of adhesion per atom of the molecule of the chemical species of the term in equation 1. They are shown in Table 1[10][11]; n is the proportion on a total of amorphous material (be it cellulose or total carbohydrates, see equation 2 and 3) present in the substrate and 1–n the corresponding balance of crystalline cellulose, which both vary according to the wood species used as a substrate. Equation 1 can then become, exclusively for adhesion to cellulose in a wood where the level of cellulose crystallinity is 70%:

$$\begin{aligned}\text{Adhesion} \\ \text{to cellulose}\end{aligned} = \frac{(\% \text{ urea})}{100} \times (0.3\ E_{amorph} + 0.7\ E_{cryst}) + \frac{(\% \text{ mm urea})}{100} \times$$

$$\times (0.3\ E_{amorph} + 0.7\ E_{cryst} + \frac{(\% \text{ dm urea})}{100} \times (0.3[0.75\ E_{amorph}(dmu\ 1) +$$

$$+ 0.25\ E_{amorph}(dmu\ 2)\} + 0.7[0.75\ E_{cryst}(dmu\ 1) + 0.25\ E_{cryst}(dmu\ 2)] +$$

$$+ \frac{(\% \text{ tm urea})}{100} \times (0.3\ E_{amorph} + 0.7\ E_{cryst}) + \frac{(\% \text{ UF polymers})}{100} \times$$

$$\times (0.3\ E_{amorph}(polymers) + 0.7\ E_{cryst}(polymers)) \qquad \text{(equation 2)}$$

By using these equations the theoretical values of adhesion of UF resins prepared as a single reacting mixture (Case 1, Table 2), as a UF resin with second urea addition (Case 2, Table 2), and as a UF resin prepared by three successive additions of both urea and formaldehyde (Case 3, Table 2) were obtained. The same three resins were used to prepare 12 mm thick particleboard under identical, and standard laboratory conditions. Table 2 reports both the relative values obtained by computational methods as well as the experimental internal bond values of the particleboard produced with the same three resins.

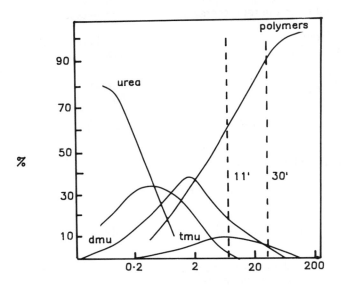

Fig.2   Percentage of chemical species formed as a function of log reaction time for a standard UF reaction in which both reagents are charged at the beginning of the reaction. mmu = monomethylolurea; dmu = dimethylol–ureas; tmu = trimethylolurea

In Table 2 both the calculated values and the experimental values appear to have a diminishing upward trend as one passes from resin 1 to resin 2 and to resin 3, indicating that the trend in the values obtained by calculation appear to be somewhat correlated to the trend in the values obtained experimentally.

In conclusion, refinements of this, still rough, approach could be used to predict the trend in adhesion of a UF formulation by calculating its relative values of adhesion.

Table 2: UF chemical species percentage distribution according to formulation and preparation process and corresponding calculated comparative relative values of total adhesion to a softwood and to amorphous + crystalline cellulose in wood at 30 minutes reaction time and at reflux (equation 1). Experimental internal bond strength of particleboard prepared with cases 1,2 and 3 resins reacted for only 11 minutes are at bottom of table.

UF Species	UF species distribution (%)		
	Case 1 Fig. 1	Case 2 Fig. 2	Case 3 Fig. 3
Urea	0.0	0.0	18.3
Monomethylolurea	0.0	5.3	12.3
Dimethylolureas	5.5	14.7	13.0
Trimethylolurea	4.5	5.3	5.7
UF polymers (trimer and up)	90.0	74.7	50.7
Total adhesion per atom of resin mixture to cellulose in kcal/mole (equation 2)	0.5543	0.6118	0.6690
Internal bond strength of laboratory particle-board in MPa	0.37	0.53	0.62

REFERENCES

1. A.Pizzi, *J. Adhesion Sci. Technol.*, **4**, 573 (1990)
2. A.Pizzi, *J. Adhesion Sci. Technol.*, **4**, 573 (1990)
3. A.Pizzi and N.J.Eaton, *J. Adhesion Sci. Technol.*, **1,3**, 191 (1987)
4. A.K.Dunker, W.E.Johns, R.Rammon, B.Farmer and S.Johns, *J. Adhesion,* **19**, 153 (1986)
5. J.J.Pratt, W E Johns, R Rammon and W I Plagemann J. Adhesion, **17**, 275 (1985)
6. R.Rammon, Ph.D. Thesis, Washington State University, Pullmann, WA (1984)
7. A.Pizzi, Wood Adhesives Chemistry and Technology, Vol.2, Chapter 4, Marcel Dekker, New York (1989)
8. A.Pizzi and N.J.Eaton, *J. Macromol. Sci., Chem. Ed.*, **A22**, 139 (1985)
9. A.Pizzi and N.J. Eaton, *J. Macromol. Sci., Chem. Ed.*, **A22**, 105 (1985)
10. A.PIZZI, *Holzforschung und Holzverwertung*, **3**, 63 (1990)

30

The water sorption isotherm of crystalline cellulose I and II and amorphous cellulose – a molecular mechanics approach

A. Pizzi - Department of Chemistry, University of the Witwatersrand, Johannesburg, South Africa.

ABSTRACT

Values of the sorption energies of single molecules of water on elementary models of amorphous cellulose and crystalline celluloses I and II by molecular mechanics methods are presented. The interference of water molecules with each other on vicinal sorption sites were obtained. Sites on which such interference can occur were identified. Sorption capacities of the water "monolayer" were calculated. Sorption capacities of the three cellulose forms were calculated and the water sorption isotherms reconstructed by theoretical means. Inflection points of the isotherms, start of water clustering and the variability of Dent's first constant for the water monolayer in relation to relative humidity for the three cellulose forms were also calculated by theoretical means and compared to applied results.

INTRODUCTION

Theoretical determination by conformational analysis of the energies of water sorption on available sites of amorphous cellulose, cellulose I and cellulose II crystallites have been carried out [1,2,3]. It is clear from comparative studies of the morphology and the chemistry of amorphous and crystalline celluloses that the two forms of cellulose behave very differently. The different crystallite structures of cellulose I and cellulose II also lead to noticeable differences in the number of sorption sites and in their capability and intensity of sorbing water. The modelling by molecular mechanics of water is a very difficult and complex problem. Notwithstanding this, it was attempted by introducing a few physical approximations. The results which were obtained [1,2,3] showed notwithstanding such approximations an acceptable level of correspondence with experimental results presented by other authors.

EXPERIMENTAL

The calculation method, and techniques used were already reported [4,5]. The detailed structure of crystalline cellulose I, amorphous cellulose and cellulose II used as a water sorption substrate in this article, and the detailed extent of the interatomic forces involved have already been reported [1–6].

The spatial positions for each molecule of water and its minimum energies (E_{tot}, E_{VdW}, E_{H-bond}, E_{elec}) of interaction with each possible sorption site on the surface of the cellulose II crystallite were calculated by rotating the molecule of water on a sphere in which the three angles, 1,2, and 3, were simultaneously rotated by 2° over 360°. As a consequence of this, all the possible positions and interactions with all atoms of the 20 glucose residues cellulose crystallite for 2° × 2° of the two hydrogens and oxygen of the water molecule on the surface of their own sphere were calculated as well as the relative position of the hydrogen of the hydroxyl group of the sorption site.

This thorough system of examination was obtained mathematically by creating an axis orthogonal to C_1C_2 and OC_2, creating an axis parallel to the previous one but centered on the hydroxyl O, positioning H_{W_1} and O_W by extending the C_2O axis, extending the $H_{W_1}O_W$ axis by 0.2404 Å, moving 0.9294 Å in the direction of the OD_1 axis and positioning H_{W_2}, and rotating angles 1, 2, and 3.

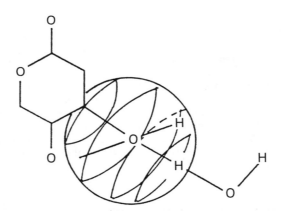

The water sorption isotherms were calculated by Dent's equations:

$$m_1 = m_0 k_1 h / (1 + k_1 h - k_2 h),$$
$$m = m_0 k_1 h / (1 - k_2 h)(1 + k_1 h - k_2 h), \qquad (1)$$

where m_1 is the monolayer water sorption, m_0 is the maximum value of water theoretically possible in monolayer sorption (which was available exactly for the first time), h is the relative humidity, and k_1 and k_2 are equilibrium constants related to the binding energies of the primary and secondary water layers. The latter were obtained from the expressions

$$E_{tot1} = -(RT/18) \ln k_1,$$

$$E_{tot2} = -(RT/18)\ln k_2, \tag{2}$$

where E_{tot1} used was the most probable average sorption energy for the monolayer calculated by conformational analysis derived from its frequency distribution curve (7.7 kcal/mol was used for E_{tot1} of cellulose II and 5.41 kcal/mol for cellulose I.) The value of k_2 was obtained in the same manner by using the E_{tot2} already obtained by conformational analysis [1,2]. As it is known that k_1 and k_2 vary with relative humidity h, the m_1 curve was also calculated for each addition to the crystal of a water molecule with the single exact energies obtained for each specific sorption site involved (cf. Fig. 2). The knowledge of the exact sorption energy for each cellulose II site allowed the determination of the dependence of k_1 on the relative humidity. All the calculations were carried out for a temperature of 300 K. It was found that k_2 had very little variation [1], and the value of k_2 in the system of reference used in the calculation was 0.934.

DISCUSSION

Figure 1 shows the water sorption isotherm (m) and the monolayer isotherm (m_1) for the schematic five–chains crystal of cellulose II built according to Dent's equations [7] by using the theoretical energy values calculated by conformational analysis [1,2,3]. The equivalent curves for the same size crystallite of cellulose I and for amorphous cellulose are reported in the same figure for comparative purposes. It must be clearly pointed out that this is only an example of how an isotherm can be built by using values calculated by theoretical means. The true isotherm will be somewhat different, as the reputed size of an elementary fibril [8] is of 17 chains and of a microfibril [9] of approximately 176 chains. Figure 1 indicates that when Dent's equations is used with exact energy values, the shape of the isotherm (m) is not a slightly sigmoid curve as when calculated with other data, and the shape of the monolayer isotherm (m_1) is a straight line. This behaviour is due to 1) the available sorption sites of the monolayer being exhausted long before reaching equilibrium moisture content, 2) a marked tendency to sorb a second water molecule ("second layer") on already sorbed sorption sites, 3) variation in slope in bigger crystals as in nature, as obtained on adding water molecules to numerous strong sorption sites of the same value, and 4) the variation of k_1 with h in Dent's equations.

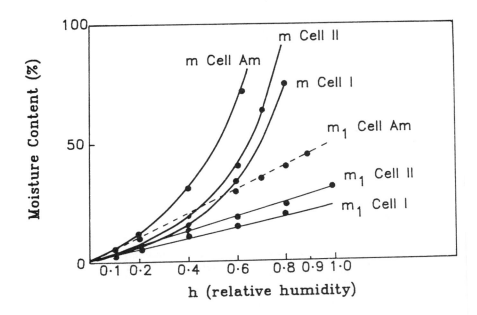

Fig. 1   Comparison of water sorption isotherms (m) and monolayer isotherms (m_1) calculated for amorphous cellulose (Cell Am), and the schematic five—chains crystals of cellulose I and cellulose II. Note the absence of inflection points when Dent's k_1 is taken as constant at different h values. These curves do not adequately describe water "clustering" and the start of further water layers.   Clusters already form at $h = 0.1$ (2.6 to 3.0%). Cellulose   II, $\Delta G_1 = 7.7$ kcal/mol, $m_0 = 33.82\%$ cellulose   I, $\Delta G_1 = 5.41$ kcal/mol, $m_0 = 26.18\%$;   cellulose   amorphous, $\Delta G_1 = 6.48$ kcal/mol.

The first reason is supported by a number of primary sorption sites being eliminated by the interference effect of water molecules with each other [3] but this is not likely to have an extensive influence.   The second reason is likely to be caused by the internal and interassociative energy characteristics of water molecules in the vapor phase.   When the associative energy of the water molecules is equal to or higher than the sorption energy of the still—available primary sites, sorption will become preferential on the first water molecule already sorbed on a site and water clusters will then start to form [10].   The monolayer sorption isotherm (m_1) will then reach a maximum at a certain h and stabilize at that level for any increasing value of h.   This will cause the appearance of the known sigmoid shape of the m curve for lignocellulosic materials.   If one considers the water clustering effect to become markedly evident at a sorption site energy of 5.5 to 6 kcal/mol, which is borne out by experimental evidence on wood isotherms [7], then the inflection point of the water sorption isotherm m can be foreseen to be at approximately 0.4 to 0.5 relative humidity for crystalline cellulose II and at 0.25 for crystalline cellulose I, in good accord with experimental evidence [7].

The calculation of the m_1 and m isotherms, molecule of water by molecule of water, presents the problem that the average energy of sorption obtained for amorphous cellulose and two crystalline forms cannot be used. The sorption energy of each single site must be used instead. As this is variable, at least the monolayer constant k_1 in Dent's equations [7] also varies. This proved to be the case. The m_1 and m isotherms for crystalline cellulose I, at very low relative humidities, were then obtained by progressive addition of single molecules of water. The values obtained are shown in Table 1. Dent's equations were never conceived for variable constants but for use with average sorption energies. That Dent's equations need revision in this respect is shown in Table 1 where the equations predict water clustering (a difference exists between m_1 and m values) even after sorption of the first molecule of water. This is not possible. The $m-m_1$ values difference must be at least equal to the m_1 value for sorption of the first molecule of water to allow sorption of a molecule of water on the second layer, hence the real start of water clustering. The values in Table 1 allow to forecast that, in the schematic cases under consideration, clustering cannot start earlier than when the 7th molecule of water is sorbed, i.e., at a relative humidity h of 0.17 and 0.14 for cellulose I and II respectively, or approximately 4% equilibrium moisture content for both types of cellulose.

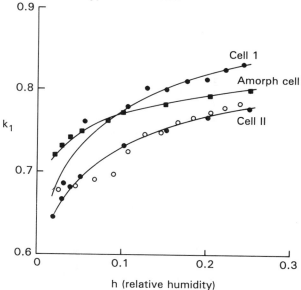

Fig.2   Comparison of the variation of Dent's constant k_1 for the water monolayer isotherm as a function of relative humidity h for amorphous cellulose (Cell Am), and the schematic five–chains cellulose I and cellulose II crystallites. Values obtained by adding one water molecule at a time on the schematic cellulose crystallites.

Table 1. Theoretically Calculated Values of Dent's Constant k_1, Water Isotherm m, and Monolayer Isotherm m_1 at Low Relative Humidities h when Adding One Water Molecule at a Time to the Schematic Five–Chains Cellulose II.

Crystalline celluose I					Crystalline cellulose II					Amorphous cellulose		
k_1	m_1	h_1	m	$m - m_1$	k_1	m_1	h_1	m	$m - m_1$	k_1	m_1	h_1
0.688	0.00545	0.0300	0.00562		0.0684	0.00545	0.0234			0.790	0.0545	0.1242
0.763	0.0109	0.0539			0.688	0.0109	0.0463			0.792	0.1090	0.2437
0.765	0.0164	0.0803			0.696	0.0164	0.0685			0.818	0.1636	0.3517
0.781	0.0218	0.1042			0.701	0.0218	0.0900			0.823	0.2182	0.4612
0.804	0.0273	0.1265			0.729	0.0273	0.1083			0.824	0.2727	0.5687
0.804	0.0327	0.1508	0.03125	-0.00145	0.751	0.0327	0.1258			0.824	0.3279	0.6742
0.812	0.0382	0.1738	0.03850	0.0003	0.753	0.0382	0.1460	0.0442	0.00600	0.845	0.3818	0.7714
0.815	0.0436	0.1969	0.04624	0.00264	0.765	0.0436	0.1639			0.848	0.4364	0.8726
0.828	0.0491	0.2180			0.770	0.0491	0.1829			0.856	0.4909	0.9716
0.833	0.0545	0.2399			0.775	0.0545	0.2013					
					0.783	0.0600	0.2191					
					0.786	0.0655	0.2377					

The dependence of Dent's k_1 on h for the schematic five—chains crystallite under consideration can then be expressed (Fig.2) as

$$k_1 = 0.06261 \ln h + 0.92388 \text{ for cellulose I,}$$
$$k_1 = 0.05178 \ln h + 0.85313 \text{ for cellulose II,} \qquad (3)$$
$$k_1 = 0.03228 \ln h + 0.03228 \text{ for amorphous cellulose}$$

The value of the Dent's k_2 was found to vary little with h, due to the very small sorption energy variation between water molecules sorbed as second layer (clustering) on different sorption sites [2].

CONCLUSIONS

1.   The results obtained now render possible the construction of theoretical water sorption isotherms for cellulose II crystallites of any known size, one molecule of water at a time. This affords considerably higher precision at lower humidities than any previous method.

2.   Water clustering appears to start very early, much earlier than previously thought, in water sorption by all three forms of cellulose.

3.   The k_1 parameter for the water monolayer in Dent's equations has been shown not to be constant when calculated for each successive water molecule sorbed. Definite relations determine its variation with humidity for crystalline celluloses I and II and amorphous cellulose. This variability is one of the main causes of the sigmoid shape of experimental water isotherms.

4.   The average sorption site energy for cellulose II crystallites is consistently higher than in cellulose I and amorphous cellulose [1,2,3]. This indicates that the water monolayer is more strongly bound in cellulose II. It may also indicate that, in cellulose II, the water monolayer adsorbs more readily and desorbs less readily than in the other two forms of cellulose.

5.   The cellulose II crystallite appears to be able to adsorb more water than cellulose I crystallites of identical molecular weight, but considerably less water than amorphous cellulose [1,2] in direct relation to the respective number of sorption sites available. This is consistent with other experimental evidence.

6.   Sorption sites interference appears to be more marked in cellulose II than in cellulose I crystallites [1].

7.   As for cellulose I crystallites, the stronger "monolayer" sorption sites of cellulose II cannot always form a second layer due to strong steric hindrance from vicinal groups.

8.   As in cellulose I, in crystalline cellulose II there appears to be favourable sorption on sites exerting high attractive forces and on sites which are exposed and protrude from the crystal surface.

REFERENCES

1.   A.Pizzi, N.J.Eaton, and M.Bariska, "Theoretical Water Sorption Energies by Conformational Analysis; Crystalline Cellulose I," *Wood Sci. Technol.*, **21**, 235–248 (1987).

2.   A.Pizzi, N.J.Eaton, and M.Bariska, "Theoretical Water Sorption Energies by Conformational Analysis; Amorphous Cellulose," *Wood Sci. Technol.*, **21**, 317–327 (1987).

3.   A.Pizzi and N.J.Eaton, "The Structure of Cellulose by Conformational Analysis. Part 5. The Cellulose II Water Sorption Isotherm, *J. Macromol. Sci., Chem. Ed.*, **A24(9)**, 1065–1084 (1987).

4.   A.Pizzi and N.J.Eaton, *J. Macromol. Sci. — Chem.*, **A21(11–12)**, 1443 (1984).

5.   A.Pizzi and N.J.Eaton, *Ibid.*, **A22(2)**, 139 (1985).

6.   A.Pizzi and N.J.Eaton, *Ibid.*, **A22(1)**, 105 (1985).

7.   C.Skaar, "Wood–Water Relationships," in *The Chemistry of Solid Wood* (*Advances in Chemistry Series*, No.207), American Chemical Society, Washington, D.C., 1984.

8.   R.B.Hanna and W.A.Coté, *Cytobiology*, **10(1)**, 102 (1974).

9.   A.J.Panshin and C.H.De Zeeuw, *Textbook of Wood Technology*, 4th ed., McGraw–Hill, New York, 1980.

10.   A.Misra, D.J.David, J.A.Snelgrove, and G.Matis, *J. Appl. Polym. Sci.*, **31**, 2387 (1986).

31

Thermal properties of water around the cross-linking networks in cellulose pseudo hydrogels

T. Hatakeyama and H. Hatakeyama* - Research Institute for Polymers and Textiles, *Industrial Products Research Institute, Tsukuba Ibaraki 305, Japan.

ABSTRACT

The dehydration process of microfibril cellulose (MFC) in the form of pseudo-hydrogels is investigated in this paper using differential scanning calorimetry (DSC). MFC pseudo-gel retains a large amount of bound water in non-freezing and freezing state. During the dehydration process, pure water was excluded from the gel and two clearly separated crystallization peaks were observed in DSC curves. It is suggested that through coagulation of cellulose molecules by hydrogen bonding, bound water separated from the gel and changed into free water.

INTRODUCTION

Various kinds of polysaccharides are known to form hydrogels [1,2]. It is reported that the cross-linking points of polysaccharide hydrogels have various types, such as double helical assemblies of two or three molecules, molecular association via divalent cations and tri-valent cations, molecular entanglement, and inter-molecular hydrogen bonding forming micro-crystallites. The structure of the junction zone, including the cross-linking points and surrounding water molecules, is considered to affect the physical properties of hydrogels.

It is reported that the physical properties of water molecules restrained in three-dimensional networks are markedly different from those restrained by linear

hydrophilic polymers [3]. Nuclear magnetic relaxation
studies indicate that the relaxation time of water
molecules in gels suggests a non-rigid solid state [4].
Thermal studies show that the amount of freezing bound
water, which is calculated based on the melting and
crystallization ethalpies of ice in the hydrogels, is higher
than that of hydrophilic linear polymers [5]. It is thought
that water molecules play an important role during junction
zone formation of polysaccharide hydrogels.

In this study, the thermal behaviour of restrained water
molecules in gel-like suspension of microfibril cellulose is
investigated during molecular association via hydrogen
bonding between cellulose molecules. The above process
accompanies the elimination of water molecules which are
bound directly to the hydroxyl group of cellulose.

EXPERIMENTAL

Micro-fibril cellulose was obtained from Daicell Chemical
Industry Co. The sample was dispersed in an excess amount
of pure water using an ultrasonic oscillator at 25°C. The
water content (W_c) of the above sample was more than 100 g/g
(W_c=grams of water/gram of dry sample, g/g).

A Perkin Elmer differential scanning calorimeter,DSC II was
used. The sample weight was ca. 2.5 mg and scanning rate
was varied from 1 to 40K/min. Transition temperatures and
heat of transition were calculated according to the method
previously reported [6,7]. Pure water was used as standard
material.

The bound water content was calculated by the method
reported previously [8].

$$W_c = W_{nf} + W_{fb} + W_f \tag{1}$$

where W_{nf} is the amount of non-freezing bound water, W_{fb} the
freezing bound water and W_f free water. The summation of
W_{fb} and W_f was calculated from the enthalpy of
crystallization and that of melting.

RESULTS AND DISCUSSION

The suspension of MFC (W_c>20,g/g) is a highly viscous
solution and looks like a homogeneous pseudo gel. When
water is squeezed by pressing or centrifugation, the pseudo
gel starts to separate into a gel-like portion and water.
This suggests that MFC in the pseudo gel form retains a
large amount of water. In order to investigate the
structural change of water in the above process of water
separation in pseudo-gel formation, phase transition

behaviour of water in the system was investigated by differential scanning calorimetry (DSC).

Figure 1 shows DSC cooling curves of water-MFC systems having various water contents. It is known that phase transition temperature of water depends on the size of the sample [9]. We maintained the sample weight at a constant value (ca. 2.5 mg), in order to eliminate the effect of weight variation. Exothermic peaks in Figure 1 are attributed to the crystallization of water in the system. As shown in Figure 1, crystallization peaks vary as a function of W_c in an unusual manner. When water content is in the range higher than 9.0, a low temperature peak (TcII) is observed. In the W_c's ranging from ca. 9.0 to 5.0, two sharp crystallization peaks are observed. In the W_c ranging from 5.0 to 1.0 crystallization of high temperature peak (TcpI) is observed.

Figure 2 shows representative DSC melting curves of MFC-water system having various water contents. The samples were heated at 10K/min after cooling from 310K to 200K at

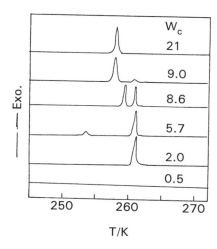

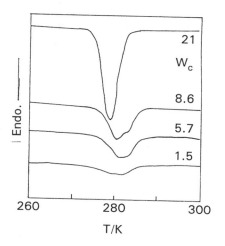

Figure 1 Representative DSC cooling curves of MFC-water systems.
Cooling rate=10K/min
Numerals in the figure show water content (W_c),(g/g).

Figure 2 Representative DSC melting curves of MFC-water systems.
Heating rate=10K/min
Numerals in the figure show W_c (g/g).

the cooling rate of 10K/min. The two peaks overlap in the melting curves and a large sub peak is observed on the the low temperature side of the main peak. All of the quenched samples were heated at various heating rates and the peak height of the main peak and the low temperature side peak was calculated. The side peak of melting decreased with decreasing heating rate, suggesting that the low temperature peak is attributed to the melting of unstable ice in the system [5].

Figure 3 shows the isothermal crystallization curves of water in MFC with W_c of 8.9. It is clearly seen that there are two types of freezing water in the system. When the crystallization temperature (T_c) ranged from 262K to 259K, a fast crystallization peak can be observed. A slow crystallization peak is observed from 258K.

Figure 4 shows the isothermal crystallization curves of the sample with W_c=5.7. In contrast to the sample with W_c=8.9, the fast crystallization peak is far larger than that of the slow crystallization peak. This indicates that the fast component is dominant when W_c decreases.

Figures 1 to 4 indicate that there are two kinds of freezing water in the system, one has a slow crystallization rate and the other a fast rate. After finishing the isothermal crystallization at each T_c, the sample was immediately heated from T_c to 310K. Figure 5 shows DSC melting curves. The sub peak observed at the low temperature side corresponded to the slow crystallization peak.

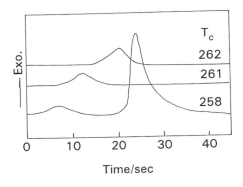

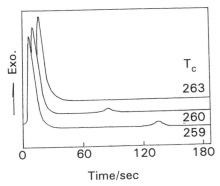

Figure 3 Isothermal crystallication curves of MFC-water system with W_c=8.9

Figure 4 Isothermal crystallication curves of MFC-water system with W_c=5.6.

Figure 6 shows the relationship between W_c and W_{nf} calculated from equation 1. W_{nf} values decreased from $W_c=ca.$ 5. It is reported that W_{nf} of ordinary natural celluloses is 0.1 to 0.2 depending on the crystallinity of the samples [10]. This fact indicates that MFC retains a far larger amount of W_{nf} compared with the other celluloses.

The above results suggested that MFC in the form of pseudo gel retains a large amount of bound water. The fact that the crystallization of water was observed on the   low temperature side suggests that water molecules in the system are affected by the presence of MFC. During the dehydration process, cellulose molecules approach the adjacent molecules and link up with each other through hydrogen bonding. At the same time, water molecules tightly bound to the hydroxyl group of cellulose are excluded from the aligned cellulose molecules. Two crystallization peaks shown in Figure 1 indicated that there are two kinds of water in the system, one is easily crystallized and the other takes   longer to crystallize. In the above reorganization stage, water is separated into the excluded free water and bound water attached to MFC.   In the last stage of dehydration, cellulose molecules are coagulated and the system separates into cellulose and water. A simple illustration of the above process is presented in Figure 7.

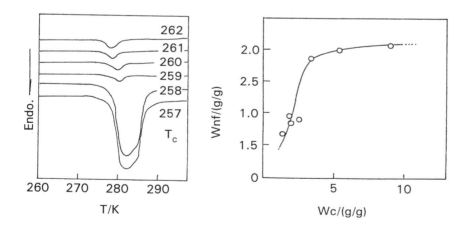

Figure 5 DSC melting curves of water in MFC after iso-thermal crystallization. $W_c=8.9$

Figure 6 Relationship between non-freezing water content (W_{nf}) and W_c.

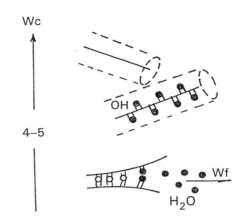

Figure 7 Schematic illustaration
of water-MFC system during dehydration

REFERENCES

[1]   Burchard,W; Ross-Murphy, S.W. (Eds.) Physical Networks
      -Polymers and Gels, Elsevier Applied Science, London
      (1990)
[2]   DeRossi D.; Kajiwara, K.; Osada, Y; Yamauchi, A. (Eds.)
      Polymer Gels - Fundamentals and Biomedical
      Applications, Plenum Press, New York and London (1991)
[3]   Hatakeyama, T; Yoshida, H; Nakamura, K.; Hatakeyama, H.
      in Maeno, N.; Hondho, T., (Eds.) Physics and Chemistry
      of Ice, Hokkaido University Press, Sapporo, (1992)
      p262.
[4]   for example, Woessner, D.E.; Snowden,Jr., B.S. J.
      Colloid and Interface Sci., **1970**, 34, 290.
[5]   Hatakeyama, T.; Yamauchi, A; Hatakeyama, H. Eur. Polym.
      J., **1987**, 23, 361.
[6]   Nakamura, S.; Todoki, M.; Nakamura, K.; Kanetsuna, H.
      Thermochimica Acta, **1998**, 136, 163.
[7]   Hatakeyema, T.; Kanetsuna, H., Thermochimica Acta, **1989**
      138, 327.
[8]   Hatakeyama, T.; Nakamura, K.; Hatakeyama, H., Thermo-
      chimica Acta, **1988**, 123, 153.
[9]   Angell, C.A., in Franks, F. (Ed.), Water, A Compre-
      hensive Treatise, Plenum Press, New York and London, 7,
      Chap. 1 (1982)
[10] Nakamura, K.; Hatakeyama, T.; Hatakeyama, H., Text.
      Res. J., **1981**, 51, 607.

32

Orientation of crystallites in the bacterial cellulose-fluorescent brightener complex membrane

Akira Kai and Ping Xu - Department of Industrial Chemistry, Tokyo Metropolitan University, Hachioji, Tokyo 192-03, Japan.

ABSTRACT

The orientation of crystallites in the membranes of the bacterial cellulose-fluorescent brightener complex, regenerated Cell I obtained from the complex by dye-extraction, and bacterial cellulose was examined through X-ray measurements. In the X-ray diffraction profile of the complex membrane by the reflection method, only a strong diffraction pattern from the $(1\bar{1}0)$ plane was observed. In contrast, in the X-ray diffraction profile by the transmitting method a strong diffraction pattern at near $2\theta = 22.1°$ was observed. The $(1\bar{1}0)$ plane and the plane observed at near $2\theta = 22.1°$ are at right angles to each other. The $(1\bar{1}0)$ plane of the complex is parallel to the surface of the membrane as the planes of bacterial cellulose and the regenerated Cell I. It is suggested that the diffraction pattern at near $2\theta = 22.1°$ is due to the (110) plane.

INTRODUCTION

It is well known that the $(1\bar{1}0)$ plane of crystallites in the bacterial cellulose membrane is parallel to the membrane surface. However, there is no report about the orientation of crystallites in the cellulose-fluorescent brightener complex membrane. In this case the dye is included between the cellulose sheets corresponding to the $(1\bar{1}0)$ plane of cellulose I (Cell I) [1,2]. It is of particular interest to understand the structure of the complex and how it is affected by the crystal orientation.

In this paper, the orientation of crystallites in the complex membrane and the structure change of the complex, by dye-extraction were examined through X-ray measurements.

EXPERIMENTAL

Preparation of complex membrane: A 60 ml cell suspension (Acetobacter xylinum IFO 13693) was added to 140 ml of the complex medium with the fluorescent brightener at 0.1 wt%, and was incubated for 24 h at 28.0°C. Afterwards the product was rinsed with 0.2 wt% aqueous NaOH solution for 48 h in order to remove the brightener not related to the dyeing; this was washed out with the distilled water until alkali-free, and the product dried on a teflon plate to obtain the complex membrane. The content of the brightener of this complex is 0.095 mol/glucose residue mol [3].
    Preparation of regenerated Cell I membrane: After the never-dried complex was extracted by boiling in a solution of 70 vol% aqueous ethanol for 18 h (aqueous ethanol was exchanged for fresh every 3 h), the regenerated Cell I refined by boiling in 1 wt% aqueous NaOH solution for 10 h under N_2 atmosphere was prepared for membrane on the teflon plate.
    Dye-extraction of the complex membrane: The dye-extraction of the complex membrane was performed by boiling in a solution of 70 vol% aqueous ethanol for a given time. The extraction time is the integrated time.
    X-ray measurements: The MXP[18] X-ray diffraction instrument (MAC SCIENCE made) was used for the X-ray diffraction measurements. The X-ray diffraction diagrams of the sample membranes were obtained with the reflection method and the transmitting method. In the case of the reflection method, the sample was set on the specimen rotation attachment to obtain the diffraction profile. On the other hand, in the transmitting method, the sample was set on the fiber specimen attachment to obtain the diffraction profile so that the X-ray could incident perpendicular to the membrane surface. The measurement condition of the X-ray diagram is as follows: X-ray: CuKα (with Ni filter); Occurrence condition: 40 kV, 200 mA; Scattering slit: 1° (in case of the transmitting method: collimator: 1 mmφ), Receiving slit: 0.15 mm, Scanning speed; 4°/min.

RESULTS AND DISCUSSION

    Fig. 1 shows the X-ray diffraction profile of each sample of bacterial cellulose, complex and regenerated Cell I obtained with the reflection method. Fig. 2 shows the X-ray profile of each sample obtained with the transmitting method. Table 1 shows 2θ value and the spacing of the diffraction peaks of bacterial cellulose and the complex.
    Takai et al. have found that when the bacterial cellulose membrane is washed with the hydrophilic solvent and is dried on the glass plate, its (1$\bar{1}$0) plane is preferentially parallel to the membrane surface, and that when it is washed with the

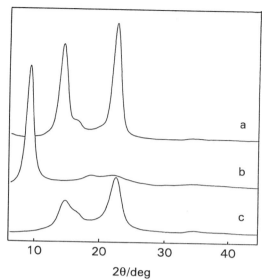

Fig. 1. X-ray diffraction profiles obtained by the reflection method: a, bacterial cellulose; b, complex; c, regenerated Cell I.

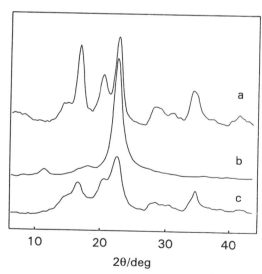

Fig. 2. X-ray diffraction profiles obtained by the transmitting method: a, bacterial cellulose; b, complex; c, regenerated Cell I.

Table 1.   Lattice spacing of bacterial
cellulose and complex

	Bacterial cellulose		Complex	
	2θ (°)	d/Å	2θ (°)	d/Å
(1$\bar{1}$0)	14.6	6.1	9.4	9.4
(110)	16.6	5.4	22.1	4.0
(020)	22.6	3.9	--	--

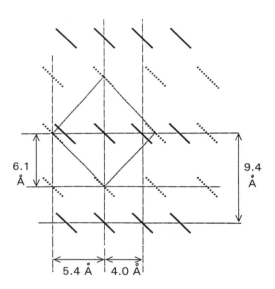

Fig. 3.   Schematic models of the reflecting planes of
bacterial cellulose and complex: Dotted line, bacterial
cellulose; solid line, complex.

hydrophobic solvent, the (1$\bar{1}$0) plane orientation drops [4]. In
this experiment, after the bacterial cellulose membrane is
washed with water, it is dried on the teflon plate. However, as
seen from the diffraction profile of bacterial cellulose in
Figs. 1 and 2, it is clear that the (1$\bar{1}$0) plane is parallel to
the membrane surface the same as in the case when dried on a
glass plate. Although the crystallinity and the orientation of
the regenerated Cell I membrane slightly drops, compared with
that of bacterial cellulose, the (1$\bar{1}$0) plane has not lost the
tendency which is parallel to the membrane surface.
 In the diffraction by the (1$\bar{1}$0) plane of the complex, since

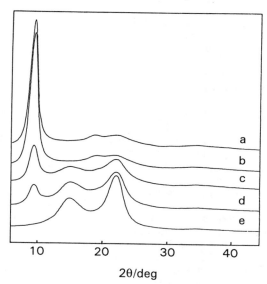

Fig. 4. Change in X-ray diffraction patterns of the
complex by dye-extraction.
Treatment time (h): a, 0; b, 0.5; c, 10; d, 18; e,35.

the brightener is included between the planes, the spacing of
the plane is expanded and appears at $2\theta = 9.4°$, and the diffraction patterns by the (1$\bar{1}$0) and (110) planes of bacterial cellulose disappear. The diffraction peak of the (1$\bar{1}$0) plane of
the complex obtained by the reflection method appears strongly,
in contrast, that obtained by the transmitting method is very
weak. These results show that the (1$\bar{1}$0) plane of the complex is
parallel to the membrane surface the same as the (1$\bar{1}$0) plane of
bacterial cellulose.
    Meanwhile, the diffraction of the complex which is observed
at near $2\theta = 22.1°(4.0$ Å) appears at a position near the diffraction angle $2\theta = 22.6°$ (3.9 Å) of the (020) plane of bacterial cellulose. However, the diffraction of the complex differs from that of the (020) plane of bacterial cellulose. Because it appears as a very weak diffraction in the reflection
diffraction profile, while it indicates strong diffraction in
the transmitting method. This indicates that the diffraction
plane observed at near $2\theta = 22.1°$ is perpendicular to the (1$\bar{1}$0)
plane. Therefore, it is suggested that this diffraction plane
is the (110) plane as shown in Fig. 3.
    The results from the X-ray measurements support the concept
that bacterial cellulose is extruded from the cell in the shape
of the cellulose sheet corresponding to the (1$\bar{1}$0) plane as
reported previously [1,2]. The hydrogen bonding is not recognized in the nascent cellulose sheet, and it is suggested that
the structure having the Cell I formative ability is maintained
by hydrophobic bonding or van der Waals forces [5,6]. When the

dye is included between the nascent cellulose sheets, hydrogen bonding among the sheets is prevented, and it is therefore suggested that the cellulose chain distance in the sheet will change from 5.4 Å to 4.0 Å.

From the study of the electron and X-ray diffraction results of the sample prepared by washing the product obtained from the acetic acid bacteria culture with the fluorescent brightener in HCl aqueous solution of pH 3, Haigler et al. reported that the brightener is dyed on the protofibril surface in the stacked state [7], and refuted our complex concept. They recognized only the plane of 4.0 Å distance, but did not recognize the existence of the plane of 9.4 Å distance. They explain that this 4.0 Å plane is the diffraction based on the dye stack (benzene plane of the dye). However, the many diffractions by the brightener crystallites are recognized on the X-ray diffraction profile of the sample prepared by washing the product with HCl water of pH 3, while these diffraction easily disappear if the sample is washed out with 0.2 wt% aqueous NaOH solution [8].

Fig. 4 shows the X-ray diffraction profile change as a function of extraction time of the complex membrane. When the extraction time was approximately 30 minutes, no change was recognized in the diffraction pattern. When the extraction time exceeds 10 h, the diffraction strength by the (1$\overline{1}$0) plane of the complex gradually becomes weak as time elapses, while the diffraction strength by the (1$\overline{1}$0), (110) and (020) planes of Cell I becomes strong. However, preferential orientation is not recognized on each plane of the regenerated Cell I. This indicates that the state of the crystal orientation will change in the process of the dye-extraction of the complex and the regeneration of Cell I. To this, as mentioned above, the (1$\overline{1}$0) plane of the membrane prepared with the regenerated Cell I is preferentially parallel to the membrane surface, the same as the bacterial cellulose membrane.

REFERENCES

1. Kai A. Makromol. Chem. Rapid Commun. 1984, 5, 307
2. Kai A. Makromol. Chem. Rapid Commun. 1984, 5, 653
3. Kai A.; Xu P. Polymer J. 1990, 11, 955
4. Takai M.; Tsuta Y.; Hayashi J.; Watanabe S. Polymer J. 1975, 7, 157
5. Kai A.; Koseki T. Makromol. Chem. 1985, 186, 2609
6. Kai A.; Kogusuri J: Koseki T; Kitamura H. "CELLULOSE and WOOD Chemistry and Technology", Schuerch C., Ed, John Wiley & Sons Ltd., New York, 1989, p 507
7. Haigler C. H.; Chanzy H. J. Ultrastruct. Mol. Struct. Res. 1988, 98, 299
8. Ishikita S.; Xu P.; Kai A. Abstracts of Papers, Part II, 90th Annual Spring Meeting of the Chemical Society of Japan, 1990, 1074

33

Characteristic hygroscopicity of bacterial cellulose-brightener complex

Akira Kai and Ping Xu - Department of Industrial Chemistry, Tokyo Metropolitan University, Hachioji, Tokyo 192-03, Japan.

ABSTRACT

The relation between the structure and hygroscopicity of cellulose-brightener complex and Cell I, which was regenerated from the complex by dye-extraction, was examined by a vapor phase deuteration-IR spectroscopy method. In 60 min, 37% of the OH groups of bacterial cellulose were deuterated, In contrast, 80-90% of the OH groups of cellulose in the complex were deuterated in 5 min. A 43% portion of the OH groups of regenerated Cell I was deuterated in 5 min and 48.6% in 60 min. The OH groups of the complex showed a broad peak near 3400 cm^{-1}. In the spectrum, however, absorptions due to intramolecular- and intermolecular-hydrogen bonding, which appeared in that of bacterial cellulose, were not observed. The OD groups of the complex also showed a broad peak at 2517 cm^{-1}, but no absorptions due to intramolecular- and intermolecular-hydrogen bonding appeared. In the OH region of the regenerated cellulose, an absorption due to intramolecular-hydrogen bonding at 3345 cm^{-1} appeared, indicating the regeneration of Cell I, but the band was broad owing to low crystallinity. The increase in hygroscopicity of the complex is thus attributable to the inclusion of the brightener between the mono-molecular cellulose sheets corresponding to the (1$\bar{1}$0) plane thereby preventing the direct binding of the sheets.

INTRODUCTION

According to the results of CP/MAS ^{13}C NMR measurements, it was suggested that the cellulose component of the bacterial cellulose-fluorescent brightener complex will have a structure similar to that of the non-crystalline component of bacterial cellulose [1,2]. Furthermore, even if the complex is dipped into the 4.3 or 5.0 wt% aqueous NaOH solution in which bacterial cellulose never transfers to Cell II, II type cellulose is regenerated [3,4]. These results can be considered due to the brightener being included between the cellulose sheets [5,6] and this then prevents the sheets hydrogen bonding. It is suggested that the structure of the complex estimated from the mercerization behavior, X-ray [7] and ^{13}C NMR measurements of the complex that the hygroscopicity differed entirely from the Cell I samples.

In this paper, the relation between the structure and hygroscopicity of the complex, its regenerated Cell I and bacterial cellulose was examined by a vapor phase deuteration-IR spectroscopy method.

EXPERIMENTAL

Preparation of the membranes of the complex and its regenerated Cell I: Each membrane was prepared to the thickness suitable for IR spectroscopy after the methods of the previous paper [7].

Deuteration method: The deuteration method of the sample was performed after the method described in the previous paper [8]. The sample membrane was set to the cell so that the IR beam was perpendicular to the membrane surface. After the inside of the cell was dried under reduced pressure (10^{-4} torr) for 48 h, the vapor phase deuteration of the sample was performed by D_2O with purity of 99.8% for a given time at room temperature. Then this was dried under reduced pressure for 60 min again, and its IR spectrum was measured. FTIR 4000 (Shimazu made) was used for the IR spectroscopy.

The degree of deuteration (D) was obtained as described below: The value of the absorbance band at 3405 cm^{-1} (intermolecular hydrogen bonding of Cell I) in the IR spectrum of the sample deuterated for the given time was divided by the absorbance of the CH band (2900 cm^{-1}) proportioning to the thickness of the cellulose membrane, was divided by the value of the never-deuterated sample (bacterial cellulose, complex and regenerated Cell I).

$$D = (1 - A_t/A_{t=0})$$

Where, $A_{t=0}$ is $A_{3405\ cm^{-1}}/A_{2900\ cm^{-1}}$ value at 0 h of the deuteration time. The deuteration time is the integrated time since one sample has been measured by repeating the deuteration - drying cycle.

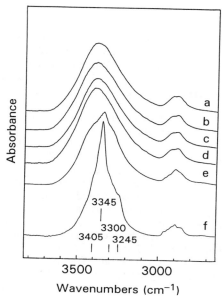

Fig. 1. IR spectra of the complex (a - d), Cell I regenerated from sample d (e), and bacterial cellulose (f). The content of brightener in the complex (mol/ glucose residue mol): a, 0.059; b, 0.083; c, 0.095; d, 0.096;

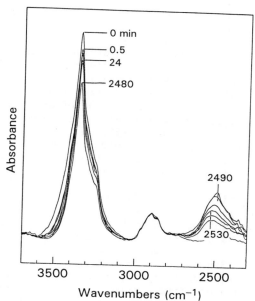

Fig. 2. Change in IR spectra of bacterial cellulose by deuteration.

RESULTS AND DISCUSSION

Fig. 1 shows the IR spectra of the complex, the regenerated Cell I and bacterial cellulose. In the spectrum of the OH groups of bacterial cellulose, the absorption based upon the intramolecular hydrogen bonding appears at 3345 cm^{-1}, and the absorption by the intermolecular hydrogen bonding appears at 3405 cm^{-1} and 3245 cm^{-1} as a shoulder. However, the spectra of the OH groups of the complex did not depend on the content of the brightener, and all complex samples showed broad spectra with a peak value near 3400 cm^{-1}. In the spectrum of the regenerated Cell I, although the absorption of the OH groups by the intramolecular hydrogen bonding indicating the characteristic of Cell I is observed at 3345 cm^{-1}, the broad absorption as well as that of the complex is observed. This indicates that the crystallinity of the regenerated Cell I is lower than that of bacterial cellulose.

The IR spectra of the deuterated bacterial cellulose is shown in Fig. 2. When bacterial cellulose is deuterated for 3.5 min, 28% of OH groups of it is deuterated. The spectrum of the OD band at this time is a non-crystalline spectrum with the peak value at 2530 cm^{-1}. However, if the deuteration time becomes long, the peak of the OD band shifts to the lower wave number side. When bacterial cellulose is deuterated for more than 60 min, its spectrum indicates that the deuteration of the OH groups of the crystalline region occurs.

The spectra of the deuterated complex sample are shown in Fig. 3. 71.5 - 87.1% of the OH groups of cellulose of the complex are deuterated in 3 min. In 75 - 87 min, 92.4 - 97.3% of the OH groups are deuterated. If experimental error is considered, all OH groups of the complex may be deuterated in about 80 min. Furthermore, the remarkable thing in the spectra of the deuterated complex is that the spectra of the OH groups always indicate broad spectra with the peak value near 3400 cm^{-1} and the absorption by intramolecular and intermolecular hydrogen bonding observed on the OH band of bacterial cellulose is not observed at all, even if the degree of deuteration is different. Furthermore, the OD band also indicates the same broad spectra with the peak value at 2517 cm^{-1}, even if its degree of deuteration is changed. It becomes clear that the OH groups of the complex are accessible in the non-crystalline region but also most of them in the crystalline region.

Judging from the deuteration results of the complex, it is suggested that the direct binding of the cellulose sheets as well as intramolecular and intermolecular hydrogen bonding in the sheet in the complex does not exist as it does in bacterial cellulose, and that the distribution will exist in the hydrogen bonding strength. Since the brightener included between sheets in the complex prevents hydrogen bonding, it is suggested that the D_2O molecules diffuse into the complex and the exchange reaction with the OH groups of the sheet surface and D_2O occurs in all regions

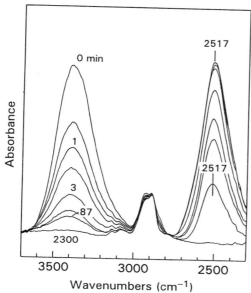

Fig. 3. Change in IR spectra of complex (sample b
in Fig. 1) by deuteration.

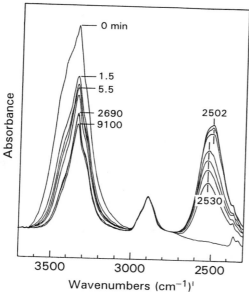

Fig. 4. Change in IR spectra of regenerated Cell I
(sample e in Fig. 1) by deuteration.

The IR spectra change of the regenerated Cell I by deuteration is shown in Fig. 4. The regenerated Cell I deuterated in a few minutes (1.5 to 5.5 min) showed spectra with the peak value based on the intramolecular hydrogen bonding at 3345 cm^{-1} and the intermolecular hydrogen bonding at 3405 cm^{-1} and 3245 cm^{-1} as a shoulder, which the spectra of the OH groups indicates as characteristic of Cell I. The degrees of deuteration were 31%, 43.9% and 50.1% in 1.5 min, 5.5 min and 77 min deuteration respectively. Since the OD band of the sample deuterated in 77 min indicates its spectrum is characteristic of the non-crystalline cellulose with the peak value at 2530 cm^{-1}, the OH groups of 50.1% is the non-crystalline OH groups. According to the deuteration time, the absorption peak of the OD band shifted to the lower wave number side. This indicates that the OH groups of the higher order region has been deuterated. Accordingly, the percent of the OH groups in the crystalline region of the regenerated Cell I is 49.9% in the 77 min deuteration. This value is almost consistent with 45% of crystallinity [3] obtained by the X-ray method. The deuteration results of the regenerated Cell I indicates that the accessible region is larger than that of bacterial cellulose, and that the crystallinity is lower. This is consistent with X-ray and ^{13}C NMR results.

From the deuteration experimental results of the complex, it became clear that all OH groups of the cellulose component become accessible OH groups due to the brightener included between sheets, and that the complex has extreme high hygroscopicity. Moreover, from the spectra of the OH and OD bands of the deuterated samples, it became clear that the crystallinity of regenerated Cell I is lower than that of bacterial cellulose and, as a result, its hygroscopicity becomes high. The conclusions from the deuteration experimental results are consistent with the conclusions from X-ray and ^{13}C NMR measurements.

REFERENCES

1. Kai A.; Horii F.; Hirai A. Makromol. Chem. Rapid Commun. 1991, 12, 15
2. Xu P.; Kai A.; Horii F.; Hu S. Polym. Preprints of Japan, 1990, 39, 3926
3. Kai A.; Xu P. Polymer J. 1990, 22, 955
4. Kai A.; Xu P. Polymer J. 1991, 23, 1
5. Kai A. Makromol. Chem. Rapid Commun. 1984, 5, 307
6. Kai A. Makromol. Chem. Rapid Commun. 1985, 5, 653
7. Kai A.; Xu P. this volume.
8. Okajima S.; Kai A. J. Polym. Sci., A-1, 1968, 6, 2801

34

Effect of Ca ions on the structural change of sodium alginate hydrogels

K. Nakamura*, T. Hatakeyama** and H. Hatakeyama*** - *Otsuma Women's University, 12, Sanbancho, Chiyoda-ku, Tokyo, 102, Japan. **Research Institute for Polymers and Textiles, 1-1-4, Higashi, Tsukuba, Ibaraki, 305, Japan. ***Industrial Products Research Institute, 1-1-4, Higashi, Tsukuba, Ibaraki, 305, Japan.

ABSTRACT

The structural change of sodium alginate (NaAlg) hydrogels having D-mannuronic acid / L-guluronic acid (M/G) ratio of 0.18 and 1.28 was investigated in the presence of Ca ions. The effect of Ca ions on the bound water content of NaAlg was measured by differential scanning calorimetry (DSC).

The minimum value of the bound water content of the G segment rich NaAlg (0.30 g/g) was observed when the Ca ion concentration was less than 0.5%. The bound water content of M segment rich NaAlg showed a linear increase with increasing Ca ion concentration. From the difference in bound water content between M and G rich NaAlgs, it was clear that Ca ions were covered by G segments and water molecules were excluded from the G segment of NaAlg.

INTRODUCTION

Alginic acid is a linear copolysaccharide consisting of D-mannuronic acid and L-guluronic acid residues extracted from brown algae [1-3]. The conformation of polymannuronic acid is a series of obtuse angles, and that of polyguluronic acid is a series of acute angles [4]. Sodium alginate (NaAlg) is a polyelectrolyte and is widely used in various fields, for example, as a gelling agent and a stabilizer in the food industry, and a dying adhesive in the textile industry.

It is known that alginate salts of monovalent alkali metals are soluble in water. It is also known that alginate salts of divalent and trivalent metals, such as Ba, Ca, Cu, Fe, Co, Al except Mg and Hg form hydrogels [5]. Hydrogels can be easily formed when Ca^{2+} ions are replaced with Na^+ ions of NaAlg in an aqueous solution. The above molecular arrangement of NaAlg gel with Ca ions exchanged for sodium is known as the "egg-box model" [4,6]. The gelation mechanism of alginate with Ca^{2+} ions was also studied by the method of intrinsic viscosity measurement [7]. We have reported that NaAlg shows liquid crystalline properties when the water content (Wc) is between ca. 1.0 and 3.0 g/g (Wc=weight of water/weight of NaAlg) [8,9].

In this study, the mechanism of formation of alginic acid hydrogel with Ca^{2+} ions was investigated by measuring the bound water content by differential scanning calorimetry (DSC). The change of bound water contents with gelation was studied using samples having certain ratios of guluronic/mannuronic acids.

EXPERIMENTAL

Sodium alginates (NaAlgs) which are used as bioreactors were obtained from Kibun Food Chemical Co. Ltd. Mannuronic to guluronic acids ratios (M/G) of NaAlg were 1.28 and 0.18. The mixing ratio of $CaCl_2$ with NaAlg was defined as $CaCl_2$/NaAlg (g/g).

The phase transition of NaAlg-water-$CaCl_2$ system was measured using a differential scanning calorimeter, Seiko Instruments Inc., SSC-220, equipped with an auto-cooling apparatus. A DSC sample was prepared in the following way. A certain amount of NaAlg was mixed with a $CaCl_2$-water solution in a sample tube and then the mixture was allowed to stand for one night. The sample was then weighed in a DSC aluminum volatile sample pan which was pre-treated in boiling water for 2 hours. DSC curves were obtained in the temperature range from $-150^{\circ}C$ to $80^{\circ}C$. The scanning rate was $10^{\circ}C/min$. Phase transition temperatures and enthalpies of transition of absorbed water were calibrated using pure water as a standard. The nonfreezing water content (Wnf) of the system was calculated using the melting enthalpy of water (334J/g). The quantitative measurement of bound water content was reported in detail in our previous papers [10-12].

RESULTS AND DISCUSSION

Fig.1 shows the schematic DSC curves of NaAlg-water (A) and NaAlg-water-CaCl$_2$ (B) systems having M/G=0.18. When the sample of NaAlg-water system of Wc=3.0 g/g without the presence of CaCl$_2$ was cooled from room temperature to -100°C, a sharp crystallization peak (T_c) was observed at about -20°C. Then the sample was heated from -100°C to 70°C and a melting peak (T_m) was observed at about -20°C. In the case of NaAlg-water-CaCl$_2$ system, Tc was observed at about -25°C, and a small exothermic peak (T_{cII}) was also observed at about -50°C in the cooling curve as shown in the figure. In the heating curve of the NaAlg-water-CaCl$_2$, the sub-peak (T_{mII}) which was not observed in the NaAlg-water system, appeared at about -20°C, and T_m was observed at -15°C. When the sample was cooled to ca. -40°C, T_{cII} was not observed. On heating to room temperature, T_m was observed as shown in this figure.

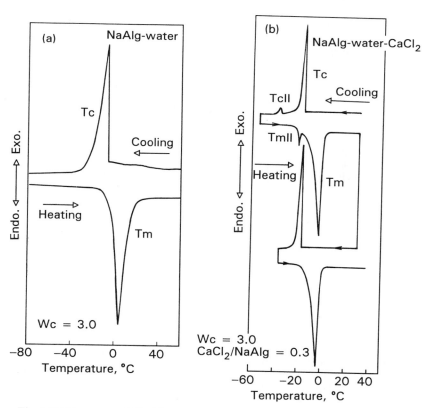

Fig.1 DSC curves of NaAlg-water (A) and NaAlg-water-CaCl2 (B) systems containing Wc=3.0.

From the difference between NaAlg-water and NaAlg-water-CaCl$_2$ systems, it was considered that the sub-peaks of T_{cII} and T_{mII} were due to the presence of Ca ions.

Fig.2 shows DSC heating curves of the NaAlg-water-CaCl$_2$ system with various concentrations of CaCl$_2$. When the sample with CaCl$_2$/NaAlg ratio higher than ca. 0.2 was heated from ca. $-100°C$ to $80°C$, T_m was observed at about $0°C$ and T_{mII} was observed at $-20°C$. T_{mII} became larger with increasing concentration of CaCl$_2$. In order to exchange all Na ions in NaAlg with Ca ions, 0.28 (111/198/2) g/g of CaCl$_2$ is required, since the molecular weights of NaAlg and CaCl$_2$ are 198 and 111 respectively.

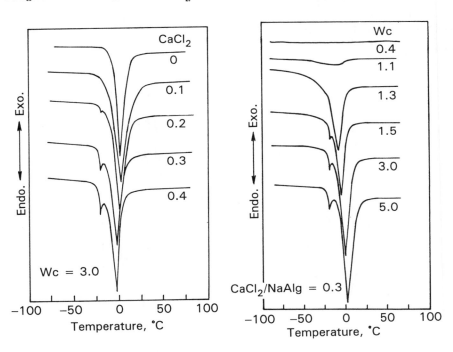

Fig.2 DSC heating curves of the NaAlg-water-CaCl2 systems with various concentrations of CaCl2. Wc=3.0 g/g.

Fig.3 DSC heating curves of the NaAlg-water-CaCl2 systems containing various Wc. CaCl2 / NaAlg = 0.3 g/g.

Fig.3 shows the DSC curves of NaAlg-water-CaCl$_2$ systems containing various Wcs. Water molecules in the system containing less than ca. 0.6 g/g of Wc did not freeze and became nonfreezing water. However, when Wc was more than 1.0, T_m was observed. When Wc was more than 1.5 g/g, T_{mII} appeared. T_{mII} became larger with increasing Wc. This means that water molecules which were restricted by alginate molecules show a T_m and that T_{mII} appears when the excess amount of water molecules are restricted by Ca ions.

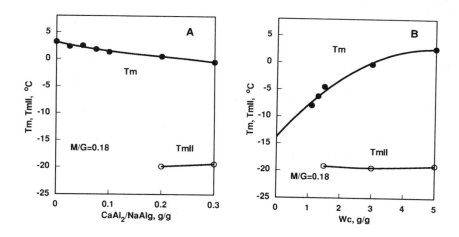

Fig.4 Phase transition temperatures of NaAlg-water-CaCl2 systems plotted against CaCl2 concentration (A) and Wc (B). M/G ratio is 0.18.

Fig.4 shows phase transition temperatures of the NaAlg-water-CaCl$_2$ system plotted against CaCl$_2$ concentration (A) and Wc (B). T_m of the NaAlg-water-CaCl$_2$ system containing a certain Wc shifted to the lower temperature side as the CaCl$_2$ concentration increased. However, T_m of the sample having a certain CaCl$_2$ concentration shifted to the higher temperature side as Wc increased. The temperature of T_{mII} was not changed by CaCl$_2$ concentration and Wc. This means that water molecules were restricted strongly with Ca ions around alginic acid molecules and that the T_{mII} was not strongly influenced by alginate molecules but depended on excess amount of Ca ions.

Fig.5 shows the relationship between nonfreezing water content (Wnf) of the NaAlg-water-CaCl$_2$ system and concentration of CaCl$_2$. In a region in which T_{mII} is not observed in Fig.5 no excess Ca ions exist. In the case of the G segment rich sample (M/G=0.18), the curve of Wnf showed a minimum point at about 0.01 g/g CaCl$_2$ concentration and then gradually increased with increasing CaCl$_2$ concentration. However, in the case of that is the M segment rich sample (M/G=1.28), Wnf increased almost linearly with increasing $CaCl_2$ concentration. This suggests that water molecules are excluded by Ca ions in G segments under the process of forming more compact molecular structures.

Moreover, when one takes into consideration that G segments of NaAlg molecules form the compact structure with Ca ions, the Ca ion concentration of 0.24g/g (0.28x0.847, in the case of M/G=0.18) is needed in order to form the structure. However, T_{mII} was observed at ca. 2.0 of $CaCl_2$/NaAlg: that is, the experimental value of Wc in which T_{mII} was observed, is smaller than the calculated value. This suggests that a part of the G segments form compact structure. This corresponds the fact that T_{mII} of NaAlg-water-$CaCl_2$ system was observed in the presence of excess amounts of Ca ions. The $CaCl_2$-water system showed an endothermic peak at about 22°C. It was reported that the melting temperature of the eutectic compound of sodium chloride and water was -21.7°C [13-16].

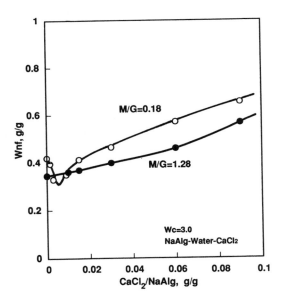

Fig.5 The relationship between nonfreezing water content (Wnf) of NaAlg-water-CaCl2 systems and CaCl2 concentration below 0.1.

Fig.6 shows the relationship between Wc and Wnf of NaAlg hydrogels having M/G=0.18 (A) and M/G=1.28 (B) systems. Wnf of NaAlg-water systems increased with increasing Wc in the region of less than ca. Wc=0.5 and then saturated at ca. Wnf=0.42 (M/G=0.18) and Wnf=0.34 (M/G=1.28). However, Wnf of the NaAlg-water-$CaCl_2$ systems increased with increasing Wc. This means that alginate hydrogel which is rich in G segments contains a higher amount of Wnf=0.42 compared with 0.32 g/g of that which is rich in M segments. Moreover, Wnf of alginate hydrogel with Ca ions is more than that of NaAlg-water system since Wnf of $CaCl_2$ is rather high at about 2.3 g/g. Therefore, the effect of Ca ions on Wnf of alginate hydrogels is very high.

Wnf is a sum of the Wnf's of the G and M segments. Therefore, Wnf is expressed as follows:

$$Wnf = W_{nfG}*G + W_{nfM}*M \qquad (1)$$

where, W_{nfG} and W_{nfM} are the amounts of nonfreezing water in G and M segments, and G and M are mole fractions of G and M segments in alginic acid molecules, respectively. In the case of the system without Ca ions, Wnf's of the systems having M/G=0.18 and 1.28 are 0.42 and 0.34 g/g, respectively. Therefore, from equation (1), W_{nfG}=0.450 g/g (4.95 mol/mol) and W_{nfM}=0.254 g/g (2.75 mol/mol) are obtained. In the case of the system having Ca ions, at the minimum point, W_{nfG}=0.305 g/g (3.36 mol/mol) and W_{nfM}=0.403 g/g (4.43 mol/mol) are obtained.

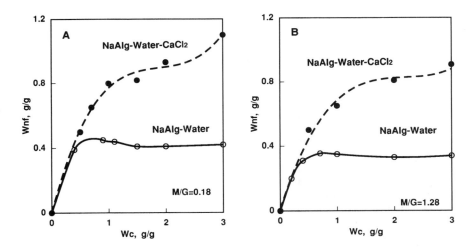

Fig.6 The relationship between Wc and Wnf of NaAlg hydrogels having M/G ratio=0.18 (A) and M/G=1.28 (B).

From the above results, it is considered that five water molecules are restricted by G segment when the system had no Ca ions and that after Ca ions are added to the system, 3.4 moles of water molecules are restricted by a G segment. However, 2.8 moles of water molecules were restricted by a M segment, and then after Ca ions were added to the system, 4.4 moles of water molecules were restricted. This means that G segments form a molecular structure of a series of acute angles and tight packing which is called the egg-box structure. In the presence of Ca ions, M segments form a molecular structure of a series of obtuse angles and loose packing. Accordingly, it may be said that water molecules exist locally in G and M segments of alginic acid molecules.

REFERENCES

1. Haug A., Larsen B., and Smidsrod O., Acta Chem. Scand., 1967, 21, 691.
2. Atkins E. D. A., Nieduszynski I. A., Mackie W., Parker K. D., and Smolko E. E., Biopolymers, 1973, 12, 1868.
3. Atkins E. D. A., Nieduszynski I. A., Mackie W., Parker K. D., and Smolko E. E, Biopolymers, 1973, 12, 1879.
4. Rees D. A. and Welsh E. J., Angew. Chem. Int. Ed. Engl., 1977, 16, 214.
5. Kasahara F., "Water Soluble Polymers", Nakamura M. Edt. Kagakukogyo-sha, Tokyo 1984, p25.
6. Rees D. A., Adv. Carbohydr. Chem. Biochem., 1969, 24, 267.
7. Gamini A., Civitarese G., Cesaro A., Delben F., and Paoletti S., Makromol. Chem., Macromol. Sympo., 1990, 39, 143.
8. Nakamura K., Hatakeyama T., and Hatakeyama H., Polym. J., 1991, 23, 253.
9. Nakamura K., Hatakeyama T., and Hatakeyama H., Sen-i, Gakkaishi, 1991, 47, 421.
10. Hatakeyama T., Nakamura K., and Hatakeyama H., Thermochemica Acta, 1988, 123, 153.
11. Yoshida H., Nakamura K., Hatakeyama T., and Hatakeyama H., Polymer, 1990, 31, 693.
12. Hatakeyama H., Nakamura K., and Hatakeyama T., "Cellulose and Wood", C. Schuerch, Ed., Wiley Interscience, New York, 1989, p419.
13. Fujiwara S. and Nishimoto Y., Anal. Sci., 1990, 6, 771.
14. Fujiwara S. and Nishimoto Y., Anal. Sci., 1990, 6, 907.
15. Fujiwara S. and Nishimoto Y., Anal. Sci., 1991, 7, 683.
16. Fujiwara S. and Nishimoto Y., Anal. Sci., 1991, 7, 687.

35

Industrial scale production of organosolv lignins: characteristics and applications

Jairo H. Lora, Albert W. Creamer, Leo C.F. Wu, Gopal C. Goyal - Repap Technologies Inc., 2650 Eisenhower Avenue, Valley Forge, PA 19482, USA. Phone: (215) 630-9630, telecopy: (215) 630-0966.

ABSTRACT

The characteristics and applications of the lignin obtained during the organosolv pulping of North American hardwoods in the industrial scale are discussed in this paper. This lignin is a polyphenol characterized by its high purity, low molecular weight, low polydispersity, low glass transition temperature and relatively high decomposition temperature. Industrially produced organosolv lignin has been successfully used as partial replacement for phenol formaldehyde resins in the manufacture of structural wood panels including plywood, oriented strand board and waferboard. Several approaches to increase substitution from current levels have shown promise in lab and pilot tests. The material has also found application in rubber compounding as a tackifier resin.

INTRODUCTION

Recently a new type of lignin has become available in industrial quantities. This lignin is recovered in the ALCELL® Process (ALCELL is a Registered Name owned by Repap Technologies Inc.), an organosolv pulping process that uses ethanol-water as the delignifying agent. This process has distinct environmental and capital cost advantages over conventional pulping technology and produces a hardwood pulp that is comparable in strength to kraft pulps (1,2). A Demonstration Plant for this technology has been in operation in Eastern Canada since the Spring of 1989 and has generated several thousand tons of pulp and lignin, and significant amounts of other co-products. The availability of large quantities of ALCELL lignin has led to a significant product development effort in several areas. In this paper the production and characteristics of ALCELL lignin will be reviewed and the results of the application of this new material in the wood adhesives and rubber industries will be discussed.

PROCESS DESCRIPTION

The ALCELL process has been described in detail in the literature (3-6) and only a brief description of the lignin recovery section will be given here. First, the spent liquor obtained as a result of pulping is flashed to atmospheric pressure and then lignin is precipitated by dilution and pH adjustment. The lignin is recovered as a cake and then is dried to obtain a powder product.

PLANT DESCRIPTION

The Demonstration Plant is located within a kraft pulping and paper-making complex in Newcastle, New Brunswick, Canada. The plant occupies approximately 3,000 square meters and is able to produce more than 6 tons of pulp and in excess of 2 tons of lignin per batch. The facility normally runs 24 hours a day, seven days a week.

Plant operation is fully automated and computerized. Control functions are performed through a Distributed Control System. More details on the characteristics of this plant have been reported elsewhere (2).

LIGNIN CHARACTERISTICS

The Demonstration Plant uses as raw material a mixture of hardwoods comprising 50% maple (Acer rubrum), 35% birch (Betula papyrifera) and 15% poplar (Populus tremuloides). A yield of about 20% lignin based on oven dried wood is normally obtained. Typical chemical and physical properties of the industrially produced ALCELL lignin are shown in Table 1. Properties for other commercially available lignins are included for comparison purposes.

Relative to other commercially available lignins, ALCELL lignin has lower content of sugars, ash and sulfur. The low content of sugars and the lack of sulfonic groups results in enhanced hydrophobicity. Compared with kraft and sulfite lignins ALCELL lignin is characterized also for a lower molecular weight, lower polydispersity and lower glass transition temperature (8,9). The low glass transition temperature of the ALCELL lignin and its high hydrophobicity partially explain why this material performs better than other lignins when tested in waferboard systems that use powder phenol formaldehyde adhesives (10). The low molecular weight and low polydispersity are expected to result in better processability when the material is used as a pre-polymer for the synthesis of other macromolecules.

Table 1
Typical Properties Of Industrially Produced Lignins

	Mixed Hardwood ALCELL Lignin	Softwood Kraft	Lignosul- fonate
Carbon, %	65-68	66*	53*
Hydrogen, %	5.5-6.5	5.9*	5.4*
Methoxyl, %	17-21	14*	12.5*
Ash, %	less than 1	3	2.5
Moisture, %	less than 3		
Wood Sugars, %	less than 0.5	low	up to 50%
Sulfur, %	less than 0.3 ppm	1.6*	6-7.9*
Specific Gravity	1.27		
Glass Transition Temperature, °C	90-100	140	not detected
Softening, ring and ball, °C	145		
Heating Value, J/kg	4.6×10^7		
Number Avg. Molecular Wt.	less than 900	2000*	400-150,000*
Weight Avg. Molecular Wt.	less than 2,000		
Median Particle Size, microns	20-40		

*Reference 7.

Proton NMR studies (9) reveal that ALCELL Lignin typically has 0.6-0.8 aromatic hydroxyl groups per C_9 unit. The material reacts with formaldehyde under alkaline conditions as evidenced by the increase in the 1.8-2.2 ppm band in the proton NMR spectrum of methylolated lignins that have been acetylated.

UTILIZATION AS A WOOD ADHESIVE

It has been reported previously that ALCELL lignin can be a direct partial replacement for powder phenolic resins used in waferboard and oriented strand board (OSB) manufacture (8-12). Routine use of the material by an OSB manufacturer started in the US in 1991.

Several approaches have been studied to increase substitution levels and compatibility, especially with resins used in the core layers of OSB panels (13). These include: increasing the platten temperature and/or extending the pressing time during panel manufacture and heat treating the lignin to achieve some polymerization prior to blending with the phenolic resin, and addition of small amounts of low molecular weight phenolic compounds to the binder formulation. The latter approach appears to be the most promising in terms of achieving higher levels of substitution without affecting the board manufacturing operation. As can be observed in Table 2 favorable results at substitution levels as high as 50% were obtained by replacing 10% of the lignin with phenol.

Table 2
Effect of Phenol Addition on Compatibility of ALCELL Lignin With Commercial Powder PF Resins Used in WB/OSB

	Modulus of Rupture % on Control	One-Cycle Durability % on Control	Internal Bond % on Control
80% Resin B+ 20% Lignin	90	86	80
80% Resin B+ 18% Lignin + 2% Phenol	104	103	97
50% Resin C+ 45% Lignin + 5% Phenol	108	109	106

Since a significant proportion of North American OSB manufacturers use liquid phenolic resins, the incorporation of ALCELL lignin in such systems has also been examined. As indicated above, ALCELL lignin reacts with formaldehyde and can be incorporated during resin synthesis as a partial substitute for phenol. Table 3 shows the properties relative to the control of 3-layer boards made using in the face a resin synthesized replacing 20% of the phenol with ALCELL lignin. As observed the boards made with the lignin containing resin were equal to or better than the control made without using any lignin.

Table 3
Performance of Lignin Containing Liquid PF Resin (20% Phenol Replacement)

Property	% Based on 100% PF Resin
Modulus of Rupture	115
Internal Bond	111
One-Cycle Durability	100
Six-Cycle Durability	109

ALCELL lignin has also been tested as component of plywood glues by directly replacing PF resin solids. Table 4 shows the results obtained in a commercial mill trial in which 15% of PF resin solids were replaced by ALCELL lignin. As observed, the panels obtained exceeded the 85% wood failure required for exterior grade applications in the USA.

Table 4
Performance of ALCELL Lignin Containing Glue Mix in Commercial Mill Trial

Construction	% Wood Failure after vacuum-pressure soak	% Wood Failure after 4-hour boil treatment
7.9 mm, 3-ply	93.2	94.7
16.7 mm, 5-ply	89.1	94.6
19.8 mm, 5-ply	93.4	93.8

UTILIZATION IN THE RUBBER INDUSTRY

During the manufacture of multilayered rubber goods such as tires, it is customary to add phenolic resins with a molecular weight of about 2,000 to enhance tack and processability. In laboratory testing (14) ALCELL lignin has shown to be an effective tackifier and to exhibit some synergism when used in conjunction with commercially available resins. In addition, since lignin is a hindered phenol, it exhibits some antioxidant properties, as illustrated in Table 5.

Table 5
Performance of ALCELL Lignin as an Antioxidant Enhancer

	Ultimate Elongation %	Tensile Strength %*
No Antioxidant	14	21
Commercial Antioxidant 3 PHR	40	50
Commercial Antioxidant 3 PHR ALCELL Lignin 5 PHR	50	76

*Based on unaged properties. Accelerated heat aging done at 100°C for 168 hours (ASTM D573).

CONCLUSIONS

A new lignin has become available in commercial quantities. This lignin is different from other lignins. It has high purity, low molecular weight, low polydispersity, low glass transition temperature and high functionality. The material is being used industrially as a partial replacement for phenolic resins used in OSB. Substitution levels can be increased by various techniques. The material can also be used as a rubber tackifier and antioxidant.

REFERENCES

1.   Pye, E.K. and Lora, J.H., 1991, The ALCELL Process: a proven alternative to kraft pulping, Tappi J., 74 (3), pp 113-118.

2. Lora J.H., Winner, S.R. and Pye, E.K., 1990, Industrial Scale Alcohol Pulping, American Institute of Chemical Engineers National Meeting, Chicago.
3. Williamson P., 1987, Pulp and Paper Canada, 88(12), pp. 47-49.
4. Lora, J.H. and Aziz, S., 1985, Tappi, 68(8) pp. 94-97.
5. Diebold V.B., Cowan W.F. and Walsh, J.K., 1978, US Pat 4,100,016.
6. Lora, J.H., Katzen, R., Cronlund, M. and Wu, C.F., 1988, US Pat 4,764,596.
7. Glasser, W.G., 1981, Forest Products J., 31 (3), pp 24-29.
8. Lora, J.H., Wu, L.C.F., Pye, E.K. and Balantinecz, J.J., 1989, Characteristics and Potential Applications of Lignin produced by an organosolv pulping process, in Lignin Properties and Materials, edited by Glasser, W.G. and Sarkanen, S., American Chemical Society Symposium Series 397, Washington, D.C.
9. Lora, J.H., Creamer, A.W., Wu, L.C.F. and Goyal, G.C. 1991, Chemicals Generated during Alcohol Pulping: Characteristics and Applications, Proceedings 6th International Symposium on Wood and Pulping Chemistry, pp 431-438, Melbourne, Australia.
10. Lora, J.H., Wu., L.C.F., Goyal, G.C., Creamer, A.W. and Ash, J.R., 1990, A progress report on the development of ALCELL lignin as a wood adhesive, Wood Adhesives Symposium, Madison, WI.
11. Lora, J.H., Wu, L.C.F. and Klein, W.R., 1991, Mill trials on the use of ALCELL® lignin as an adhesive for structural panels, 25th International Particleboard/Composite Materials Symposium, Pullman, WA.
12. Wu, L.C.F., Lora, J.H. and Edwardson, C.F., 1989, ALCELL Lignin: a new adhesive for waferboard, 23rd International Particleboard/Composite Materials Symposium, Pullman, WA.
13. Lora, J.H., 1991 Adhesives and Bonded Wood Symposium, Seattle, WA.
14. Trojan, M. and Lora J.H., 1990, American Chemical Society Rubber Division Meeting, Las Vegas, Nevada, Paper No. 68.

36

Mechanical spectroscopy – a tool for lignin structure studies

Anne-Mari Olsson and Lennart Salmén - STFI, Swedish Pulp and Paper Research Institute, P.O. Box 5604, S-114 86 Stockholm, Sweden.

ABSTRACT

In dynamic mechanical testing, i.e. mechanical spectroscopy, the sample is subjected to cyclic loads of various frequencies. The temperature at which a maximum in energy dissipation occurs denotes the onset of main movement of the polymer chain i.e. its glass transition temperature, Tg. For water saturated wood the main transition is that of lignin in situ, a transition which is greatly affected by the lignin structure.

The flexibility of the lignin macromolecule in hardwood species with different syringyl/guaiacyl ratios, in softwoods with compression wood lignin and in sulphonated wood have been evaluated by this method.

POLYMER BACKGROUND

Amorphous polymers such as most wood components are viscoelastic, i.e. they dissipate energy during deformation. At the glass transition temperature, Tg, the material softens, which is reflected in the fact that the energy dissipated reaches a maximum and mechanical properties change drastically [1].

Wood is a polymer blend consisting of crystalline cellulose, disordered cellulose, hemicelluloses, and lignin. For wood soaked in water the amorphous disordered cellulose and hemicellulose soften below room temperature [2]. Lignin with its branched crosslinked structure has a limited water uptake and is thus first softened at a temperature of 50°C to 100°C [3]. Therefore the glass transition temperature measured on wood in water in the temperature range of 10°C to 100°C is that of lignin, as illustrated in figure 1. Under these conditions, the other polymers are either in their glassy state, i.e. crystalline cellulose, or far above their softening temperature, i.e. hemicelluloses and disordered cellulose [2].

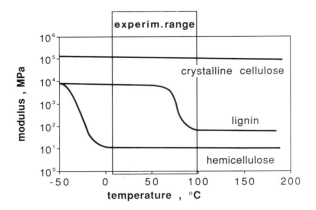

Figure 1. The storage moduli versus temperature for the water-saturated wood polymers show different glass transition temperatures. Lignin is the only component which passes through a transition in the experimental range from 10°C to 100°C.

Polymers can be visualized as entangled macromolecules. Their rigidity depends on the amount and type of bondings between the chains, and the degree of entanglement and type and number of groups that are bonded into the network. For a rigid network, more energy (i.e. a higher temperature) is needed to make the whole macromolecule move and reach its glass transition temperature, Tg, than in a polymer with a flexible structure and a lower Tg. A branched and crosslinked network like lignin is more rigid and has a higher Tg than a network with low crosslink density [4].

In dynamic mechanical testing, i.e. mechanical spectroscopy, the material is subjected to cyclic loads of different frequencies, where a higher load frequency gives a higher glass transition temperature. Flexible networks are more sensitive to changes in frequency than rigid ones. This is shown in an Arrhenius plot, figure 2, of log frequency versus the inverse of Tg in Kelvin (note that higher Tg is to the left on the abscissa). The slope in this diagram can be used to calculate the apparent activation energy of the softening process, which increases with increasing network rigidity.

On the basis of this knowledge, mechanical spectroscopy on wet wood has been used to study differences in lignin structure.

MECHANICAL SPECTROSCOPY ON WET WOOD

The lignin glass transition temperature has been measured on small samples of wet wood, 3 x 3 x 15 mm, in three point bending, the sample being kept in water throughout the test. The samples are subjected to a sinusoidal load with a set of frequencies from 0.03 Hz to 5 Hz with low stress amplitudes, in the

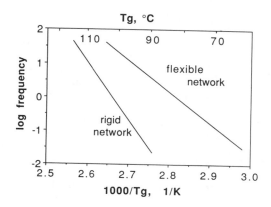

Figure 2. The Arrhenius plot for networks with different flexibilities shows a lower Tg and a greater sensitivity to stress frequency in a flexible network.

viscoelastic region i.e where the extension is reversible. The temperature has been scanned at a rate of 2°C/min from 10°C to 98°C, and the stress, strain and phase angle recorded. Values such as storage modulus and tanδ are then calculated.

The glass transition temperature, Tg, is most easily defined as the temperature at which the energy dissipation, tanδ, reaches its maximum, as indicated in figure 3.

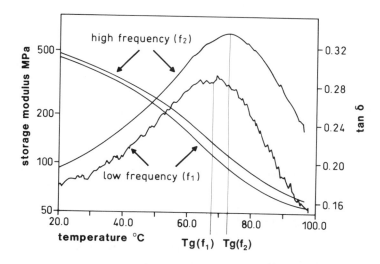

Figure 3. The calculated storage modulus and tanδ for black oak tested at load frequencies of 0.1 Hz and 0.8 Hz. The maximum in tanδ defines the lignin glass transition temperature which increases with increasing frequency.

EFFECTS OF NATIVE LIGNIN STRUCTURE

The monomers in lignin are the phenylpropane units, which are linked together to a branched crosslinked polymer network. Free groups such as methoxyl groups and free phenolic hydroxyl groups hinder crosslinking, and thus make the structure more flexible. Of the lignin units, syringyl contain two methoxyl groups, guaiacyl one, and parahydroxyphenylpropane no methoxyl groups.

Lignins from different hardwood species have different syringyl/guaiacyl ratios i.e. they contain varying amounts of methoxyl groups. This difference in lignin structure was evaluated by mechanical spectroscopy. Species with syringyl/-guaiacyl ratios from 0.0 (softwood) to 1.2 were tested.

Figure 4 shows the glass transition temperature at 0.1 Hz load frequency as a function of the syringyl/guaiacyl ratio. A high content of syringyl units give a low glass transition temperature as expected from the fact that lignin with a higher methoxyl content has a less crosslinked structure.

Softwood compression wood lignin contains parahydroxyphenylpropane. These units can form crosslinks without hindrance from any methoxyl groups. In figure 5, an Arrhenius plot for samples from the same tree of compression wood and normal wood shows that the parahydroxyphenylpropane of the compression wood gives lignin with a higher Tg and a lower sensitivity to load frequency. This confirms that the methoxyl content in the lignin is an important factor for the network flexibility.

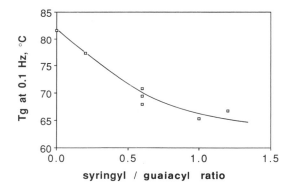

Figure 4. The softening temperature at 0.1 Hz for species with different syringyl/guaiacyl ratios.

EFFECTS OF SULPHONATION

In chemimechanical pulping processes, lignin is sulphonated in order that fibers can be separated to make good pulp. Wood is treated at high temperatures with sulphite solution under slightly basic conditions. By varying the reaction time, different pulp properties are achieved, since the reactions give differences in lignin structure [5]. Structural differences affecting the rigidity of this sulphonated lignin may also be studied by mechanical spectroscopy.

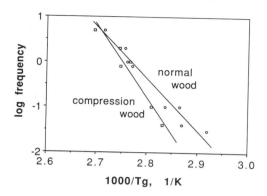

Figure 5. Arrhenius plot for compression and normal wood from red pine showing that compression wood lignin with a low content of methoxyl groups is a comparably rigid structure with high Tg.

Sulphonation lowers the glass transition temperature of lignin due to breakage of ether bonds [6]. The Arrhenius plot in figure 6 for sprucewood sulphonated to 0.3 % sulphur content shows that, after a long reaction time, the lignin is more rigid than lignin having the same sulphur content achieved by a shorter reaction time. This indicates the effect of additional slower crosslink reactions which become evident at long reaction times.

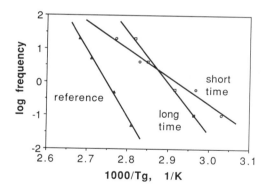

Figure 6. Arrhenius plot for sprucewood sulphonated to 0.3 % sulphur. The reaction times were 5 and 60 min respectively at 135°C. The difference in slope is interpreted as being an effect of additional crosslink reactions occurring at long reaction times.

CONCLUSIONS

Mechanical spectroscopy on wet wood has been shown to be very sensitive to structural differences in the in situ lignin. It is shown that the content of methoxyl groups in the lignin is important for the network flexibility affecting the glass transition temperature. This method may be a useful tool in investigating lignin structure, for example as it is affected by chemical treatments.

ACKNOWLEDGEMENTS

The authors thank Dr. John Obst, FPL, Madison for kindly providing hardwood samples as well as analysis of syringyl/guaiacyl ratios, and Dr. Anthony Bristow for the linguistic revision.

REFERENCES

1.   Sperling, L.H.; *Introduction to Physical Polymer Science* John Whiley & Sons, New York **1986.**
2.   Back, E.; Salmén, L.;, *Tappi*, **1982,** 65, 7:107.
3.   Salmén, L.; *J. Material Science*, **1984,** 19, 9:3090.
4.   Nielsen, L. E.; *Mechanical Properties of Polymers and Composites* Marcel Dekker, New York **1974.**
5   Ölander, K; Salmén, L.; Htun, M; *Nordic Pulp and Pap. Res. J.* **1990,** 5,.2:60.
6.   Atack, D.; Heitner, C.; *Trans. Tech. Sect. (Can Pulp Pap Assoc)* **1979,** 5, 4: TR 99.

37

Comparison of some molecular characteristics of star-block copolymers with lignin

Willer de Oliverira and **Wolfgang G. Glasser** - Department of Wood Science and Forest Products, and Biobased Materials Center, Virginia Tech, Blacksburg, Virginia, USA.

ABSTRACT

The influence of several architectural parameters of two types of star-block copolymers containing lignin as soft (core) segment on selected properties was examined. Linear (hard) arm segments were either polycaprolactone or fully substituted cellulose propionate. Total hard segment arm length ranged from approximately 5 to 50 repeat units (degree of polymerization). The properties examined included the Mark-Houwink-Sakurada exponential factor (a); intrinsic viscosity (IV); melting point (T_m); and heat of fusion (ΔH_f). The results suggest that copolymer behavior is dominated by the character of the linear hard segment component if arm length exceeds 10 repeat units for caprolactone and 20 to 30 units for cellulose propionate. This is interpreted in terms of superior compatibility between lignin and the cellulose derivative.

INTRODUCTION

Block copolymers are designed to combine the properties of two polymers with different chemistry, different morphology, and different (physical) properties. By coupling two types of polymers together, copolymers may emerge that have characteristics superior to those of the parent molecules [1,2]. In addition to creating molecules with unique properties as materials by themselves, block copolymers are also often used as compatibilizers for two otherwise incompatible polymers in mixture [3].

Star-block copolymers usually consist of copolymers representing two different chemistries whereby the inner (soft) core is multi-functional (i.e., branched) and the arm segment (like the prongs of a star) are represented by monofunctional linear (hard) polymer segments. These may have a high T_g or be capable of crystallization. Star-block copolymers have recently become interesting as thermoplastic elastomers and as three dimensional stabilizers for liquid crystalline polymers [4].

This study deals with the use of lignin, an aromatic biopolymer present in all woody plants [5], as inner (soft) core of two-types of star-block copolymers. The normally glassy nature of lignin is thereby altered, and a rubbery character created, by modification with propylene oxide [6,7]. The relationship between lignin's glass transition temperature and the degree of propoxylation has been investigated by Kelley et al. [8]. This study uses a hydroxypropyl lignin derivative with T_g below ambient as multi-functional (rubbery) core in star-block copolymers with polycaprolactone and cellulose propionate as crystallizable hard segments. The specific objective is to compare selected solution and thermal properties of the two types of block copolymers in relation to their molecular architecture.

EXPERIMENTAL

SYNTHESIS
(A) Hydroxypropyl Lignin: Kraft lignin was reacted with propylene oxide under conditions described elsewhere [8,9,10]. Chain extended hydroxypropyl lignin with T_g of -7°C was isolated by exhaustive liquid-liquid extraction followed by molecular fractionation in accordance with the procedure described by Kelley et al. [11].

(B) Copolymer with Cellulose Propionate: Commercial cellulose propionate (Aldrich Chemical Comp.) with a degree of substitution (DS) of >2.8 was converted into a mono-functional segment by depolymerization with HBr according to Mezger and Cantow [12,13]. Cellulose propionate segments with arm lengths ranging between 5 and 50 repeat units (degree of polymerization, DP) and with hydroxy functionality at a single terminus, were reacted with toluene 2,4-diisocyanate (TDI) in accordance with earlier work [14]. NCO-functional cellulose propionate segments were reacted with hydroxy- propyl lignin in THF in accordance with work reported elsewhere [14].

(C) Copolymers with Polycaprolactone: Commercially available (Aldrich Chemical Comp.) mono-functional polycaprolactone segments (OH functionality) were NCO capped in accordance with the procedure described for cellulose propionate.   Mono-functional caprolactone segments were available in two different sizes (4,400 and 11,300 daltons). In parallel work, monomeric ε-caprolactone was reacted with hydroxypropyl lignin [9] thereby creating a multi-functional star-block copolymer with grafted caprolactone arms.

ANALYSIS

(A) The solution properties of lignin based star-block copolymers were characterized by high pressure liquid chromatography in THF solution using a differential viscometer (DV) (Viscotek model 110-01) for analysis. Mark-Houwink-Sakaruda exponential factor and intrinsic viscosity were determined using the Unical program.

(B)   The thermal properties of the copolymers were determined by differential scanning calorimetry (DSC) (Perkin-Elmer Model DSC-7). Melting points (T_m) and heats of fusion (ΔH_f) were determined in the usual manner [15].

RESULTS AND DISCUSSION

The Mark-Houwink-Sakaruda (MHS) exponential factor (a) is a material constant that is sensitive to the molecular architecture of polymers, especially to branching [16]. The MHS factor relates the intrinsic viscosity of the polymer in solution (IV) to the absolute molecular weight by the relationship of

$$[IV] = KM^a$$

where K is a constant and M is molecular weight. The MHS exponential factor (a) is usually high for rigid rod-type linear polymers ($0.7 < a < 1.0$); it is of medium magnitude for freely draining coil type polymers ($0.5 < a < 0.7$); and it is low for spherical, highly branched polymers ($a = 0.5$).

The results of MHS constant determinations using lignin containing star-block copolymers with caprolactone and with cellulose propionate in relation to hard segment arm length are illustrated in Figure 1. The data reveal that there is very little effect of architecture on the a-factor for the caprolactone copolymer series.   All copolymers have a-factors of approximately 0.6 to 0.7. By contrast, a dramatic rise, from 0.2 to >0.8 is observed for the corresponding cellulose propionate copolymer series as arm length rises from 5 to 50 DP (corresponding to molecular weights between 2,000 and 15,000 daltons). Above a DP of about 35, the MHS

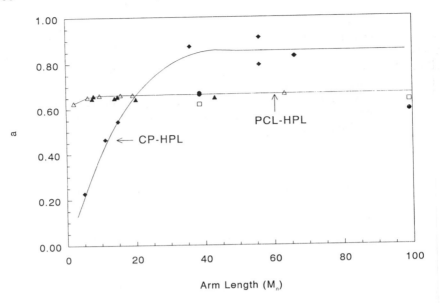

Figure 1.   Relationship betwen Mark-Houwink-Sakurada exponential factor (a) and hard segment size (arm length in degree of polymerization DP_n). ($\triangle$) and ($\blacktriangle$) represent data obtained by anionic polymerization of ε-caprolactone with low (2,100 daltons) and high (6,400 daltons) no. average molecular weight hydroxypropyl lignin, respectively; ($\bullet$) represents data obtained by attachment of preformed polycaprolactone segments of M_n 4,400 and 11,300 to lignin derivative; ($\square$) represents data obtained from ungrafted polycaprolactone segments of M_n 4,400 and 11,300; and ($\blacklozenge$) represents data obtained by attaching cellulose propionate segments to hydroxypropyl lignin.

factor remains approximately constant. The rise in MHS factor in relation to arm length is expected. The difference in the rate of rise is reflective of the overall flexibility of the hard segment component. Whereas the rigidity of the cellulose propionate intermonomer bond results in a low MHS factor until a critical hard segment size is reached, the flexible polyester bond of polycaprolactone begins to affect the MHS factor virtually immediately.

The intrinsic viscosity (Figure 2) reveals a gradual rise for the block copolymer with caprolactone as arm length rises from (M_n 250 to M_n 7,200). The rate of increase is the same for both series obtained by anionic polymerization of ε-caprolactone with low and high molecular weight hydroxypropyl lignins. Even when the average arm length is nearly as long as that of the linear polymer, the viscosities of the star-like polymers are still below that of the linear polymer of equivalent arm

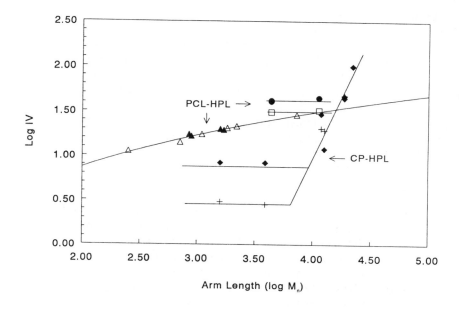

Figure 2. Relationship between intrinsic viscosity (IV) and hard segment size. Note: Symbols as in Figure 1 legend, and (+) represents data obtained from ungrafted cellulose propionate segments.

length. Thus, the presence of short and intermediate-size arms has the effect of lowering the intrinsic viscosity below the value that should be expected for linear polymers with equivalent molecular weight. For the copolymers obtained by attachment of preformed polycaprolactone sements, the surprising finding is that the overall increase in intrinsic viscosity is greater than that of the linear polycaprolactone segments that gave rise to the star-like copolymers. A possible explanation for this behavior can be attributed to the incorporation of a foreign group at the end of the polycaprolactone segments, as result of end-group capping with diisocyanate. It has been confirmed that the presence of even a very small segment of a co-monomer in the polymer chain can alter the solution behavior of the chain [17]. By contrast, cellulose propionate shows a different behavior. The viscosity curve rises sharply at about M_n 6,000 daltons. This is approximately the size of the hydroxypropyl lignin (HPL) used. When the arms become bigger than the central HPL core, the intrinsic viscosity rises abruptly, which suggests a change in copolymer composition. Similar to the block copolymer with polycaprolactone, the intrinsic viscosities of the copolymers with short arm length were higher than the linear cellulose propionate oligomers of equivalent molecular

weight. As discussed earlier, the rise in intrinsic viscosity may be related
to an increase in incompatibility between the segments, as a result of the
incorporation of urethane groups during the TDI end capping of cellulose
propionate oligomers.   As the arm length increases beyond the HPL
molecular weight, the difference in intrinsic viscosity between copolymer
and the corresponding homo-oligomer disappears. Both copolymer and
oligomer behave as rod-like molecules in solution. When HPL molecular
weight predominates in the copolymer composition, the intrinsic viscosity
is low, with values/typical of a branched polymer, whereas on copolymers
where cellulose propionate molecular weight is greater than HPL, the
intrinsic viscosity has the value of the corresponding linear cellulose
propionate oligomer.

The thermal properties of lignin based star-block copolymers are examined
in terms of melting point temperature (T_m) (Figure 3) and heat of fusion
(ΔH_f) (Figure 4). Both types of block copolymers reveal an expected rise
in T_m to the level of the respective homopolymer within approximately 10
to 20 repeat units (DP). No significant difference is established between
the cellulose vs. the polyester type copolymer. However, there appears to
be a significant difference in regard to heat of fusion (Figure 4). Whereas
ΔH_f remains low (below the ΔH_f value of the parent homopolymer
segment) forcellulose propionate- containing copolymers having arm
lengths in excess of 10 to 20, the ΔH_f value approaches that of the parent

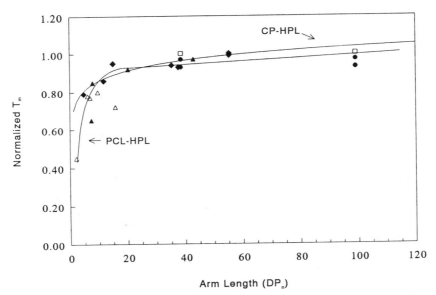

Figure 3. Relationship between melting point (T_m) (normalized) and hard
segment size (arm length in degree of polymerization, DP_n). Note: Symbols
as in Figure 1 legend.

linear polymer segment at arm lengths below 10 DP for the polycaprolactone-containing copolymer. This suggests that the cellulose propionate segments of the cellulose-based copolymer experience a greater resistance to crystallization as compared to the corresponding caprolactone copolymer. This resistance to molecular organization in crystalline regions must be explained with the greater compatibility of the cellulose derivative with lignin as compared to that of caprolactone. This preferential interaction between cellulose (derivatives) and lignin (derivatives) is consistent with work by Rials and Glasser [18,19].

CONCLUSIONS

The star-block architecture of lignin derivative-based copolymers with polycaprolactone and cellulose propionate has a noticeable influence on solution and thermal copolymer properties. Whereas the characteristic lignin properties are more influential on the copolymer properties if segment size is below 10 repeat units in the case of caprolactone, the lignin character prevails until the hard segment size is greater than the lignin size in the case of cellulose propionate. At higher segment sizes, the copolymer property is determined primarily by the character of the respective linear homopolymer segment.

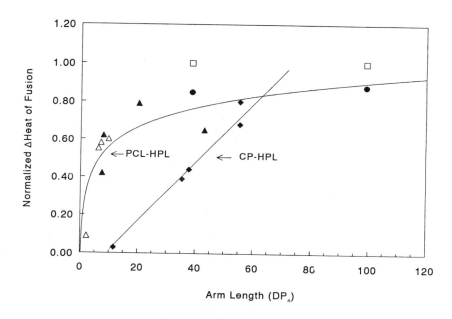

Figure 4. Relationship between Heat of Fusion (ΔH_f) (normalized) and hard segment size (arm length in degree of polymerization, DP_n). Note: Symbols as in Figure 1 legend.

The results are consistent with the hypothesis that lignin (derivative) is more compatible with cellulose (derivative) than with caprolactone.

ACKNOWLEDGEMENT

This study was financially supported by a grant from the National Science Foundation Science and Technology Center on High Performance Polymeric Adhesives and Composites at Virginia Tech.

REFERENCES

1.   Meier, D.J., in Walsh, D.J., Higgins, J.S., Maconnachie, A., Ed., *Polymer Blends and Mixtures*, NATO ASI Series, Series E No. 89, **1985**.

2.   Agarwal, S.L., in Folkes, M.J. Ed., *Processing, Structure and Properties of Block Copolymers*, Chapter 1, **1985**.

3.   Paul, D.R., in D.R. Paul and S. Newman, Eds., Polymer Blends, Vol. I, Academic Press, Inc., New York, 1978.

4.   Dickstein, W.H. *Rigid Rod Star-Block Copolymers*, Technomic Publishing Comp., Inc. Lancaster-Basel, **1990**, 106 pg.

5.   Glasser, W.G., Kelley, S.S., *Lignin*, in *Enc. Polym. Sci. Eng.*, **1987**, Vol. 8, 2nd ed., 795.

6.   Glasser, W.G., Wu, L.C.-F., Selin, J.-F. *Wood and Agricultural Residues-Research on Use for Feed, Fuels and Chemicals*, E.J. Soltes, ed., Academic Press, New York, **1983**, 149.

7.   Wu, L.C.-F., Glasser, W.G., *J. Appl. Polym. Sci.*, **1984**, 29, 1111.

8.   Kelley, S.S., Glasser, W.G., Ward, T.C., *J. Wood Chem. Technol.*, **1988**, 8(3), 341.

9.   de Oliveira, W., *Multiphase Star-Like Copolymers Containing Lignin: Synthesis, Properties and Applications*, Ph.D. Dissertation, Virginia Tech, **1991**, 250 pg.

10.   de Oliveira, W., Glasser, W.G., unpublished manuscript.

11.   Kelley, S.S., Ward, T.C., Glasser, W.G., *J. Appl. Polym. Sci.*, **1989**, 37, 2961-71.

12.   Mezger, T., Cantow, H.-J., *Angew. Makromol. Chem.*, **1983**, 116, 17.

13.   Mezger, T., Cantow, H.-J., *Polymer Photochemistry*, **1984**, 5, 49.

14.   de Oliveira, W., Glasser, W.G., *J. Appl. Polym. Sci.*, **1989**, 37, 3119.

15.   Turi, E.A., *Thermal Characterization of Polymeric Materials*, Academic Press, New York, **1981**, 972 pg.

16.   Rudin, A., *The Elements of Polymer Science and Engineering*, Academic Press, New York, **1982**, 485 pg.

17.   Bohdanecky, M., Kovar, J., "Viscosity of Polymer Solutions", Jenkins, A.D., Ed., Polymer Science Library 2, Elsevier Scientific, Amsterdam, 1982.

18.   Rials, T.G., Glasser, W.G., *J. Appl. Polym. Sci.*, **1989**, 37, 2399.
19.   Rials, T.G., Glasser, W.G., *Wood and Fiber Sci.,* **1989**, 21 (1), 80.

38

Molecular characteristcs of the residual soda-additive pulp lignins

B. Košiková and J. Mlynár - Slovak Academy of Sciences Bratislava, Czecho-Slovakia.

ABSTRACT

The sulfur-free modifications of pine wood delignification including soda pulping in the presence of additives such as anthraquinone, ethylendiamine, and methanol, respectively were investigated from the view point of molecular parameters and condensation factors of the pulp residual lignins. The effect of selected additives on the lignin condensation reactions has been also studied by model experiments with vanillyl alcohol and syringyl alcohol.

INTRODUCTION

The soda pulping of wood involves fragmentation reactions causing the dissolving of lignin and also condensation reactions which contribute to difficulties in the removal of residual lignin during pulping (1,2). The objective of the present paper was to identify the differences in branching parameters and degree of condensation between the residual lignin of soda and soda-additive pine pulps cooked approximately to the same lignin content. In addition, condensation reactions during soda pulping in the presence or absence of AHQ were investigated by using simple lignin models i.e. vanillyl alcohol and syringyl alcohol.

EXPERIMENTAL

The soda pulps were prepared from pine chips (Pinus silvestris L.) using the conditions described earlier (3). GPC analysis and molecular characterisation of the dioxan - lignin preparations isolated from the pulps under investigation were accomplished similarly as described in (4). For determination of condensation factors the method described by Glasser was followed in principle (5). Mono- and di-meric degradation products were separated quantitatively by gas chromatography as methyl esters (Fig. 1).

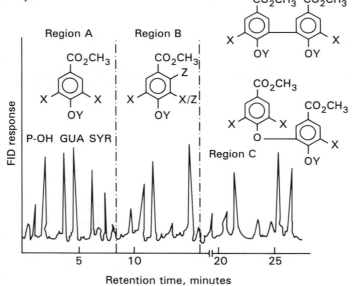

$Y_1 = CH_3$ and/or CH_2CH_3
$Y_2 = CH_3$
$Y_3 = CH_2CH_3$
$Y_4 = CH_3$ or CH_2CH_3
$X = H$ or OCH_3
$Z = CO_2CH_3$ or H

$$KF_1 = \frac{\Sigma C_{Y1, X}}{\Sigma A_{Y2, X}} \qquad KF_2 = \frac{\Sigma B_{X, Y1, Z}}{\Sigma A_{Y2, X}}$$

Fig. 1 Typical gas chromatogram of a permanganate oxidation product mixture of lignin

Vanillyl alcohol (VA) and/or syringyl alcohol (SA) (54mmol) were treated with 30mL 0.1M NaOH in the absence or presence of AHQ (4.8mml). GC-MS analyses of methylated products were done on a mass spectrometer JMS-D100 with gas chromatograph JGC-20 (JEOL).

RESULTS AND DISCUSSION

As indicated by the values summarized in Table 1 all modifications of soda pulping at a given degree of delignification caused a decrease of degree of polymerization DP_w, degree of branching Q, degree of cross linking density Q, as well as content of condensed biphenyl and diaryl structures K_1 and alkyl aryl C-C structures K_2. Mutual differences in those parameters show, in part, the various effectivity of used additived in soda pulping.

Pulping method	DP	Q	Q,	K_1	K_2
Soda	74.2	0.3614	0.0410	3.9113	1.0324
Soda-AQ	49.2	0.3033	0.0303	0.6667	0.7090
Soda-EDA	9.4	0.3149	0.0372	1.3403	0.1969
Soda-MeOH	27.6	0.3634	0.0299	0.1086	0.0778
Pine wood	29.5	0.4555	0.0530	0.0103	0.0019

Table 1 Some molecular parameters of soda-additive pine pulp residual lignin

The determined branching parameters indicate, that AHQ improves the selectivity of cooking, probably by reducing the content of trifunctional and tetrafunctional units in the lignin macromolecule, resulting in its linearization without a significant degradation of lignin. Regarding values of K_1 and K_2, AHQ seems to be very effective in retarding the condensation reactions. The latter effect of AHQ was also investigated by model experiments.

GPC analyses of the chloroform - soluble portion of vanillyl alcohol (VA) and/or syringyl alcohol (SA) reaction mixtures (methylated with dimethylsulphate) resulted from hot sodium hydroxide treatment at a temperature 130°C in the presence or absence of anthrahydroquinone (AHQ) are illustrated in Fig. 2 and Fig. 3, respectively.

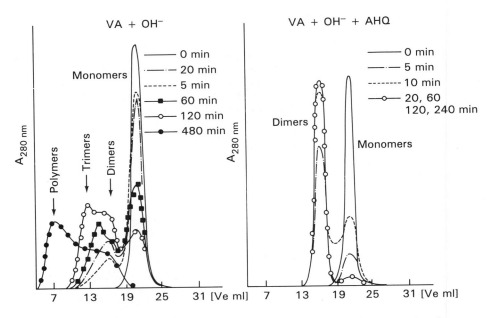

Fig. 2 Gel permeation chromatograms of the products obtained from vanillyl alcohol cooks with and without AHQ AT 130°c

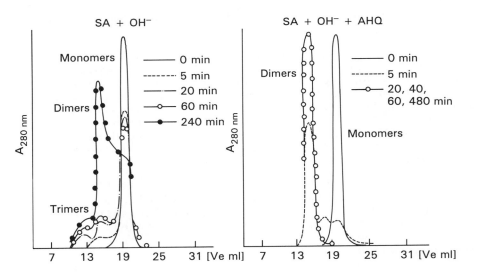

Fig. 3 Gel permeation chromatograms of the products obtained from syringyl alcohol cooks with and without AHQ at 130°C

The cooked sample which contained AHQ showed 90%
conversion of VA and SA after 20 minutes into dimers
possessing high stability in further cooking. In
contrast, both VA and SA treated with sodium
hydroxide in the absence of AHQ yielded besides
dimers also trimers and oligomers with DP 8-10,
respectively, in the case of VA. The differences in
the decrease of monomers as a function of cooking
time are shown in Fig. 4.

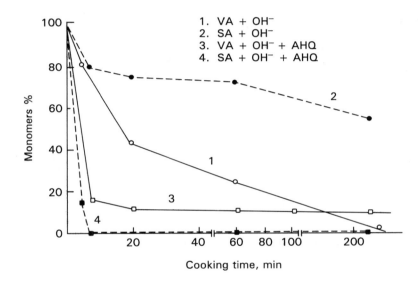

Fig. 4 Decrease of monomers as a function of cooking
time (% of the original vanillyl alcohol and syringyl
alcohol concentration)

Gas chromatography - mass spectroscopy (GC/MS)
characterization of the vanillyl alcohol and syringyl
alcohol condensation products indicates formation of
bivanillyl and bisyringyl structures on cooking with
AHQ. Based on the results obtained, which agree with
the observation of Dimmel et al. (6), we assumed that
lignin condensation includes the formation of benzyl
radicals through a quinone methide intermediate
which recombines to dimers according to the following
scheme. The presence of AHQ greatly depressed the
formation of diphenylmethane structures.

R = H or OCH$_3$

CONCLUSIONS

1. A comparison of the molecular characteristics and condensation factors of soda-additive pulp lignins indicates, that all tested additives significantly reduced the cross-linking density of lignin and depressed its condensation reactions in soda pulping.

2. The model experiments with vanillyl alcohol and syringyl alcohol confirmed that AHQ was very effective in retarding of condensation reactions by formation of stable dimers of the diphenylethane type.

REFERENCES

1. Gierer,J.: Svensk Papperstidn., 73(18),571 (1970).

2. Gierer,J., Lindberg,O.: Acta Chem.Scand., 33 (8), 580 (1979).

3. Košíková,B., Ebringerová,A., Pekkala,O.: Papír a celulóza, 39 (7-8), V70 (1984).

4. Košíková,B., Mlynár,J., Joniak,D.: Holzforschung, 44, 47 (1990).

5. Glasser,W.G., Barnett,C.A., Sano.Y.: J.Appl. Polymer Sci,m 34, 441 (1983).

6. Dimmel,D.R., Shepard,D., Brown,T.A.: J.Wood Chem. Technol., 1 (2), 123 (1981).

39

Molecular weights of organosolv lignins

D.T. Balogh, A.A.S. Curvelo and R.A.M.C. De Groote - Institute de Fisica e Quimica de São Carlos, USP C.P. 369, 13560, São Carlos, S.P. Brasil.

ABSTRACT

Organosolv lignins from *Pinus caribaea hondurensis*, extracted with nine different organic solvents, were analyzed by high performance size exclusion chromatography (HPSEC) and vapor phase osmometry (VPO). The chromatographic analysis were performed with two different column sets, using in both cases THF as eluent and polystyrene standards for calibration. In accordance to the patterns of the elution curves, the samples were classified in three groups. The molecular weight distributions (MWD) obtained from both column sets were different mainly in the higher molecular weight region. Consequently, the weight molecular weight averages (M_w) were also different. The number average molecular weight (M_n) obtained by HPSEC were in the same range for the both column sets (700 - 1500 g/mol) and are in agreement with the values obtained by VPO, demonstrating that the chromatographic systems used are adequate for the determination of M_n values for organosolv lignins.

INTRODUCTION

Organosolv extractions have been performed with a great variety of solvents and acid catalysts (mineral, organic or Lewis acids) and also together with alkaline process (based in sodium hydroxide or ammonia). The lignin obtained by different processes have different characteristics.

The lignins are analyzed mainly by elemental analysis, functional group distribution, oxidation reactions, spectroscopic methods and by molecular weight determinations.

The molecular weight determinations can be made by vapor pressure osmometry [1] (for number average molecular weight, $\overline{Mn}$), by light scattering [2] (for weight average molecular weight, $\overline{Mw}$) and high performance size exclusion chromatography [3-5] (for relative values of $\overline{Mn}$, $\overline{Mw}$, and molecular weight distribution, MWD, determination).

In this paper, the molecular weights of organosolv lignins, obtained by different solvents with hydrochloric acid as catalyst, were determined by high performance size exclusion chromatography (HPSEC) with two different column sets and by vapor pressure osmometry (VPO) and the results correlated with the extraction solvents.

EXPERIMENTAL

Organosolv lignins from *Pinus caribaea hondurensis* sawdust were obtained using nine different organic solvents in mixtures (9:1) with aqueous 2.0N HCl (final acid concentration: 0.2N), at 125°C for 6 hours [6].The solvents used were: chloroform, acetone, 1,4-dioxane, tetrahydrofuran (THF), 2-butanol, 1-butanol, 1-propanol, ethanol and methanol.

Number average molecular weights were determined by vapor phase osmometry (VPO) using a Knauer mod.N-11.00 vapor pressure osmometer, with THF as solvent at 45C° and benzil as calibration standard.

The molecular weight distributions were obtained by high performance size exclusion chromatography (HPSEC) using THF as solvent at a

flow rate of 1 ml/min. The chromatographic system
used consisted of a Waters pump model 6000A,
injector U6K, UV detector model 440 (254 nm). The
determinations were performed with two different
column sets: Plgel (10 μm, 30 cm length, 7.8"
i.d.)10^4 + 500 + 100 Å (I), and 10^5 + 10^4 + 500 + 100 Å
(II) using, in both cases, polystyrene standards and
ethyl benzene for calibration. For the first column
set, the polystyrene nominal molecular weights were:
195,000; 68,000; 28,000; 12,500; 7,600; 3,770;
1,800; 1,050 and 580. For the second set, the
standards of molecular weight 2,300,000 and 450,000
were also used for calibration.

RESULTS AND DISCUSSION

MOLECULAR WEIGHT DISTRIBUTIONS

The molecular weight distributions
obtained with the two column sets, as well as the
calibration curves are shown in figures 1 and 2.
The theoretical exclusion limits
(500,000 - 100g/mol for set I, and
5,000,000 - 100 g/mol for set II) were determined
with polystyrene standards and thus cannot be the
same for the lignins.

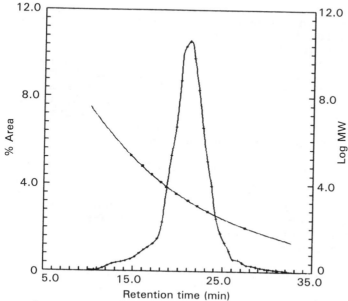

Figure 1 - Molecular weight distribution
resulting from column set I. Lignin sample obtained
by extraction of sawdust with 1,4-dioxan.

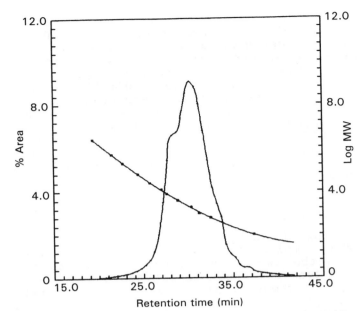

Figure 2 - Molecular weight distribution resulting from column set II. Lignin sample obtained by extraction of sawdust with 1,4-dioxan.

As can be seen in the figures, the major differences in the molecular weight distribution are in the higher molecular weight region. A fraction of some samples eluted beyond the theoretical exclusion limits. In set I, this fraction is higher than in the second set. Also the resolution of the second set seems to be better than the first, mainly in the higher molecular weight range. These observations are related to the higher exclusion limits and to the higher length of the second column set.

The fact that a very small fraction elutes beyond the columns exclusion limits does not mean that this fraction has a molecular weight higher than 5,000,000, in the case of set I, but that this fraction seems to have different structures from that of the other fractions and, thus different hydrodynamic behavior.

The molecular weight distribution data for the samples analyzed in the two column sets are shown in Table 1.

The lignins can be classified in three major groups, in accordance with the MWD obtained

in the two column sets: first group: formed by the
samples obtained with chloroform and acetone,
showing a shoulder at lower molecular weight range;
second group: formed by the samples obtained with
THF and 2-butanol, showing only unimodal
distribution; third group: formed by the remaining
five samples and having a shoulder at higher
molecular weight. This difference between the
samples is related to the extraction solvent, as all
other extraction system parameters were kept
constant, and indicates that some solvents are
better for the solubilization of high molecular
weight fragments , or are more selective for the
extraction of high molecular weight fragments [6].

Table 1. Molecular weight distribution data for the
         organosolv lignins

SAMPLE	SET I		SET II	
	M_i max.	SHOULDER	M_i max.	SHOULDER
CHLOROFORM	1260	490	1100	430
ACETONE	1260	490	1100	430
DIOXAN	1640	5420*	1850	5730
THF	1640	——	1420	——
2-BUTANOL	1640	——	1420	——
1-BUTANOL	2160	5420*	1850	5730
1-PROPANOL	2160	5420*	1850	5730
ETHANOL	2160	5420*	1850	5730
METHANOL	2160	5420*	1850	5730

*Samples that presented a fraction beyond the column
 exclusion limit

WEIGHT AND NUMBER AVERAGE MOLECULAR WEIGHTS

The $\overline{Mw}$ and $\overline{Mn}$ values, shown in Table 2,
obtained from the MWD of the two column sets were
calculated choosing the integration limits [7] in
such a way that less than 5% of the total
chromatogram area were discarded in the
calculations. The integration limits used were
14.00-28.00 min. for set I and 24.00-38.00 for set
II.

The $\overline{Mw}$ values calculated from the two
column sets were very different due to the
differences in the MWD in the higher molecular
weight region.

Derivation, Characterisation and Physiochemical [Pt.2

Table 2. Weight and number average molecular weights of the organosolv lignins

SAMPLE	SET I		SET II		$\overline{Mn}_{VPO}$
	$\overline{Mn}$	$\overline{Mw}$	$\overline{Mn}$	$\overline{Mw}$	
CHLOROFORM	810	1420	720	1340	800
ACETONE	970	4070	860	2150	780
DIOXAN	1050	8970	1000	3780	1450
THF	980	3380	860	2070	980
2-BUTANOL	1130	6430	970	2490	1040
1-BUTANOL	1330	10040	1210	4080	910
1-PROPANOL	1400	11750	1330	4050	1820
ETHANOL	1610	10430	1410	4580	1520
METHANOL	1680	5790	1570	3670	1470

As can be seen from Table 2, the $\overline{Mn}$ values show little difference between the column sets. Also, these values are in the same range as that obtained by VPO experiments, indicating that the chromatographic systems used are adequate for determinations of $\overline{Mn}$ values of organosolv lignins.

REFERENCES

[1]Froment, P.; Pla, F.-" Determinations of average molecular weights and molecular weight distributions of lignin" in Lignin Properties and Materials, ACS Symposium Series 397, 1989.
[2]Pla, F.- Etude de la structure macromoleculaire des lignines- Universite Scientifique & Medicale PhD.Thesis- Grenoble-1980.
[3]Chum, H.L.;Johansson,D.K.; Tucker, M.P.; Himmell, M.E.- Holzforschung, 41, 97 (1987).
[4]Faix, O.;Lange,W.; Besold,G. - Holzforschung,35, 137 (1981).
[5]Faix, O.; Lange,W.; Beinhoff, O.- Holzforschung, 34,174 (1980).
[6]Balogh,D.T.;Curvelo, A.A.S.;De Groote, R.A.M.C.- Holzsforschung, (1992) (in press).
[7]Yau, W.W.; Kirkland, J.J. & Bly,D.D.- Modern Size Exclusion Chromatography. J. Wiley and Sons- New York, 1979.

40

Role of phenolic hydroxyl groups in sulfonation

Y.-Z. Lai, X.-P. Guo, T. Suzuki, Y. Kojima dn H.-T. Chen - Empire State Paper Research Institute, SUNY College of Environmental Science and Forestry Syracuse, New York 13210, USA.

ABSTRACT

The extent of sulfonation conducted at pH 7.5 and 140°C for a variety of softwood and hardwood species was shown to be directly proportional to the phenolic hydroxyl group content of the wood lignin. Sulfonation of hardwoods, in contrast to softwood, displayed a significant variation among different species and appears to be largely associated with guaiacyl propane units, because syringyl units contain relatively few free-phenolic structures.

INTRODUCTION

Lignin sulfonation, the dominant reaction of sulfite-based treatments or pulping of wood [1,2], is a major factor in determining the strength properties of chemimechanical or chemithermomechanical pulps, especially from softwood [3-7]. Although the rate of sulfonation can be accelerated by modifying the sulfite treatment conditions [8-10], e.g. SO_2 concentration and temperature, the extent of fiber sulfonation under preferred neutral or slightly alkaline conditions is generally thought of as being limited to the phenolic units of lignin [1]. The conceptual role of phenolic hydroxyl groups in sulfonation has been largely established through studies of lignin model compounds, and it has not been quantitatively determined for wood samples.

Recently we have shown [11,12] that the phenolic hydroxyl group content of wood lignin can be conveniently determined *in situ* by a periodate oxidation method. For softwoods [12], the phenolic hydroxyl group content (12-13%) based on the percentage of C_9 units, like the methoxyl group content [13], were quite similar among different species. There is considerable variation in the phenolic hydroxyl group contents among hardwood species ranging from 11% (red oak) to 7% (white birch). This paper discusses the extent to which the sulfonation of softwood and hardwood can be correlated to their phenolic hydroxyl group contents.

EXPERIMENTAL

The wood species used in sulfonation studies included Norway spruce (*Picea abies*), loblolly pine (*Pinus taeda*), red pine (*Pinus resinosa* Ait.), Balsam fir (*Abies balsamea* Mill), aspen (*Populus tremuloides*), eastern cottonwood (*Populus deltoides* Bartr.), American beech (*Fagus grandifolia* Ehrh.), red aod (*Quercus borealis* L.), sweetgum (*Liquidambar styraciflua* L.), and white birch (*Betula papyrifera* March). Extractive-free woodmeals (40-60 mesh), prepared by extraction with acetone were used in all the experiments.

Sulfite treatments, as described previously [14], were conducted in small autoclaves using a 6% SO_2 concentration, prepared by adjusting sodium bisulfite solution to pH 7.5 with alkali, at a liquor-to-wood ratio of 10, and then heated in an M & K digester at 140°C for 1 h.

Lignin contents (Klason plus acid-soluble lignin) were determined by the Tappi Standard Methods. The sulfonate groups were measured by a conductometric method [15] while the phenolic hydroxyl groups was determined by a periodate oxidation method [11,12].

INFLUENCE OF SULFITE TREATMENTS

The yield of sulfite treatments among softwoods was quite uniform being around 93%, and the yield loss (7%) was attributed to the dissolution of both carbohydrate and lignin, roughly in equal proportion. The yield loss was higher in the case of hardwoods, ranging from 10% for birch to 15% for beech and red oak. The proportion of lignin components in the dissolved fraction varied from 30% (for cottonwood) to 46% (for red oak). On the average, the amount of lignin removed during sulfite treatments of hardwood was about 4% of the wood for most species.

Also, it was shown that for most species, the apparent phenolic hydroxyl group content of the sulfonated sample was significantly higher than that of the untreated wood. Analyses of the phenolic hydroxyl groups on the basis of residual lignin content indicate an excellent correlation (0.979 correlation coefficient) between the phenolic hydroxyl group content of lignin before and after sulfite treatments. The linear regression line had a positive slope of 1.324. It extended nearly through the origin with a negative intercept equivalent to about 0.6 phenolic hydroxyl group per 100 C_9 unit. The magnitude of increase among four softwoods was quite uniform with an average of about 30%. Interestingly, Yang *et al.* [16], using the ultraviolet microscopic technique, observed a 40% increase in phenolic hydroxyl groups for the secondary wall lignin of a neutral sulfite-treated spruce wood obtained at 93% yield.

NATURE OF SULFITOLYTIC CLEAVAGES OF
ARYL ETHER LINKAGES

Two major mechanisms, as revealed from model compound reactions [1], may account for the formation of phenolic hydroxyl groups under neutral or slightly alkaline sulfite conditions. These are the sulfitolytic cleavages of the β-aryl ether linkages of (a) phenolic units and (b) etherified units containing an α-carbonyl group. Separate experiments on neutral sulfite treatments of thermomechanical pulp samples from Norway spruce indicated that borohydride reduction prior to the sulfite treatment has virtually no effect on the phenolic hydroxyl group content of the resulting sulfonated pulp. This observation seems to discount any significant contribution of mechanism (b) to the phenolic hydroxyl group content of residual sulfonated lignin.

Thus, it is evident that the phenolic hydroxyl groups present in the wood lignin play two important roles under neutral sulfite treatment conditions by being able to facilitate: a) the sulfonation reaction and b) the sulfitolytic cleavage of phenolic β-aryl ether structures. It is anticipated that the sulfitolytic cleavage reaction is strongly dependent upon temperature. Therefore, the observed increase in fiber sulfonation at a higher temperature [8] may be readily explained in terms of a more extensive sulfitolytic cleavage reaction.

VARIATION OF WOOD SPECIES IN SULFONATION

Under otherwise identical sulfite treatment conditions, the sulfonate content expressed as mmol/100 g sample was considerably higher for softwoods (22.7-26.0) than for hardwoods (9.9-12.4). Since fiber sulfonation is virtually confined to lignin components, the variation of wood species in sulfonation should be compared on the basis of lignin.

Figure 1 illustrates an excellent relationship between the sulfonate and phenolic hydroxyl group content of sulfonated samples for a variety of wood species. The linear regression line had a high correlation coefficient (0.977) and a positive slope of 1.07 which confirms the conceptual role of phenolic lignin units in neutral sulfonation.

Although it is unclear on the extent to which the observed sulfonation was attributed to the reaction of the cinnamaldehyde end groups and other carbonyl groups, the observed high correlation between the sulfonate and phenolic hydroxyl group content strongly supports a previous finding of Gellerstedt and Gierer [17] that the quinone methides of lignin model compounds in neutral sulfite solution at ambient temperature gave the sulfonated products in nearly quantitative yield.

SULFONATION-METHOXYL GROUP CONTENT RELATIONSHIP

Figure 2 illustrates the relationship between the sulfonate and methoxyl group content of lignin in the wood. The methoxy group content was taken

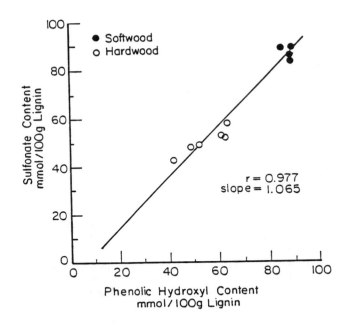

Figure 1. Relationship between the sulfonate and phenolic hydroxyl group content of sulfonated lignin.

from the literature reported for milled-wood-lignin (MWL) or thioglycolic acid preparations [18,19]. Both the data of softwood and hardwood seem to fit a linear regression line reasonably well (-0.952 correlation coefficient). The linear relationship indicates that the extent of sulfonation decreases with an increase in the methoxyl group content (or the quantity of syringyl units) of the lignin. This finding supports the contention that the neutral sulfonation of hardwood lignin is largely limited to the guaiacyl propane units [20], and the syringyl units are primarily of etherified type [2,21-23].

CONCLUSIONS

The present study clearly demonstrates that the extent of neutral sulfonation for a variety of woods is directly related to their lignin phenolic hydroxyl group contents. It appears that sulfonation of hardwoods is primarily associated with the guaiacyl propane units, and thus it varied significantly among different species. Since many lignin reactions proceed through quinone methide intermediates resulting from the ionization of phenolic hydroxyl groups, the capacity to form such intermediates may serve as a relative measure of lignin reactivity. Consequently, the response of lignocellulosic materials to neutral sulfite treatments may be suitable for measuring the relative reactivity of lignin.

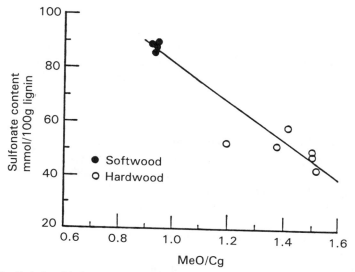

Figure 2. Relationship between the sulfonate and methoxyl group contents of wood lignins.

ACKNOWLEDGEMENT

Financial support from the Empire State Paper Research Associates (ESPRA) is greatly appreciated.

REFERENCES

1.   Gellerstedt, G. *Svensk Papperstidn.* **1976**, *79*, 537.
2.   Adler, E. *Wood Sci. Technol.* **1977**, *11*, 169.
3.   Iwamida, T.; Sumi, Y.; Nakano, J. *Cellulose Chem. Technol.* **1980**, *14*, 497.
4.   Sinkey, J.D. *Appita*, **1983**, *36*, 301.
5.   Heitner, C.; Hattula, T. *J. Pulp Paper Sci.* **1988**, *14*, J6.
6.   Mutton, D. B., Tombler, G.; Gardner, P. E.; Ford, M. J. *Pulp Paper Can.* **1982**, *83*, T189.
7.   Lai, Y.-Z.; Situ, W. *J. Wood Chem. Technol.* in press.
8.   Heitner, C.; Beatson, R. P.; Atack, D. *J. Wood Chem. Technol.*, **1982**, *2*, 169.
9.   Beatson, R. P.; Heitner, C.; Atack, D. *J. Wood Chem. Technol.* **1984**, *4*, 439.
10.   Beatson, R. P.; Heitner, C.; Rivest, M.; Atack, D. *Pap. ja Puu* **1985**, *67*, 702.

11. Lai, Y.-Z.; Guo, X.-P.; Situ, W. J.*Wood Chem.Technol.* **1990**, *10*, 365.

12. Lai, Y.-Z.; Guo, X.-P. *Wood Sci. Technol.* **1991**, *25*, 467.

13. Sarkanen, K. V.; Chang, H.-M.; Allan, G. G. *Tappi* **1967**, *50*, 583.

14. Lai, Y.-Z.; Guo, X.-P. *Holzforschung* in press.

15. Katz, S.; Beatson, P.; Scallan, A. M. *Svensk Papperstid.* **1984**, *87*, R48.

16. Yang, J.-M.; Yean, W. Q.; Goring, D.A.I. *Cellulose Chem. Technol.* **1981**, *15*, 337.

17. Gellerstedt, G.; Gierer *J. Acta Chem. Scand.* **1970**, *24*, 1645.

18. Musha, Y.; Goring, D.A.I. *Wood Sci. Technol.* **1975**, *9*, 45.

19. Sarkanen, K. V.; Hergert, H. L. in Sarkanen, K. V.; Ludwig, C. H. (Eds.) *Lignins* Wiley-Interscience, New York, **1971**, p. 43.

20. Beatson, R. P.; Heitner, C.; Rivest, M.; Atack, D. *Pap. ja Puu* **1985**, *67*, 702.

21. Chang, H.-H.; Cowling, E. B.; Brown, W.; Adler, E.; Miksche, G. *Holzforschung* **1975**, *29*, 153.

22. Nimz, H. H.; Robert, D.; Faix, O.; Nemr, M. *Holzforschung* **1981**, *35*, 16.

23. Lapierre, C.; Monties, B., in *Proceedings of Fifth International Symposium on Wood and Pulping Chemistry* **1989**, *Vol. 1*, 615.

41

Analysis of lignin phenolic units by a combined phenyl nucleus exchange — periodate oxidation method

Y.-Z. Lai and M. Funaoka - Empire State Paper Research Institute, SUNY College of Environmental Science and Forestry, Syracuse, New York 13210, USA.

ABSTRACT

The proportion of phenolic units among uncondensed structures of aspen lignin has been estimated *in situ* by a combination of phenyl nucleus exchange and periodate oxidation techniques. The data indicate that the syringyl units, in contrast to guaiacyl units, are primarily of the etherified type.

INTRODUCTION

Despite the prominent role of lignin phenolic units and condensed structures [1,2] in pulping and bleaching reactions [3,4], available methods for measuring these functionalities generally do not reveal quantitatively the structural environment in which they occur. Among the reported methods [5-11], the phenyl nucleus exchange (NE) [10,11] and periodate oxidation (PO) techniques [6,7] appear to be the most direct procedures for determining the uncondensed and phenolic units of lignin, respectively. In addition, the NE method when combined with a nitrobenzene oxidation (NO) technique can be used to estimate the proportion of diphenylmethane-type structures [10-12] which may be formed under acidic and alkaline pulping conditions [1-4]. This report estimates the proportion of phenolic units among uncondensed guaiacyl and syringyl units for aspen lignin *in situ* by a combination of NE, NO, and PO methods.

EXPERIMENTAL

A sample of extractive-free aspen (*Populus tremuloides*) wood meal (passed through 80 mesh) was prepared by extracting with ethanol-benzene (1:2 by volume) in a Soxhlet extractor for 48 h.

The procedures of NE and NO methods used are similar to those previously described by Funaoka *et al.* [10,11]. The NE reaction employed a mixture of phenol, boron trifluoride-phenol complex, and xylene in a ratio of 19:4:10 by volume, respectively, at 180°C for 4 h. The NO was conducted in sodium hydroxide (2 M) solution at 170°C for 3 h.

The PO was carried out by treating samples with a saturated sodium periodate solution at 4°C with occasional stirring for 48 h.

Lignin contents (Klason plus acid-soluble lignin) were determined by Tappi Standard Methods. A Hewlett Packard 5890 A gas chromatograph equipped with flame ionization detectors, a computerized integrator, and a Quadrex fused silica capillary column (50 m x 0.25 mm x 0.25 μm) packed with the 007 series methyl silicone was used for analyses of reaction products. Nitrogen was used as the carrier gas.

For NO experiments, the molar yields of vanillin and vanillic acid were determined for the reaction of guaiacyl units while the yields of syringaldehyde and syringic acid were measured for syringyl units. For NE experiments, the molar yields of guaiacol and catechol were determined for guaiacyl units while the yields of pyrogallol-1,3-dimethyl ether, pyrogallol-1-methyl ether, and pyrogallol were determined for syringyl units.

UNCONDENSED UNITS

Table 1 summarizes the yields of nitrobenzene oxidation products (NOP) from untreated and periodate-treated aspen samples. The periodate treatment which was conducted at 4°C for 48 h should have completely degraded the phenolic units based on the behavior of Norway spruce lignin. It is evident that the yield reduction in the periodate-treated sample was substantially higher for guaiacyl units (26%) than for syringyl moieties (10%) indicating a significant difference in phenolic hydroxyl group content among these uncondensed structures.

Table 2 shows the yield of nucleus exchange products (NEP) and the ratio of NOP and NEP for the guaiacyl units of aspen. The data indicate that the periodate treatment resulted in a 22% reduction in NEP yield which is slightly lower than that observed in NOP yield (26%). The NOP/NEP ratio also showed a slight decrease (from 0.84 to 0.79) which may have resulted from the periodate oxidation of lignin side chain structures.

Table 1. Nitrobenzene oxidation products from untreated and periodate-treated aspen wood meal

Sample	NOP, mole % of Lignin			
	Guaiacyl Unit		Syringyl Unit	
	Vanillin	Vanillic acid	Syring-aldehyde	Syringic acid
Untreated	15.3	1.6	33.1	4.3
Periodate-treated	11.2	1.3	30.1	3.4

The proportion of syringyl units in aspen lignin was estimated to be 45%, which, as discussed by Funaoka *et al.* [10,11], was obtained by dividing the NOP yield (37.4%) with a correction factor of 0.84. The correction factor used was the ratio of NOP and NEP for guaiacyl units which was assumed also to be applicable to syringyl units in view of their similarity in response to the NE reaction [10,11]. Similarly, the proportion of syringyl units present in etherified structures was estimated to be 43%. This was obtained by dividing the NOP yield of the periodate-treated sample (34%) by a correction factor of 0.79 (Table 2). It follows that only about 4% of syringyl units in aspen possess a phenolic hydroxyl group.

The proportion of total guaiacyl units in aspen lignin obtained by difference was 55% while the proportion of guaiacyl units of the uncondensed type was estimated directly from the NEP yields to be 20%. Similarly, the yield of NEP from the guaiacyl units of periodate-treated sample (16%) represents the etherified guaiacyl units of the uncondensed type.

Table 2. Nucleus exchange products and the ratio of NOP and NEP from guaiacyl units of untreated and periodate-treated aspen.

Sample	NEP Yield, mole % of Lignin			$(NOP/NEP)_G$
	Guaiacol	Catechol	Total	
Untreated	5.5	14.6	20.1	0.84
Periodate-treated	4.0	11.8	15.8	0.79

Table 3. Distribution of phenolic units in aspen lignin

Type of Unit	% of C_9 Units
Guaiacyl	
uncondensed	20
phenolic	4
etherified	16
condensed	35
Syringyl	
phenolic	2
etherified	43
Total phenolic[*]	10

[*] Measured by a periodate oxidation method [6].

DISTRIBUTION OF STRUCTURAL UNITS

Table 3 summarizes the structural information of aspen lignin obtained by the combined NO, NE, and periodate oxidation techniques. Also included is the total phenolic hydroxyl group content which was analyzed by a periodate oxidation method based on the maximum methanol formation [6].

It is clearly shown that the proportion of phenolic units among syringyl structures (4%) is considerably lower than that of uncondensed guaiacyl units (20%). A similar finding was reported previously by Lapierre et al. based on thioacidolysis of poplar wood meal [13]. They estimated the proportion of phenolic units among uncondensed guaiacyl and syringyl units of β-0-4 structures to be about 29 and 3%, respectively. These quantitative data strongly support the general contention that the syringyl units present in hardwood lignins are primarily of the etherified type [13-16]. The data also indicate that about 40% of the phenolic units are associated with the condensed type structures.

CONCLUSIONS

This study shows that periodate oxidation can be used along with the NE method to estimate the proportion of phenolic and etherified structures of lignin. Among the uncondensed structures of aspen lignin, the percentage of syringyl units having a phenolic hydroxyl group was quite small (4%) being approximately one- fourth that of guaiacyl moieties (20%). This finding fully confirms a previous observation that the phenolic hydroxyl group content of wood lignin decreased proportionally with an increase in syringyl units.

ACKNOWLEDGEMENT

This work was supported by a grant from the Empire State Paper Research Associates (ESPRA).

REFERENCES

1.   Adler, E. *Wood Sci. Technol.* **1977**, *11*, 169.
2.   Lai, Y.-Z.; Sarkanen, K. V., in Sarkanen, K. V.; Ludwig, C. (Eds.), *Lignins*, Interscience publishers, 1971, p. 165.
3.   Gierer, J. *Wood Sci. Technol.* **1985**, *19*, 289.
4.   Gierer, J. *Wood Sci. Technol.* **1986**, *20*, 1.
5.   Gellerstedt, G.; Lindfors, E. *Svensk Papperstidn.* **1984**, *87*, R115.
6.   Lai, Y.-Z.; Guo, X.-P.; Situ, W. *J. Wood Chem. Technol.* **1990**, *10*, 365.
7.   Lai, Y.-Z., in Lin, S.; Dence, C. (Eds.) *Methods in Lignin Chemistry*, Springer, in press.
8.   Erickson, M.; Larsson, S.; Miksche, G. E. *Acta Chem. Scand.* **1973**, *27*, 903.
9.   Kirk, T. K.; Obst, J. R. in Wood, W. A.; Kellogg, S. T. (Eds.) *Methods in Enzymology - Biomass*, Academic Press. **1988**, *Vol. 161*, 87.
10. Funaoka, M.; Abe, I. *Mokuzai Gakkaishi*, **1983**, *29*,   781.
11. Funaoka, M.; Abe, I.; and Chiang, V., in Lin, S.; and Dence, C. (Eds.) *Methods in Lignin Chemistry*,   Springer, in press.
12. Chang, V. L.; Funaoka, M. *Holzforschung* **1988**, *42*,   385.
13. Lapierre, C.; Monties, B. in *Proceedings of Fifth   International Symposium on Wood and Pulping Chemistry* **1989**, *Vol. 1*, 615.
14. Lai, Y.-Z.; Guo, X.-P. *Wood Sci. Technol.* **1991**, *25*, 467.
15. Chang, H.-M.; Cowling, E. B.; Brown, W.; Adler E.; Miksche, G. *Holzforschung* **1975**, *29*, 153.
16. Nimz, H. H.; Robert, D.; Faix, O.; Nemr, M. *Holzforschung* **1981**, *35*, 16.

42

A spin probe study of the interaction of hydrophobically modified hydroxyethyl cellulose with sodium dodecyl sulphate

R. Tanaka, J. Meadow, P.A. Williams and G.O. Phillips - Polymer and Colloid Chemistry Group, Faculty of Science, Health and Medical Studies, The North East Wales Institute, Connah's Quay, Deeside, Clwyd, CH5 4BR, UK.

Abstract

The aqueous solution interaction of the anionic surfactant sodium dodecyl sulphate (SDS) with hydrophobically modified hydroxyethyl cellulose (HMHEC) has been investigated using small deformation oscillatory measurements and electron spin resonance spectroscopy.

Both techniques indicated that significant polymer / surfactant interaction occurred at concentrations of SDS considerably lower than its expected cmc. Electron spin resonance (e.s.r.) spin probe studies, using the amphiphilic spin probe 5-doxyl stearic acid (5-DSA), indicated that the level of incorporation of SDS molecules within the mixed HMHEC / SDS 'aggregates' increased with increasing levels of surfactant addition. From this, and previously reported data, a mechanism is outlined to explain the complex rheological behaviour of HMHEC in the presence of SDS.

Introduction

Hydrophobically associating water-soluble polymers such as HMHEC are a relatively new class of polymers which are gaining increasing importance due to their ability to impart improved rheological behaviour to a variety of industrial formulations (1). Their unique rheological properties arise from the intermolecular association of their constituent hydrophobic groups, which leads to the formation of reversible three dimensional network structures (2). Due to their hydrophobic character, such regions of intermolecular association represent areas of high affinity for surfactant interaction. Consequently, the addition of surfactants has been found to have a dramatic effect on the rheological properties of solutions containing such polymers. However, the observed rheological effects have been found to be strongly dependent upon both the chemical nature of the surfactant, and the actual levels of its addition (3-5). It is evident that the explanation of such complex rheological behaviour is therefore dependent upon a detailed understanding of the nature of the polymer / surfactant interaction in the system.

In this paper we report on the use of a free radical nitroxide spin probe in conjunction with electron spin resonance spectroscopy to study the interaction of HMHEC with SDS.

Materials

The water-soluble HMHEC was kindly supplied by Aqualon (U.K.) Ltd., under the trade name Natrosol Plus Grade 330. The manufacturer reports it to have a molecular mass of approximatey 250,000 and a molar substitution of 3.3. In addition the HMHEC also contains approximately 1% w/w of chemically grafted C_{12} - C_{18} alkyl side chains (6).

SDS [specially pure grade] was obtained from BDH Chemicals Ltd., and was used as supplied.

The amphiphilic spin probe 5-doxyl stearic acid (5-DSA) was obtained from Sigma Chemicals Ltd., and was used as supplied.

Methods

(i)Rheological measurements

Equal amounts of various aqueous SDS solutions of known concentration were added to approximately 5g of 2% aqueous solutions of HMHEC. The mixtures were then tumbled overnight to ensure complete homogeneity. The storage (G') and loss (G") moduli of the various solutions were determined over the frequency range 0.01 - 10 Hz at a fixed amplitude of 6 x 10^{-3} radians using a Carrimed CS100 controlled stress rheometer (Carri-Med Ltd., U.K.) fitted with either a 4cm 2 degree or a 2cm 2 degree cone and plate measuring system.

The values of both moduli at a frequency of oscillation of 1 Hz were recorded.

(ii) Spin probe experiments

All solutions were prepared by dissolving the appropriate amount of HMHEC and SDS in a slightly alkaline (pH9) aqueous solution of 5×10^{-6} mol dm^{-3} 5-DSA. The polymer was completely solubilised by stirring continuously for at least 18hr before the addition of the appropriate amount of SDS.

The e.s.r spectra were recorded at 20°C on a Bruker ESP300 Electron Spin Resonance Spectrometer (Bruker Spectrospin Ltd.,U.K.) using a quartz cell suitable for aqueous solutions.

The use of e.s.r. spectroscopy to study polymer - surfactant interaction.

Nitroxide free radicals give rise to well characterised three-lined e.s.r spectra. The relative shapes and intensities of the lines are a reflection of the mobility of the nitroxide moiety. If the motion of the nitroxide radical is unrestricted, then the three lines are relatively narrow and are of similar intensities. However, as the mobility of the free radical is reduced, line broadening occurs due to anisotropic effects.

Specifically selected nitroxide spin probes can be used to monitor the formation of regions of microheterogeneity within bulk systems. In this instance, the spin probe experiments were carried out using 5-DSA which is amphiphilic in nature and, thus, can be expected to preferentially reside close to any microregions of hydrophobicity present within a bulk aqueous environment. The decrease in rotational mobility experienced by the probe as a consequence of any such preferential association can be monitored through the resultant changes in the shape of its e.s.r. spectrum. This is illustrated in Figure 1 which shows the e.s.r spectra of 5-DSA (a) in water and (b) in the presence of SDS micelles.

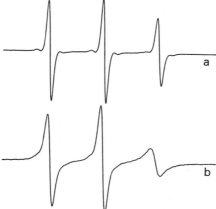

Figure 1. E.s.r. spectra of 5×10^{-6} mol dm^{-3} 5-DSA in (a) water and (b) 10^{-2} mol dm^{-3} aqueous SDS solution.

Results and Discussion

The effect of the addition of SDS on G' and G" of 1% aqueous solutions of HMHEC at a frequency of oscillation of 1 Hz are given in Figure 2. It is evident that the interaction of SDS with HMHEC is relatively complex, since both moduli at first increase with increasing SDS concentration up to maximum values approximately one order of magnitude greater than those observed for the HMHEC solution in the absence of surfactant. At this 'optimum' concentration (C_{max}) of added SDS, the originally viscous fluids exhibited distinctly gel-like character -istics. Further additions of SDS above the level of C_{max} produced a marked decrease in both moduli. At sufficiently high concentrations of added SDS, the samples became fluid-like again. The actual magnitudes of both moduli at such high surfactant additions were significantly less than in the absence of surfactant, suggesting a complete disruption of the network structure of HMHEC had occurred.

The rheological curves illustrated in Figure 2 are of similar shape to those observed for the addition of various other surfactants to aqueous solutions of HMHEC. For homologous series of sodium alkyl sulphate and sodium alkanoate surfactants, we have found (5) that, above an apparently critical alkyl chain length of six carbon atoms, the maximum value of G'_{max} and $G"_{max}$ achievable through the addition of surfactant to a 2% aqueous solution of HMHEC increased exponentially with increasing alkyl chain length of the added surfactant. In contrast, the differences in the anionic headgroups of these two series of surfactants appeared to have no influence on the observed values of G'_{max} and $G"_{max}$.

Figure 3a gives the e.s.r spectra of 5-DSA in a 1% aqueous solution of HMHEC. The observed spectrum is clearly 'composite' in nature, in that it contains both isotropic and anisotropic components. This is indicative of a partitioning of the spin probe between the bulk aqueous solution and regions of intermolecular hydrophobic association. Computer analysis of the composite spectra given in Figure 3a indicates it comprises of 26% of the narrow isotropic spectra (Figure 3c) arising from those 5-DSA molecules residing in the bulk aqueous solution, and 74% of the distinctly anisotropic spectrum given in Figure 3b. This latter spectrum can be attributed to 5-DSA molecules whose mobility has been significantly reduced due to their interaction with the regions of intermolecular hydrophobic association.

The significantly broader, more anisotropic nature of spectrum 3b compared to spectrum 1b indicates that the 5-DSA probe molecules have a considerably lower rotational mobility when associated with the regions of intermolecular hydrophobic associations of HMHEC than when incorporated within an SDS micelle. This suggests a higher internal freedom of movement within an SDS micelle, possibly due to the higher number of constituent SDS molecules compared to the number of bound hydrophobes which constitute a region of intermolecular hydrophobic association.

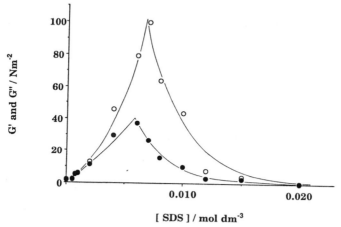

Figure 2. The effect of the addition of SDS on (O) G' and (●) G",(ω = 1Hz), of 1% aqueous solutions of HMHEC.

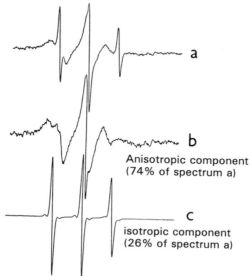

Figure 3. E.s.r. spectra of a) 5 x 10^{-6} mol dm^{-3} 5-DSA in a 1% aqueous solution of HMHEC, and its resolved (b) anisotropic and (c) isotropic components.

Resolution of the 'composite' spectra, such as that shown in Figure 3a, by subtraction of the 'interfering' isotropic component, enables the interaction of the SDS molecules with the HMHEC - bound hydrophobes to be studied through monitoring changes in the e.s.r. spectra of only those 5-DSA molecules which are closely associated with the regions of intermolecular hydrophobic association.

The e.s.r spectra given in Figure 4 are the resolved anisotropic components of the composite spectra obtained for 5×10^{-6} mol dm^{-3} 5-DSA in 1.0% aqueous solutions of HMHEC containing various amounts of SDS. The percentage values given alongside each spectrum represent the fraction of the total number of 5-DSA molecules closely associated with regions of intermolecular hydrophobic association at that level of SDS addition. From Figure 4, it can be seen that this value increases as the SDS concentration increases. In addition, the resolved spectra become less anisotropic with increasing SDS addition, indicating that the mobilty of the 5-DSA molecules associated with the regions of intermolecular hydrophobic association increases. These observations suggest that the HMHEC / SDS aggregates are continually growing in number and/or size with increasing levels of surfactant addition. This is consistent with our own specific ion electrode investigations (5) and the steady-state fluorescence probe studies of Dualeh and Steiner (3).

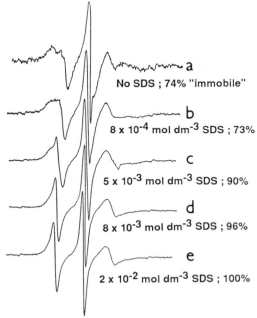

a

No SDS ; 74% "immobile"

b

8×10^{-4} mol dm^{-3} SDS ; 73%

c

5×10^{-3} mol dm^{-3} SDS ; 90%

d

8×10^{-3} mol dm^{-3} SDS ; 96%

e

2×10^{-2} mol dm^{-3} SDS ; 100%

Figure 4. E.s.r. spectra of 5×10^{-6} mol dm^{-3} 5-DSA in a 1.0% aqueous solution of HMHEC containing varying amounts of added SDS.

At an added SDS concentration of 2×10^{-2} mol dm^{-3}, the initial unresolved e.s.r. spectrum of the 5-DSA show no composite character, with all of the 5-DSA molecules having a rotational mobility similar to that when incorporated in pure SDS micelles (compare spectra 1b and 4e). This indicates that, at such surfactant concentrations, the incorporation of any HMHEC - bound hydrophobe(s) into an SDS micelle has a negligible effect on the internal structure / mobility of that aggregate.

From the above, and our previously reported (5) data, the complex rheological behaviour of HMHEC in the presence of SDS may be explained as follows :

At levels of surfactant addition considerably lower than its expected cmc, 'micellar-type' aggregates of SDS molecules form around the intermolecular hydrophobic 'bridges' of the polymer. The number of SDS molecules associated with each hydrophobic 'bridge' increases with increasing surfactant addition. The exact composition of the HMHEC / SDS aggregates at C_{max} is still a matter for speculation but, ideally, it may correspond to a situation where each surfactant aggregate incorporates a single grafted alkyl side chain from each of two neighbouring polymer molecules, thereby maximising the number of interpolymer associations.

At concentrations of added surfactant above C_{max}, additional surfactant aggregates are available for the 'solubilisation' of the HMHEC-bound hydrophobes. Consequently the average number of bound hydrophobes per surfactant aggregate would decrease, thereby reducing the number of effective micellar-type 'bridges', and thus producing the decrease in G' and G" with increasing surfactant addition observed in the latter portions of the curve given in Figure 2.

At sufficiently high levels of SDS addition, each bound hydrophobe would be encapsulated in its own individual surfactant micelle, thereby breaking all interpolymer 'bridges' and causing the complete disruption of the network structure, as indicated by the very low values of G' and G" observed at high SDS additions.

References

1. Glass, J.E. Ed. *Polymers in Aqueous Media : Performance Through Association*; American Chemical Society : Washington D.C.; *Advances in Chemistry Series*, **1989**, Volume 223.
2. Shaw, K.G.; Leipold, D.P. *J. Coatings Technol.* **1985**, 57, 63.
3. Dualeh, A.J.; Steiner, C.A. *Macromolecules.* **1991**, 24, 112.
4. Sivadasan, K.; Somasundaran, P. *Colloids Surf.* **1990**, 49, 229.
5. Tanaka, R.; Meadows, J.; Phillips, G. O.; Williams, P.A. *Macromolecules* **1992**, 25, 1304.
6. Landoll, L.M. *J. Polym. Sci. Polym. Chem.* **1982**, 20, 443.

43

Hydrophobically modified hydroxyethyl cellulose – a study of hydrophobe length and reagent reactivity

H. Hofman - Aqualon B.V., Zwijndrecht, The Netherlands.

INTRODUCTION

Water soluble cellulose ethers are traditionally made using alkyl halides (Williamson etherification) or epoxides (alkali catalyzed oxyalkylation).
These methods of etherification have been carried out for some fifty years now. Basically, all cellulose ethers are products in which the etherifying agents are attached directly to the cellulose backbone. In recent years, novel modifications of hydroxyalkylated cellulose have been investigated, resulting in unique properties. Landoll (1,2) discovered that hydroxyethyl cellulose (HEC) modified with long chain aliphatic hydrocarbons showed so called associative thickening behavior.
The modification is carried out not at the cellulose backbone, but at the hydrophobic groups attached at the hydroxyethyl side chains. With the use of longer alkyl chains to modify cellulosics, two questions arise. First, what is the minimum chain length of the alkylating reagent needed for noticeable association in the water phase and, secondly, what is the effect on the overall reaction efficiency of the alkyl chain length.

EXPERIMENTAL

All products were made according to the procedure described below. The base material in all cases was a previously made hydroxyethyl cellulose with an HE-MS of about 4.0 and constant molecular weight.

The following reagents were loaded into a stainless steel reactor: 80 grams of Hydroxyethyl cellulose, 800 grams of tert.-butanol, 80 grams of water and 2.78 grams of sodium hydroxide. Oxygen was removed and the mixture was allowed to swell for 30 minutes.

Following swelling in alkali, the alkylating reagent was added and the mixture heated to 90 deg. centigrade. Samples of the reaction diluent were taken periodically and analyzed by gas chromatography, to follow the decrease of alkylating agent with time.

Alkylhalides and epoxides were used as the alkylating reagents. Differentiation between real epoxides and glycidyl ethers is needed. The glycidyl ethers used contain either alkyl groups alone, or have a phenyl group between the glycidyl moiety and the alkyl group. The latter are referred to in this discussion as alkylaryl glycidyl ethers. (Fig 2).

For a homologous series of alkylaryl glycidyl ether modified HEC, the relationship between the aqueous viscosity and the Degree of Substitution (DS) was determined. The DS is the average number of hydrophobic molecules attached to one monomeric cellulose unit.

RESULTS AND DISCUSSION

I. The effect of the alkyl chain length on associative behavior in aqueous solution. Association of polymers in the water phase occurs through interaction of the alkyl groups attached to different cellulose molecules. The hydrophobes tend to cluster into aggregates, as depicted in Fig 1. The driving force for this cluster formation arise from the tendency to minimize the disruption of the water structure around the polymer by the hydrophobe. (3)

Association between alkyl groups can occur above a minimum chain length. If the alkyl chain length is too short, the chains attached to different cellulose molecules simply cannot come close enough to interact. Furthermore, the degree of substitution of the hydrophobe has to exceed a certain critical limit to give noticeable interaction. A given number of alkyl chains per cellulose molecule have to be present to create the possibility for interaction.

Fig. 1. Structure of hydrophobically
modified HEC in solution

Fig. 2. Hydrophobic reagents used

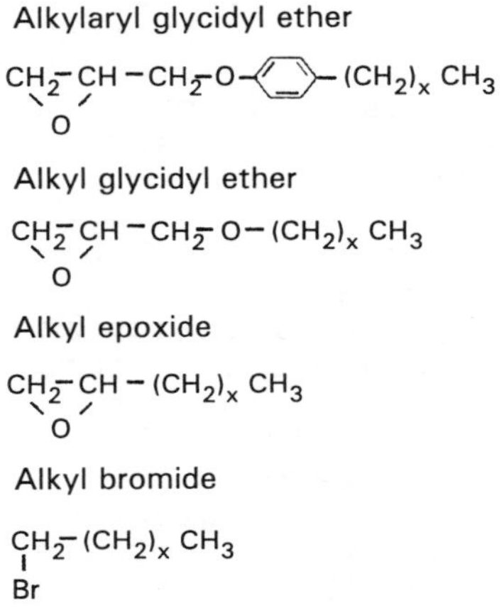

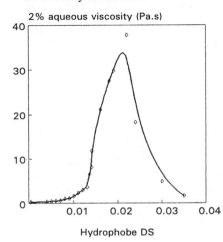

Polymer backbone

Hydrophoba

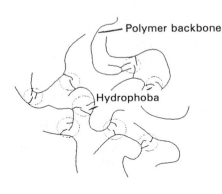

Alkylaryl glycidyl ether

$CH_2-CH-CH_2-O-\langle\ \rangle-(CH_2)_x\ CH_3$
O

Alkyl glycidyl ether

$CH_2-CH-CH_2-O-(CH_2)_x\ CH_3$
O

Alkyl epoxide

$CH_2-CH-(CH_2)_x\ CH_3$
O

Alkyl bromide

$CH_2-(CH_2)_x\ CH_3$
Br

I.1 Influence of hydrophobe DS on associative behavior.

In Fig 3, the degree of substitution of a hexadecyl glycidyl ether modified HEC is plotted against the viscosity of a 2 weight% aqueous polymer solution.

The viscosity increases as the DS increases, up to a maximum. After that maximum is reached, the viscosity decreases again. As the number of alkyl chains increases, the possibility for interaction of these chains attached to different cellulose molecules increases. The viscosity increases as a result.

Fig. 3. Influence of hydrophobe DS
on viscosity of HMHEC

Fig. 4. Association as function
of alkyl chain length

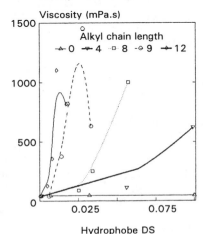

2% aqueous viscosity (Pa.s)

Hydrophobe DS

Viscosity (mPa.s)

Alkyl chain length
0   4   8   9   12

Hydrophobe DS

Increasing the number of hydrophobe chains, increases the possibility of intramolecular hydrophobic interaction. The result is a decline in viscosity due to decreased intermolecular association. As the hydrophobe DS increases even further, the intermolecular interaction predominates the intramolecular interaction, resulting in further decrease of viscosity and ultimately water insolubility.

I.2. Influence of hydrophobe chain length on associative behavior.

In Fig 4, the viscosity at as shear rate of 30 rpm of 1 weight% aqueous solutions of various hydrophobicallly modified HEC's are plotted. The hydrophobes are alkylaryl glycidyl ethers, with alkyl chain length varying from zero to 12.

With only a phenyl group present, no viscosity increase is noticed with an increase in hydrophobe DS. For a butyl group, the viscosity increases slightly at higher hydrophobe DS. The viscosity increase with higher hydrophobe DS becomes very clear as the alkyl chain length increases from 8 through 9 to 12.

Thus, for the class of hydrophobes studied, the chain length needed for noticeable interaction is 4.

I.3. Influence of hydrophobicity on associative behavior.

A second observation that can be made from graph 2, is that the point at which a maximum viscosity is achieved, moves toward lower hydrophobe DS, as the length of the hydrophobe chain increases.

At short alkyl chain lengths, the hydrophobicity of the polymer is close to the hydrophobicity of the HEC backbone. As the alkyl chain length increases, the hydrophobicity of the polymer increases. With increased hydrophobicity, the water structure surrounding the polymer is disrupted to a greater extent. This results in an increased tendency toward cluster formation between neighboring polymer molecules. As a consequence, the maximum viscosity is reached at lower hydrophobe DS.

II. Reaction kinetics

The reactions between the alkylating agents and HEC can best be described assuming first order reaction kinetics. This is exemplified by the reaction between HEC and hexyl glycidyl ether (Fig 5). A semi logarithmic plot of the decrease in concentration of the glycidyl ether with time results in a straight line, as expected in first order reaction kinetics. Under this assumption, the relative reaction constants

for all reagents used were determined. They are listed
in Table I.

Fig. 5. Decrease of hexyl glycidyl
ether concentration in time

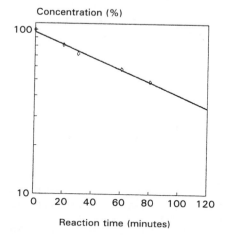

Fig. 6. First order rate constant
of various hydrophobic agents

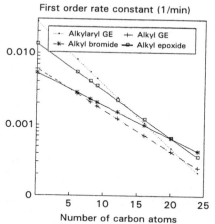

Table 1: Reaction rate constant for various hydrophobic agents

Number of carbon atoms	k (1/min)			
	Alkylaryl glycidyl ether	Alkyl glycidyl ether	Alkyl bromide	Alkyl epoxide
0	0.0290			
4	0.0110	0.0035		
6		0.0013		0.0067
8			0.0023	
9	0.0073			
10				0.0021
12	0.0025	0.0013	0.0015	
14		0.0009		
16			0.0010	0.0014

A plot of the first order rate constants of the various
alkylating reagents versus the alkyl chain length
(Fig 6) shows that:
- alkylaryl glycidyl ethers are most reactive;
- alkyl epoxides are more reactive than alkyl glycidyl
ethers;
- alkyl bromides and alkyl glycidyl ethers seem to
exhibit similar reactivity toward HEC;

The reason for the differences in reactivity can be explained by a combination of the following factors:
- the hydrophobicity of the alkylating reagent;
- the increasing steric hindrance with increasing alkyl group length.

II.1 The hydrophobicity of the alkylating reagent.
The alkylating agent must approach the reactive sites in the HEC for reaction to occur. To do so, the molecule has to migrate from the hydrophobic reaction diluent through a possible water phase to the strongly hydrophilic alkoxide ions on the side chains of the activated HEC.
The migration is a function of the solubility of the reagent. Solubility is a function of hydrophobicity of the reagent. Hydrophobicity is a function of the nature of the reagent, and of the length of the alkyl chain, in case of similar reagents.
In Fig 6, the reaction rate constants for the reaction between HEC and alkyl bromides, alkyl glycidyl ethers and alkyl epoxides are plotted against the length of the alkyl chain. Below an alkyl chain length of 5, both epoxides are more reactive than the halide, as can be concluded from the slope of the lines. Apparently, at these relatively short chain lengths, the influence of the oxygen atom on hydrophilicity is significant. Above an alkyl chain length of 5, the situation alters.

II.3 Steric Hindrance.
The slope of the lines in Fig 6 is much steeper for the alkylaryl species than for the alkyl glycidyl ether. Due to the increasing size of the molecule, phase partitioning of the reactants will favor the reaction diluent instead of the alkalized HEC. This is partly due to the increasing hydrophobicity, but can also be partly attributed to steric hindrance.
The same effect of steric hindrance can be noticed for epoxides and glycidyl ethers. In essence, the glycidyl ether functionality is nothing more than an extension to the alkyl chain, as compared to the epoxide. The glycidyl ethers exhibit lower reactivity than the epoxide, but the slope of the lines are almost equal. As indicated earlier, at short alkyl chain lengths, glycidyl ethers are more reactive than alkyl bromides. Possibly, this can be explained by steric hindrance effects, as well. The alkyl bromide first has to approach the reactive alkoxide ion. Then, after reaction, the relatively bulky bromide ion has to migrate from the reaction surface to the TBA phase, thus obstructing incoming molecules.

CONCLUSIONS

Associative behavior of hydrophobically modified hydroxyethyl celluloses depends on the length of the hydrophobe chain, the nature of the hydrophobe and the amount of hydrophobe attached to the polymer.

For alkylaryl glycidyl ethers, the minimum chain length of the alkyl group needed to give noticeable associative behavior is 4.

The degree of substitution needed for noticeable associative behavior also depends on the length of the hydrophobe. The longer the alkyl chain, the lower the required DS for association.

The difference in reactivity between alkyl epoxides, alkyl glycidyl ethers, alkyl halides and alkylaryl glycidyl ethers can be attributed to a combination of steric hindrance and hydrophobicity.

REFERENCES
1. Landoll, L.M. US Patent 4 243 802, 1981
2. Landoll, L.M. US Patent 4 352 916, 1982
3. Sau A.C., Landoll, L.M. Polymers in Aqueous Media; Glass, J.E. Ed.; Advances in Chemistry Series 223; American Chemical Society: Washington, DC 1989; pp 343-364.

44

Thermogelation in cellulose ethers

K. Philp and R. Ibbett - Courtaulds Research, 101 Lockhurst Lane, Coventry, CV6 5RS, UK.

ABSTRACT

Evidence about the mechanism of the thermal phase separation seen in several non ionic cellulose derivatives is derived from a study of the polymers solution properties in the presence of salts and the use of NMR relaxation studies to show the polymers phase separation. Evidence is provided to substantiate previous theories about the gelation of methyl cellulose and a theory is put forward to explain the difference in thermal properties shown by methylcellulose and hydroxypropyl cellulose.

INTRODUCTION

Aqueous solutions of many non-ionic cellulose ethers exhibit the unusual phenomena of reversible thermal precipitation. In those polymers where the major substituent is methoxyl this phenomena is modified to one of reversible thermal gelation. The precipitation/gelation temperature and the type of gel/precipitate formed has been shown to be a function of the polymers molecular weight, concentration, molecular structure and the presence of additives [1]. The overall mechanism of gelation within methyl cellulose is now widely recognised as the result of hydrophobic interactions between the trimethoxy anhydroglucose segments within the polymer chains [2]. However the mechanism for the thermal precipitation of hydroxypropyl cellulose is not so clearly understood. Evidence is discussed that helps us understand, at the molecular level, the phase separation phenomena.

EXPERIMENTAL

Samples

Samples of methylcellulose, hydroxypropyl methylcellulose and methyl ethylcellulose were all provided by Courtaulds PLC. Samples of hydroxypropyl cellulose were provided by Hercules Corp'.

Thermal and Physical Analysis

Gel points, syneresis points and cloud points were all measured on 2 w/w solutions. The gel point is defined as the temperature at which an increase in the resistance to manual stirring is noted. The cloud point is the point at which visual clarity of the solution is lost and the syneresis point is the temperature at which syneresis around the gel is first detected visually.

NMR Spectroscopy

Samples were dissolved at 10%w/v in deutero water in-situ in 10mm od high resolution NMR tubes. Measurements were carried out using a Bruker AC300 pulsed spectrometer at a carbon frequency of 75MHz. A capillary tube containing a solution of 3 mg/ml/Cr(acac)3 relaxation agent in proteo-dimethylsulphoxide was held coaxially in the NMR tube as an external reference.

PHYSICAL PROPERTIES OF CELLULOSE ETHERS

Reverse solubility is seen in a variety of modified cellulosics which contain a substantial hydrophobic character. Those modified cellulosics that possess methoxyl groups are also likely to form thermoreversible gels.

The type and level of substitution on the cellulose chain affects the type of phase separation behaviour with regard to gel point, gel strength and the level of syneresis detected. From table 1 it can be seen that the greater the level of methyl substitution the lower the gel point and the stronger the gel formed. There is also very little syneresis seen in purely methyl substituted products. However in products containing solely hydroxypropyl substitution a different behaviour is seen, the solution clouds, the polymer forms a floc and effectively precipitates out of solution. It is interesting to note that the mixed ethers show both types of behaviour in the same solution. This shows itself as a gel point, a separate cloud point and finally a point where syneresis is observed with a subsequent contraction of the gel.

TABLE 1

Effect of Substitution Un Gel

Substituent	Gel Point	Type of Gel
Methyl	45	Firm
High Methyl	40	Firm
Methyl Ethyl	50	Very Weak
Hydroxypropyl Methyl	61	Firm, Syneresis
Low Methyl HPM	75	Weak, Bad Floc
Hydroxypropyl	40-60	Floc

From this observation alone it appears that the thermal
gelation is intrinsically linked to the methyl content of the
polymer and the solid/liquid phase separation seen as syneresis
is closely associated with the polymer's hydroxypropyl content.

Looking closely at the actual thermally induced transition from
a free flowing, viscous liquid to a gel in hydroxypropyl
methylcellulose there is a pronounced hysteresis loop. It is
interesting that the hysteresis effect seen can be shown to be
a thermodynamic effect as opposed to a kinetic effect. Taking
a hydroxypropyl methylcellulose gel and cooling to below its
gel point, but not below its solutioning temperature, does not
cause the polymer to redissolve. The polymer gel can be held
below its gel point for several hours without any resolutioning
occurring. One would therefore surmise that the gel could be
held at this point indefinitely without resolutioning
occurring.

As the concentration of a thermally gelling polymer is
increased so the gel point of the polymer solution is reduced.
This fact will help in the consideration of the salt effects to
be discussed. It was decided to investigate the hydrophobic
nature of the interaction with the use of monovalent salts
taken from the Hofmeister lyotropic series. The lyotropic
series has been used to probe the type of interaction in
several gelling systems. The addition of a salt may cause the
gel point of an HPM sample to deviate way from its value in
distilled water. The greater the concentration the greater the
deviation. The use of a wide range of monovalent salts with
different anions all followed the Hofmeister series with regard
to the fact that the large anions, structure breakers, cause
the gel point to increase whereas the smaller anions, which
would be structure formers, caused the gel point to decrease.
This behaviour can be explained in terms of available water.
The structure breaker disrupts the hydrogen bonding within
water itself and hence free water is available to solvate the
polymer. This effectively dilutes the polymer and hence
reduces its gel point.

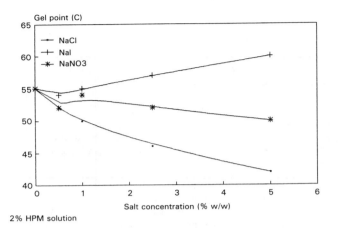

figure 1. How the concentration of added salts affects the gel
point of hydroxypropyl methylcellulose

When the salt is a structure former the opposite is true and
the gel point is reduced. By plotting the gel point of
solutions containing anions against the lyotropic series a
steady decrease in the gel point is seen across the lyotropic
series as one would expect for a product which gels by
hydrophobic interactions.

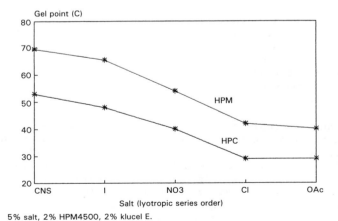

figure 2. How salts in the lyotropic series affect the gel
point or cloud points of modified cellulosics.

NMR RELAXATION STUDIES

It was decided to investigate the gelation of cellulose ethers using NMR. By observing the changes in the relaxation of the carbon spectra of a cellulose ether as its temperature is raised to its gel point and beyond it was hoped to determine which parts of the molecule lose their mobility first during the gelation or phase separation process. Methyl cellulose was chosen first being the system that is best understood. It can be seen that as the temperature is raised so, initially, the intensity of the spectra, compared to the standard increases. This is just an increase in the overall mobility of the solution due to a lowering of the viscosity of the solution. However as the temperature is increased to the gel point so the intensity of both the methyl carbon and also the C-1 carbon on the backbone decreases significantly. According to the theory the crosslinking loci are due to the tri-methyl substituted residues of the polymer chain and these only make up a small percentage of the overall methyl substitution on the polymer. What we are seeing in the methyl cellulose case is a gradual phase separation taking place in the system as the polymer chain desolvates.

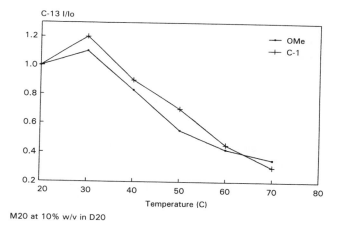

M20 at 10% w/v in D20

figure 3.   13C intensities and their change with temperature

In the hydroxypropylcellulose case a very different intensity change is seen as the temperature is increased. The C-1 on the carbon backbone quite rapidly loses its mobility and ceases to present an NMR signal. However the terminal methyl group on the hydroxypropyl chain maintains practically all of its mobility and hence has not phase separated but has remained mobile and in solution. The methyl groups situated along the hydroxypropyl chain show an intermediate behaviour typical of carbon atoms along a chain that are anchored at one end of the chain and have the other end mobile. So what we are seeing here is the formation of a rigid backbone surrounded by mobile chains of poly-hydroxypropyl units which are mobile and still fully solvated.

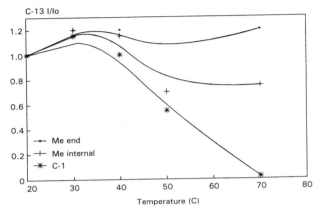

figure 4. Hydroxypropylcellulose 13C intensities and their change with temperature.

The effect of ions on the phase separation of modified cellulosics can be predicted using the Hofmeister series. The modification of gel point is a function of the salt and water structuring and is polymer independent.

From the NMR studies the mechanism of phase separation of methylcellulose is different from the mechanism for hydroxypropylcellulose. The methylcellulose dehydrates after forming crosslinks between its most hydrophobic sections. However the crosslinked network forms an extended cage like structure which once formed acts like a sponge for the water and does not collapse. The hydroxypropylcellulose cannot form crosslinks due to the hydrated layer formed by the sidechains, what we effectively see is a dehydration of the backbone whilst the sidechains remain solvated.

REFERENCES

1   Sarkar, N,; Journal of Applied Polymer Science, Vol 24, 1073-1087, 1979.

2   Takahashi, S.; Fujimoto, T.; Miyamoto, T.; Inagaki, H.; Journal of Polymer Science, Part A, Polymer Chemistry, 1987, 25, 187.

45

Pharmaceutical applications of thermogels composed of nonionic cellulose ether and ionic surfactant

B. Lindman*, J. Tomlin** and A. Carlsson*+ - *Physical Chemistry 1, Chemical Center, Lund University, P.O.B. 124, S-221 00 Lund, Sweden. **Centre for Human Nutrition, Northern General Hospital, Herries Rd., Sheffield S5 7AU, UK.

ABSTRACT

Polymer-surfactant interactions are discussed with particular reference to thermal gelation and drug delivery. A novel gel with some interesting properties are described for systems of a nonionic polymer (ethyl(hydroxyethyl)cellulose) and an ionic surfactant and some possible applications are discussed. Special emphasis is made on the effects of the gel on gastric emptying in man.

INTRODUCTION

Many practically used nonionic polymers, like cellulose ethers and copolymers of ethylene oxide and propylene oxide, show a phase separation at elevated temperature in aqueous solution. This manifests itself in a strong turbidity of a solution in a certain temperature interval and the phenomenon is usually referred to as clouding, occurring at the so called the cloud point (CP) temperature. The clouding is influenced more or less strongly by different cosolutes. Particularly strong effects result from the addition of ionic surfactants, which cause a dramatic increase of the CP even for low concentrations. These nonionic polymers are extensively used in different formulations, such as paints, cosmetics and pharmaceuticals, and often they occur together with ionic surfactants. The resulting inhibition of clouding due to the ionic surfactants can be very significant for many applications, because it allows the use of more hydrophobic and more surface-active polymers without phase separation. However, the approach requires careful attention since it was found that even very small amounts of

---

+) Present address: Karlshamns LipidTeknik AB, P.O. Box 15200, S-104 65 Stockholm, Sweden.

electrolyte (by themselves having no influence on the clouding), in the presence of ionic surfactant, causes a dramatic lowering of the CP [1].

Nonionic polymers are known to interact very generally with ionic surfactants. The systems of nonionic polymer and ionic surfactant were considered to be particularly relevant for the attempt to produce novel structures and formulations with novel rheological properties since (physical) cross-linking effects can be modulated both by an ionic surfactant and by temperature. Thus, as these polymers become less polar at a higher temperature, they will offer better nuclei for surfactant self-assembly at a higher temperature. In the semi-dilute regime there should be a possibility for the surfactant micelles of cross-linking polymer chains resulting in thermogelation [2,3].

Principles for using a water-soluble, amphiphilic cellulose ether (ethyl(hydroxyethyl)cellulose; EHEC) in combination with an ionic surfactant as a basis for *in situ* activated thermogelation are described in the following. In connection with this, the concepts 'liquid carrier' and 'liquid fibre' are introduced, the latter because EHEC may be included in the definition of dietary fibre.

EXPERIMENTAL

Ethyl(hydroxyethyl)cellulose (EHEC) of medical grade was provided by Berol Nobel AB, Stenungsund, Sweden, and used as received. The viscosity average molecular weight is 150 kg/mol. The degree of substitution of ethyl groups is equal to 1.9 and the molar substitution of oxyethylene groups is equal to 1.3. The cloud point of this particular sample is *ca.* 31 °C on cooling and is thus more hydrophobic than other water-soluble cellulose ethers commercially available. Sodium dodecyl sulphate (SDS) was obtained from BDH, Poole, UK. The model drug riboflavine was of pharmaceutical grade.

In vitro dissolution tests of riboflavine and SDS were performed on gels initially containing 1.0 wt% EHEC, 3.0 mmolal SDS and 0.004 wt% riboflavine. The diffusion medium was 0.9 wt% NaCl (aq.) equilibrated at 37 °C. The diffusion cell (a simple steel net basket) containing *ca.* 8 g gel (preheated to 37 °C) was immersed into the flask of a USP rotating paddle apparatus (Prolabo Dissolutest, Paris, France) with a paddle rotating speed of 100 rpm. Aliquots of medium were withdrawn at selected times and analysed by HPLC with post-column ion-pair extraction with UV detection (SDS) and fluorimetric detection (riboflavine).

Ten healthy male volunteers aged 21 to 31 years were recruited for the gastric emptying study. Each volunteer had two studies performed in a random order; a test in which they took an EHEC-SDS solution ('liquid fibre'), and a control in which they took a placebo drink. The studies began at 10.00 a.m. after an overnight fast from 8 p.m. the previous day. They were instructed to eat the same foods and follow the same routine the day before each study. The volunteers sat in front of a gamma camera (Pho Camer Model 1201, Nuclear Chicago, Amsterdam, The Netherlands) and drank 250 ml of the appropriate liquid which had been labelled with 1.85 MBq of the radioactive marker Tc-tin colloid (Amersham, Bucks, U.K.).

A dorsal image was taken every 2 min for the first 30 min and every 10 subsequently. The images were stored on a computer. A lateral image of the stomach region was taken at the end of the study after 200 ml water labelled with

1.0 MBq Tc-tin colloid, to allow correction for tissue attenuation and lateral movement of the marker. The movement of marker out of the stomach was analysed by outlining an area of interest of the stomach region and plotting the %-age of radioactive counts present against time. Statistical analysis was performed using the Student's paired t test and the results are expressed as means and standard errors.

RESULTS AND DISCUSSION

Figure 1 shows a partial phase diagram for a system consisting of 1 wt% EHEC and varying concentrations of SDS in water. At low concentrations of surfactant, the EHEC solution phase separates (the clouding phenomenon) into a white gel. However, if the surfactant concentration reaches a certain point, the phase separation ceases to occur, and a clear gel is formed when the temperature increases from room to body temperature. The process is reversible. At too high concentrations of the surfactant, the solution stays clear upon heating but the gel does not form. Normally, the optimal surfactant concentration varies with the amount of drug, so the phase behaviour has to be determined in every case.

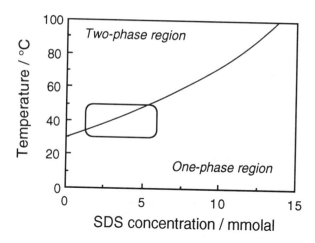

Figure 1. Phase diagram for 1.0 wt% EHEC and SDS in water. The marked area defines the condition of very high viscosity.

Figure 2 shows the release of riboflavine and SDS from an EHEC-SDS gel. Both curves show that the gel gives rise to a slow release of the model drug. Moreover, the release of SDS is slower than riboflavine, which is necessary for the gel to remain, since there is strong evidence that surfactant takes part in the gel structure [4,5].

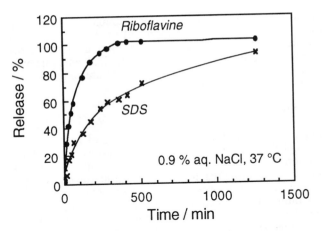

Figure 2. *In vitro* release of riboflavine and SDS from a gel initially containing 1.0 wt% EHEC, 3.0 mM SDS and 0.004 wt% riboflavine.

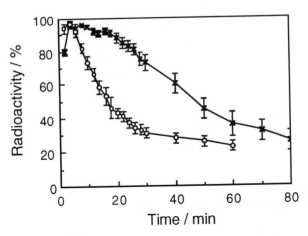

Figure 3. The mean gastric emptying curves for ten volunteers on 'liquid fibre' (crosses) and placebo (open circles). The 'liquid fibre' was composed of 0.85 wt% EHEC of medical grade and 3 mM SDS plus orange flavour and preservatives in water. The placebo drink consisted of orange flavour and preservatives in water.

The EHEC-ionic surfactant combination, which is a drinkable liquid at room temperature, has shown to form a gel in simulated (warm) gastric juice [6]. At present the concept of the EHEC-ionic surfactant mixture as a 'liquid fibre' is evaluated. The mean gastric emptying curves in Figure 3 show that significantly more radioactivity was present in the stomach region after 'liquid fibre' for all times between 8 and 50 min but there was less present in the initial 2 min frame compared to the placebo ($p<0.05$). This implies that 'liquid fibre' has begun to thicken even before it reaches the stomach and some has coated the mouth and oesophagus, emphasising its bioadhesive properties. The reduced rate of emptying is seen in the half-time ($t_{1/2}$) measurement which was extended from 17.7 min to 55.8 min ($p<0.001$). This was made up of both an increase in the lag time (time for 10 % to empty) from 7.0 min to 19.4 min, and also a decrease in the rate of emptying at $t_{1/2}$ from 3.00 % per min to 1.91 % per min. The mean curves also show that 'liquid fibre' empties quite linearly, in a manner similar to solids, rather than the typical curve as seen with the placebo.

EHEC-SDS dramatically retards gastric emptying in man (as well as in the rat [7]) and so may be expected to alter eating patterns in man and to modify the absorption characteristics of other foods. The 'liquid fibre' system may therefore prove useful in the treatment of obesity (a full stomach should retard eating), as well as dumping syndrome and diabetes where the absorption of nutrients needs to be slowed down. The liquidity of the preparation may be beneficial in improving patient compliance as it should be more palatable than the usual treatments which often have to be taken for prolonged periods of time. As EHEC is a polysaccharide with a cellulose backbone it will not be broken down by human digestive enzymes and therefore should act on the colon in a similar manner to dietary fibre [8]. *In vitro* incubations have confirmed that it would be expected to be an efficient laxative and may even reduce flatulence [7].

The thermoreversible gel based on EHEC and ionic surfactant may also have a broad applicability as a 'liquid carrier' for delivery of drugs. The system has a viscosity that allows spraying, instilling, pouring, drinking, or spreading the dosage form and should therefore facilitate the delivery of a drug to the intended biological cavity or part of the body. Upon administration the 'liquid carrier' will adhere to the mucus membrane [9] and form a high viscosity gel or layer. Our results here suggest that on oral administration of a drug in the carrier a gel will form in the gastrointestinal tract (*cf.* Figure 3) giving a slow release of the active substance (*cf.* Figure 2). Once the gel is formed it is very resistant to salt and mechanical rupture.

In principle any water-soluble substance can be incorporated in the 'liquid carrier' regardless if it is charged or not. However, if the drug is charged the maximum loading is limited since small amounts of electrolyte in the presence of ionic surfactant may cause a dramatic lowering of the CP *(vide supra)*. Macromolecules such as insulin may also be incorporated and presently the effect of the carrier on the nasal absorption of insulin in the rat is evaluated [10]. It is also possible to disperse lipophilic substances such as corticosteroids in the carrier - the resulting suspension is stable and is easily transformed to a stiff gel upon heating.

Thus, the gelling is largely independent of the presence of other species in the physiological medium or at the biological surface and requires low concentrations of polymer and surfactant (98-99 % water). Furthermore, the system is compatible with different ways of administration and has a long-term stability. Other aqueous polymer systems known to gel *in situ* require either high salt concentrations (e.g. low-acetyl gellan gum [11]), low pH (e.g. poly(acrylic acid) [12]) or a high polymer content (poloxamers [13]). Such demands may not always be possible to fulfil when using the system for delivery of drugs, for example to the eye.

CONCLUSION

As pharmacologically active substances can be rather generally dissolved or dispersed in an EHEC-surfactant vehicle, it appears that this system, which is low-viscous at room temperature but spontaneously forms a stiff gel at body temperature, can offer improvements in drug administration and drug delivery. The *in vivo* study in man has shown that the system gels in the stomach so that the possibilities of drug targeting and delivery are multiplied. The delaying effect on gastric emptying may be beneficial in its own right.

ACKNOWLEDGEMENT

Dr. Conny Bogentoft is gratefully thanked for valuable discussions. This work was financially supported by Kabi Invent AB.

REFERENCES

1. Karlström, G.; Carlsson, A.; Lindman, B. *J. Phys. Chem.* **1990**, *94*, 5005.
2. Carlsson, A.; Lindman, B.; Karlström, G.; Malmsten, M. in Kennedy, J.F.; Phillips, G.O.; Williams, P.A. (Eds.) *Cellulose Sources and Exploitation. Industrial Utilisation, Biotechnology and Physico-Chemical Properties,* Ellis Horwood, Chichester, 1990; p. 317.
3. Carlsson, A.; Karlström, G.; Lindman, B. *Colloids Surf.* **1990**, *47*, 147.
4. Nyström, B.; Roots, J., Carlsson, A.; Lindman, B. *Polymer,* in press.
5. Zana, R.; Binana-Limbele, W.; Kamenka, N.; Lindman, B. *J. Phys. Chem.,* submitted.
6. Lindman, B.; Carlsson, A.; Thalberg, K.; Bogentoft, C. *L'actualité Chimique* **1991**, Mai-Juin, 181.
7. Tomlin, J.; Read, N.W., personal communication.
8. Tomlin, J. *Scand. J. Gastroenterol.* **1987**, *22*, 100.
9. Kellaway, I.W., personal communication.
10. Rydén L.; Edman, P. *Int. J. Pharm.,* in press.
11. Rozier, A.; Mazuel, C.; Grove J.; Plazonnet, B. *Int. J. Pharm.* **1989**, *57*, 163.
12. Park, H.; Robinson, J. R. *Pharm. Res.* **1987**, *4*, 457.
13. Lenaerts, V.; Triqueneaux, C.; Quarton, M.; Rieg-Falson, F.; Couvreur, P. *Int. J. Pharm.* **1987**, *39*, 121.

46

Highly effective thickening systems with cellulose ether combinations

J.L. Mondt*, W.E. Schermann* and G. Ebert** - *HOECHST AG, Werk Kalle-Albert, D-6000 Wiesbaden 1. **CASSELLA AG, d-6000 Frankfurt/Main 61.

ABSTRACT

As a result of the maximum average degree of polymerization (DP) of about 2,000 industrial cellulose ethers give a viscosity in water of about 150000 mPa s (as a 2 % solution measured in the Höppler-falling ball rheometer). By combining methylhydroxy-ethyl-(MHEC)-cellulose and Super Absorbent Polymer (SAP) the thickening power can be increased synergistically. The thickening mechanism was investigated by means of viscoelastic measurements. It is discussed on the basis of hydrodynamic packing effects of the SAP microgel and entanglement with the MHEC molecules. Also electrostatic effects are considered.

INTRODUCTION

Water soluble cellulose ethers are used as thickeners to improve the properties of aqueous formulations. For a given amount of cellulose ether the degree of thickening in water depends directly on the average degree of polymerization (DP). In the case of commercial products this is between 150 and 2.000. It is not possible to obtain much higher DP-values, for example in the region of the original cellulose, because during pulp production and the ether formation the molecular weight is reduced to a certain extent [1]. This paper will discuss possibilities of using synergistic thickening effects with combinations of high molecular non-ionic cellulose ethers and ionic cross-linked polyacrylic acid (sodium salt).

EXPERIMENTAL

The model investigations were carried out in distilled water at 20 °C with MHEC (Tylose FD 60000, Hoechst AG). The molecular weight of this CE is about 350000. The SAP product (Sanwet IM 1500 F, Cassella AG) has a degree of swelling of about 500 g H_2O/g SAP based on a swelling time of 24 hours. The anionic charge of the polymer is in the region of 500 to 1000 µmol/g SAP.

Viscoelastic measurements in the linear region were carried out using the Bohlin-VOR-Rheometer (Bohlin Rheologi GmbH, D-7130 Mühlacker). This was also used for the dynamic investigations. The streaming current detector (Muetek GmbH, D-8036 Herrsching) was used to measure the anionic charge of the SAP [2].

STEADY FLOW VISCOSITY

The shear thinning properties of a 1 % MHEC solution are shown in Figure 1. The viscosity curve corresponds to the maximum level of thickening that can be obtained by using cellulose ethers. As a result the higher molecular weight SAP suspensions (0,5 and 1,0 %) thicken even more. With a combination of cellulose ether and SAP the thickening level for the same solid concentration can be increased by about a factor of ten.

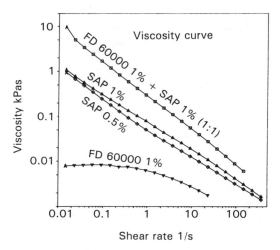

Figure 1: Viscosity development of CE and SAP and the (1 : 1) mixture thereof in water at 20 °C

The SAP suspension which consists of individual swollen particles indicates the typical characteristics for a microgel system as the amount of solid is increased. This is shown in Figure 2/curve SAP. Above a critical concentration, c*, which depends on the degree of cross-linking and/or the degree of swelling, the thickening values do not increase as much as the concentration is increased.

The (1 : 1) mixture of MHEC and SAP, which shows a clear synergistic thickening (curve a), has a similar behaviour. Curve b shows the theoretically expected result if there were no synergistic effects [3]. For comparison the results with MHEC are also included.

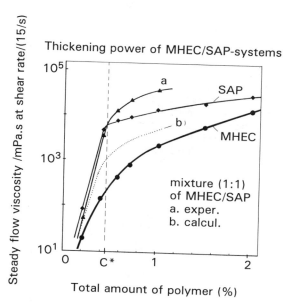

Figure 2: Degree of thickening of MHEC/SAP-systems in water at the shear rate of 15/s and 20 °C

VISCOELASTIC PROPERTIES

The 1 % MHEC solution shows the typical behaviour for highly viscous hydrocolloid solutions, Figure 3. At lower frequency the loss modulus, G'', is dominant and shows a slope of 1 in the log-log plot. The storage modulus, G', asymptotically approaches a straight line with a slope of 2 [4]. At a frequency of 20/s both moduli are equal corresponding to a loss angle of 1. The complex (oscillating) viscosity is comparable to the dynamic viscosity shown in Figure 1.

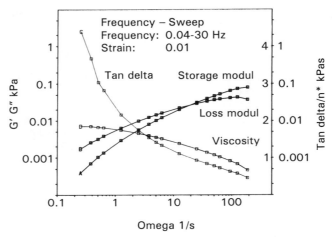

<u>Figure 3:</u> Viscoelastic properties for 1,0 wt% MHEC in water at 20 °C

The SAP suspension shows a completely different viscoelastic behaviour (Figure 4). The rheogram is characteristic for polymer cross-linked structures; there are plateau-values for both loss and storage moduli. The elastic component dominates [4]. Accordingly the complex viscosity at low frequency is about 2,5 decades higher than that of the MHEC sample.

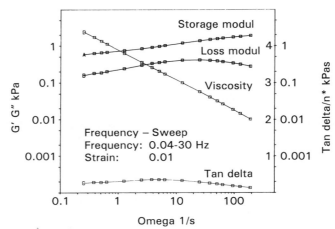

<u>Figure 4:</u> Viscoelastic properties for 1,0 wt% SAP in water at 20 °C

The viscoelastic spectrum of the (1 : 1) MHEC/SAP mixture shown
in Figure 5 has a similar behaviour. The MHEC component
increases the viscosity in general. The microgel contribution
remains the most important factor.

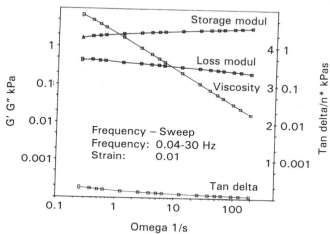

Figure 5: Viscoelastic properties for (1 : 1) mixture of MHEC
          and SAP in water at 20 °C

MECHANISM

The rheological experiments indicate that the SAP suspension has
a polymeric cross-linked structure with a randomly close packed
microgel. This is shown by the following two observations:

a) In the linear viscoelastic region storage and loss moduli are
   almost independent of the frequency (plateau) with the
   elastic component dominating.

b) At a critical concentration, c^*, microgel particles start to
   interact as a result of the increased packing density. With
   the SAP product investigated in this report we found a
   critical concentration, c^*, of 0,4 %. This correspondes to a
   reduced concentration $c.Q = 2$ as expected on the basis of
   Taylor's theory [5]. Q is the degree of swelling of the SAP
   at the concentration c.

The synergistic thickening mechanism of the MHEC/SAP⁻ system seems to be a combination of various effects. Water is immobilized by the SAP microgel. Above the critical packing density the mobility of the swollen particles are further reduced by steric hindrance. If water as solvent is replaced by MHEC solution a further thickening results. This is generally observed when a filler is added to a viscous solution (emulsion paint). Further effects are assumed: the formation of physically cross-linked structures by entanglement of the MHEC molecules with loops and chain segments of the SAP molecule and H-bonding mechanism. This mechanism is supported by experiments with SAP/anionic carboxymethylcellulose mixtures (NaCMC) where the thickening is much less than with the SAP/non ionic cellulose ether. Therefore electrostatic effects probably play a role in the mixture of anionic SAP-gel particles and nonionic cellulose.

[1] R. L. Whistler, Industrial Gums,
    Academic Press, Inc., New York, London (1973)

[2] L. Hong, B. Borrmeister, H. Dautzenberg, B. Philipp
    Zellstoff und Papier 5 (1977), p. 20 - 210

[3] W.-M. Kulicke, Fließverhalten von Stoffen und Stoffgemi-
    schen, Hüthig & Wepf, Basel, Heidelberg, New York (1987)

[4] Praktische Rheologie der Kunststoffe und Elastomeren,
    VDI-Gesellschaft Kunststofftechnik (1990)

[5] N. W. Taylor, S. H. Gordon
    Journal of Applied Polymer Sci. 27 (1982), p. 4377 - 4386

47

Cationic cellulosic derivatives

E.D. Goddard - Union Carbide Chemicals and Plastics Company Inc., Specialty Chemicals Division, Tarrytown, NY 10591, USA.

ABSTRACT

Adsorption properties of the Polymer JR family of cationic cellulosic polymers onto the natural keratins, hair and skin, which they are known to condition, are reviewed. The interaction patterns of the polymer with anionic surfactants are outlined. A new, hydrophobically modified member of the polymer family is described.

INTRODUCTION

The Polymer JR family of cationically substituted cellulosic polymers was developed by Union Carbide Corporation in the early 1970's. Though initially targeted (and shown to be effective) as flocculants it was soon realized that their ability to topically condition the natural keratins, hair and skin, made them outstanding candidates for the personal care industry. It is their properties, with relevance to this field, that will receive emphasis in this review.

STRUCTURE

The polymers are derivatized products of hydroxyethyl cellulose (HEC) containing quaternary ammonium groups [1]. The molecular weights of the three grades of Polymer JR viz., JR30M, JR400 and JR125, are approximately 700,000, 400,000 and 125,000 daltons, respectively. Other members of the "family" are Polymer LR 30M and LR 400 in which the cationic substitution level is about half that of Polymer JR.

SORPTION STUDIED BY RADIOTAGGING

Most of our early work utilized a ^{14}C tagged polymer and radiocounting of the keratin substrates, viz. hair fibers or stratum corneum membranes [2,3]. Three points are noteworthy:

1.   The uptake was appreciable--far exceeding that anticipated for monolayer coverage of the substrates.

2.   Adsorption was "slow" in the sense that after the initial adsorption segment, uptake continued for several hours.

3.   The adsorption levels were highest for the lowest MW (JR125) homolog, and vice versa.

The three points above all point to the uptake being more appropriately termed a sorption, rather than an adsorption process. Indeed, square root of time plots confirmed linearity of uptake, Q(t), over a several hour period -- in conformity with Hill's equation for diffusion into an infinite slab or cylinder [2-6],

$$Q(t) = 2C_o(Dt/\pi)^{\frac{1}{2}}$$

where C_o is bulk concentration , D the diffusion coefficient, and t is time.

Effect of Additives on Sorption: Keratins have a low isoelectric point (see below), and are thus negatively charged at normal pH. The addition of NaCl at 0.01 M concentration reduced the uptake of Polymer JR400 on hair by as much as 50% [7]. The addition of $CaCl_2$ had an even larger effect, and $LaCl_3$ larger again. The reduction is ascribed to shielding of the negative charges of the keratin by the cations of the added salt, efficiency increasing with the valency of the cation. Effects observed on incorporating surfactants in the polymer solution were, likewise, explicable on electrical grounds [4]. Another manifestation of the electrostatic effect was seen in sorption experiments carried out as a function of pH [8]. HEC, lacking any charge, showed a finite but low adsorption value, independent of pH. The sorption value of Polymer JR, on the other hand, was constant and high. As the pH was reduced towards 3.3, however, sorption dropped sharply towards the value observed for uncharged HEC.

SORPTION BY SURFACE TECHNIQUES

Streaming Potential: It was established that the iso-electric point of hair is 3.3. [8]. The zeta potential of hair, measured at natural pH (~ 6) in

10^{-4}M KNO_3 is in the range of -30 to -35 mV. Exposure to dilute solutions (0.01%) of Polymer JR resulted in an immediate increase in zeta potential to +20 to 30mV [8]. Removal of the contacting polymer solution and replacement with a 10^{-4} KNO_3 rinsing solution yielded only a slight decay of the potential; it is obvious that these polymers, once adsorbed, are strongly retained. A point of interest is that the highest molecular weight homolog, viz. Polymer JR 30M, is the one most strongly adsorbed and retained, as would be expected from polymer adsorption principles.

Colloidal Particle Deposition: This approach [9,10] was used to confirm the reversal of charge, from negative to positive, of hair fibers before and after exposure to solutions of Polymer JR. After experimenting with various latex preparations we found that 0.2μ size particles of silica, prepared by the Stober method [11], were most convenient. Little to no deposition occurred on control hair fibers, while a close packed layer of these negatively charged particles adhered to the fibers which had been exposed to a polymer solution of concentration as low as 0.01% [12].

XPS or ESCA: Clear evidence of the presence of a surface layer of polymer was obtained [13] from several observed parameters, e.g. an increased level of C-O carbon, the emergence of a quaternary nitrogen peak, and a reduction of sulfur (cystine, cysteine) due to concealment of the hair surface by adsorbed polymer. The surface layer was very resistant to aqueous rinse-off, but XPS showed that treatment with SDS solution ("shampooing") was effective in removing it. The fractional coverage of the hair surface was estimated from the spectroscopic data to be about 0.25, i.e. 25%.

Wetting: Natural keratin is hydrophobic -- at least as assessed by measurements of the advancing angle of water on it. Work by Weigman and Kamath [14] (and confirmed by us), in which single fibers of hair were the sensor in a "Wilhelmy plate" surface tension experiment, showed that the wetting angle, θ, of hair advancing slowly into water was around $100°$. After exposure to a 1% Polymer JR solution, rinsing and drying, the fibers became wettable in the sense that θ became $<90°$. These data indicate a surface coverage of the fiber surface by the polymer of $\sim 23\%$, which agrees quite well with the value estimated from ESCA [13].

CONDITIONING AND PROTECTIVE PROPERTIES OF POLYMER JR

Early subjective reports of improved "condition" and "manageability" of hair or skin treated with formulations containing Polymer JR were widespread. The lubrication (ease-of-combing) attributes of Polymer JR are amenable to quantitative assessment by using a "mechanical comb" attached to an Instron tester [15]. Typical values of the forces P_3 (the maximum recorded near the end of travel through the hair), P_2 (intermediate), and P_1 (initial maximum) for a control shampooed hair, are 100, 20 and 10g. versus 20, 10 and 5 g., respectively, for a similar hair tress subsequently treated with 0.1% Polymer JR solution and then rinsed with water. There are several quantitative in-vitro assessments of skin protection which can be carried out on separated stratum corneum membranes. Aggressive treatments by warm detergent solutions, for example, are known to reduce the membranes' resistance to water vapor transmission and to cause serious swelling, and concomitant weight loss, of the membranes. The presence of Polymer JR in all cases reduced these effects [6]. Correspondingly, in-vivo tests on humans [16] showed that the presence of Polymer JR could protect the skin from such aggressive agents as harsh surfactants, thioglycollate depilatory, and poison ivy catechols.

INTERACTION WITH ANIONIC SURFACTANTS

Surface Tension: Polymer JR, per se, is only weakly surface active at the air/water interface. The surface tension of its 0.1% solution in water is ~70 dyne/cm. On the other hand, when present together with SDS in solution it leads to a profound, synergistic increase in surface activity of their combination [17]. We have ascribed the phenomenon to the formation of a highly surface active complex, whose composition changes with the ratio of added SDS. Even at the stoichiometric equivalence point, where most of the incorporated solutes are out of solution, the surface tension is low showing that the sparingly soluble complex is extremely surface active. At higher relative concentrations of SDS, at which the complex is solubilized, the surface tension of the mixed solution coincides with that of micellar SDS.

Solubility Characteristics: Crude solubility diagrams of Polymer JR and a number of anionic surfactants have been obtained [17,18]. They are characterized by precipitation regions, and resolubilization regions at higher surfactant levels. The intercepts (C_e) of the maximum insolubility lines on the surfactant log concentration axis were found to be linear with chain length, n, in a homologous series of Na alkyl sulfates (C_8 to C_{14}).

Mathematically, the result could be expressed as
$$\log C_e = n \Delta G/kT$$
where ΔG represents the contribution of the CH_2 groups to the free energy of formation of the complex [18]. The value of ΔG found, ~ $1kT/CH_2$ group, is comparable to the value for the free energy of micelle formation

of surfactants, suggesting that aggregates of surfactant molecules are present in the complexes which form.

Small Angle Neutron Scattering: Preliminary measurements with SANS [19] showed that the presence of small amounts of SDS, i.e. in the pre-precipitation zone, substantially changed the scattering pattern of Polymer JR; and, likewise, that Polymer JR had a substantial influence on the scattering pattern of SDS micelles in the post-precipitation zone. These data are again consistent with the formation of polymer/surfactant aggregates in both of these zones.

Rheology: When SDS was added to a 1% solution of Polymer JR 400 in the pre-precipitation zone, increases in viscosity of 100-fold were observed [19]. This suggests that a network is formed by association of surfactant ions bound to separate polymer molecules. The corresponding SDS/Polymer JR system with the highest molecular weight grade of polymer, viz. JR 30M, was investigated. In this case strong gels resulted in the pre-precipitation range when the polymer concentration was quite low (1%), and the SDS level about one-tenth this level. The gels were characterized by oscillatory rheological measurements; the storage modulus, G', was found to dominate over the loss modulus, G", over the entire frequency range examined [20], and the derived viscosity of the gel at low frequency (0.001 Hz) was some 10 million cps!

HYDROPHOBICALLY MODIFIED POLYMER

A polymer with dodecyldimethylammonium groups linked to the HEC parent molecule, rather than with trimethylammonium groups as in Polymer JR [1], has been introduced by Union Carbide under the name of QUATRISOFT® LM200. This material differs from Polymer JR in being markedly surface active in water and it generates very stable foam [21]. Also, it adsorbs strongly onto hair keratin. ESCA measurements show, if anything, that it is adsorbed more strongly and more uniformly than is Polymer JR.

In common with other polycations it has precipitation ranges with added anionic surfactants. As with Polymer JR, adding anionic surfactants like SDS, causes increases in viscosity, not only in the preprecipitation zone, but in this case in the postprecipitation range as well. Oscillatory measurements were used to characterize compositions in this range rheologically [22]. Evidence was also adduced that the polymer interacts with cationic and nonionic surfactants, presumably through a process of hydrophobic association.

REFERENCES

1.   Brode, G.L.; Goddard, E.D.; Harris, W.C.; Salensky, G.A. *Polymeric Science and Engineering Proc.* American Chemical Society, **1990**, <u>63</u>, 696.
2.   Goddard, E.D.; Hannan, R.B.; Faucher, J.A. *Proc. VII Int'l Congress on Surface Active Agents*, Moscow **1977**, <u>2</u> 834.
3.   Faucher, J.A.; Goddard, E.D. *J. Soc. Cosmetic Chem.* **1976**, <u>27</u>, 543.
4.   Faucher, J.A.; Goddard, E.D. *J. Colloid Interface Sci.* **1976**, <u>55</u>, 313.
5.   Hill, A.V. *Proc. Roy. Soc. London* **1938**, <u>104B</u>, 39.
6.   Goddard, E.D., Leung, P.S. *Cosmetics and Toiletries* **1982**, <u>97</u>, 55.
7.   Faucher, J.A.; Goddard, E.D.; Hannan, R.B. *Textile Res.J.* **1977**, <u>47</u>, 616.
8.   Goddard, E.D. *Cosmetics and Toiletries* **1987**, <u>102</u>. 71.
9.   Harley, S.; Thompson, D.W.; Vincent, B. *Colloids and Surfaces* **1992**, <u>62</u>, 163.
10.   Alince, B.; Robinson, A.A.; Inoue, M.J. *J. Colloid Interface Sci.* **1978**, <u>65</u>, 98.
11.   Stober, W.; Fink, A.; Bohn, E. J. *Colloid Interface Sci.* **1968**, <u>26</u>, 62.
12.   Goddard, E.D.; Chandar, P. *Colloids and Surfaces* **1989**, <u>34</u>, 295.
13.   Goddard, E.D.; Harris, W.C. J. *Soc. Cosmetic Chem.* **1987**, <u>38</u>, 233.
14.   Weigman, H.D.; Kamath, Y.K. *Cosmetics and Toiletries* **1986**, <u>101</u>, 37.
15.   Garcia, M.L.; Diaz, J. *J.Soc. Cosmetic Chem.* **1976**, <u>27</u>, 379.
16.   Faucher, J.A.; Goddard, E.D.; Hannan, R. B.; Kligman, A.M. *Cosmetics and Toiletries* **1977**, <u>92</u>, 39.
17.   Goddard, E.D.; Phillips, T.S.; Hannan, R.B. *J. Soc. Cosmetic Chem.* **1975**, <u>26</u>, 461.
18.   Goddard, E.D.; Hannan, R.B. *J. Am. Oil Chem. Soc.* **1977**, <u>54</u>, 561.
19.   Leung, P.S.; Goddard, E.D.; Han, C.; Glinka, C.A. *Colloids and Surfaces* **1985**, <u>13</u>, 47.
20.   Leung, P.S.; Goddard, E.D. *Langmuir* **1991**, <u>7</u>, 608.
21.   Goddard, E.D.; Braun, D.B. *Cosmetics and Toiletries* **1985**, <u>100</u>, 41.
22.   Goddard, E.D.; Leung, P.S. *Colloids and Surfaces* In press.

48

Compatibility of water-soluble cellulose ethers with lignosulphonates

G. Telysheva*, M. Akim, G. Shulga and T. Dizhbite - Institute of Wood Chemistry, Latvian Academy of Sciences, Riga, Latvia.
*Present address: Institute of Wood Chemistry, 27 Akademijas st., 226006 Riga, Latvia.

ABSTRACT

Aqueous solutions of cellulose polymers (CP) and sodium lignosulphonates (LS) as well as polymer blends prepared from the same solutions were investigated. The mechanism of CP and LS interaction is proposed. We assume that the main role in this interaction belongs to phenolic OH groups of the LS and OH groups of CP at the C2-atom of the glucose units. Methylcellulose (MC-100) and LS macromolecules interact in aqueous solutions, forming polymer associates with stoichiometric composition, corresponding to LS:MC ratio 1:6.

Introduction of LS into CP blends facilitates purposeful change to their physical and mechanical properties. Combination of these polymers enabled us to obtain high-performance products and/or diminish the consumption of the most valuable cellulosic components in practical applications. For example, surface treatment of paper with aqueous solutions of LS/CP gives an increase in its tensile strength and modulus of elasticity of more than 30%.

INTRODUCTION

Water-soluble CP and LS are widely used as structure formers, binders and stabilizers of various disperse systems.

The aim of the present study was the investigation of CP and LS mixtures in aqueous solution and in the solid phase. First, we wished to find the possibility of diminishing the consumption of valuable cellulosic components in practical applications, and second - to improve the properties of LS to widen its application area.

EXPERIMENTAL

Compatibility of LS with MC in aqueous solutions was studied by the Dobry and Boyer-Kawenoki method [1]. Mixtures were prepared by mixing at 293 K LS and MC solutions with concentrations from 0.1 to 1.0% and in mass ratio range from 1:10 to 10:1. The viscosity of mixtures was determined at 298 K using a Ubbelohde viscometer for which kinetic energy correction was negligible. The pH titrations were carried out with a "Radiometer" pH meter. The conductometric titrations were carried out with a conductometer model "OK-102/1". The associates formation in polymers solutions was investigated by light scattering performed using a multiparameter flow cytofluorimeter EPIC.

The surface tension of the aqueous solutions was measured by the Wilhelmi method [2].

Mechanical properties of blends, obtained from mixed solution at various mass ratios MC:LS, as well as the same of paper surface sized with solutions of MC and LS were investigated by the method of active linear deformation at the constant deformation rate using universal apparatus "Instron-1121". ESR spectra of blends were registered using RE-1306 spectrometer at 293 K with DPPH as standard.

RESULTS AND DISCUSSION

LS:MC mass ratio and polymer concentrations in solution were found to be the major factors, influencing compatibility. Mixtures with LS:MC ratio < 1:6 are compatible within the all studied concentration intervals and stay monophase over long-time storage (Fig. 1). The mixtures, containing more LS in the concentration range from 0.1 to 1.0% start to form two phases after 24 h. storage.

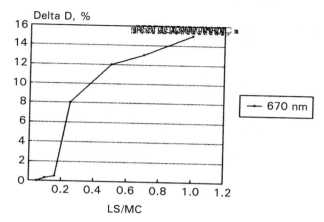

Figure 1. Alteration of optical density of 0.3% LS:MC mixtures after 72 h., 293 K.

Compatibility of LS and MC in aqueous solutions is caused by specific intermacromolecular binding accompanying with non-additive increase of specific acidity and conductance of mixtures (Fig.2). For example, pH decrease in the case of a 0.3% mixture of LS:MC=1:6 was 1...2 pH units relative to initial polymer solutions.

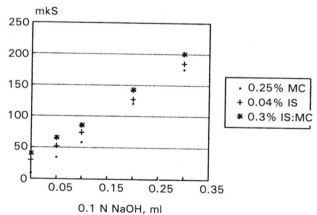

Fig. 2. Specific conductance of LS, MC and their mixture (1:6).

The data obtained by ESR (Fig. 3) indicates the increase of stable phenoxy-radical concentration in the blends formed from LS:MC solutions relative to the initial LS. The reason for radicals formation is supposed to be the donor-acceptor interaction between MC and LS, maximal at LS:MC=1:6.

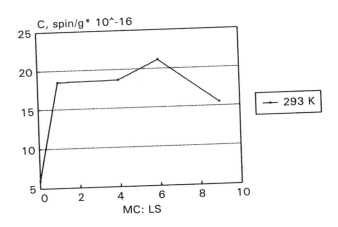

Fig. 3. Concentration of lignin stable radicals in blends from LS:MC mixtures.

It may be supposed that the main role in the donor-acceptor interaction with phenolic OH groups of LS belongs to the OH groups of MC at C2, having the least electron density because of the induction effect of two oxygen atoms at C1. Evidently, the interaction of polymers in diluted solutions leads to formation of associates, differing from the initial LS and MC in surface activity and viscosity. The polymer mixture LS:MC=1:6 gives the minimal surface tension, confirmed by formation of stoichiometric molecular associates at this polymer ratio. Increase of MC content in polymer mixtures to 6:1 leads to regular increase of kinematic viscosity. Further increase of MC content stabilizes the rheological properties of the polymer mixture. It also corresponds to light-scattering data: in diluted solutions with the same LS:MC ratio, the associates are smaller and have higher structural density, compared with LS:MC mixtures of ratio > 1:6.

Time dependences of phase division (Fig.1) and surface tension reflect the kinetic character of LS and MC associate formation. It is also confirmed by light-scattering data, namely by a two fold increase of the number of associates 48 h after the introduction of small quantities of MC into an LS solution. Just after mixing LS and MC at a stoichiometric ratio 1:6, the different types of submolecular structures are formed (Fig. 4). It reflects the interaction of LS with heterogeneous polymer. 48 h later as a result of LS binding with preferable sites of MC the formation of an average structure is observed.

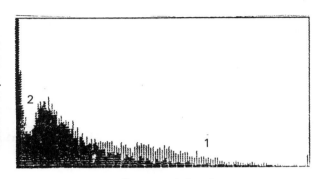

Fig. 4. The compared histograms of orthogonal light-scattering of 0.02% LS:MC=1:6 solution just after mixing (1) and 48 h later (2).

LS addition into MC aqueous solutions simultaneously diminish their gelation temperature by 15 K and the mechanical strength of the formed gel (more than twice). It is proposed that LS takes part in the phase formation of the gel network.

Modification of LS with small amounts (4...10%) of MC give products with high blend-forming properties. In the case of CP, application of LS (up to 16%) did not reduce their mechanical properties, and, furthermore, gave an opportunity to regulate the elasticity of blends. The properties of blends, obtained from solutions with constant total concentration (1.33%) of (LS+MC) and with constant content of blend-forming polymer - MC were investigated. Tensile strength of blends increases up to 16% of LS in composition, then slowly decreases and became less than the value for blends from MC only at LS content 50%. At the same time the failure energy for the blends with 16% LS does not change in comparison with 100% MC, while for LS content 66% this value decreases 20 times. In the interval of LS content 0...50% the modulus of elasticity increase more than twice.

Compositions of LS and water soluble CP may find application in the pulp and paper industry, for example, on sulfite pulp and paper mills in the production of different kinds of paper.

For example, LS modified with MC at a 9:1 ratio, is very suitable for surface sizing. In this case deformation characteristics of surface layers forming under the action of MC:LS became very near to those of the base paper, thus increasing its overall strength. Comparison of strength-elongation curves confirm the possibility of realization of overall strength of both inner (unsized) and external (sized) layers.

When LS modified with small amounts of MC (4%) is present in paper mass, it became possible to improve the properties of paper. As a result the modulus of paper elasticity increased by 45%. Failure energy increased to the same value, as achieved after the action of the maximal amount of MC. Failure deformation increased, too.

REFERENCES

1. Dobry A., Boyer-Kawenoki F. J. Polymer. Sci., 1947, v. 2, p. 90-100.
2. Adamson A. Physical Chemistry of Surfaces. Moscow, 1979, 568 p.

49

Reaction mechanism for metal crosslinking carboxymethyl hydroxypropyl cellulose

En-Pu Wang and Kai Xu - Guangzhou Institute of Chemistry, Academia Sinica, P.O. Box 1122, Guangzhou, 510650, P.R. China.

ABSTRACT

The carboxymethyl hydroxypropyl cellulose (CMHPC) crosslinked with metal ion (e.g. Cr^{3+}, Al^{3+}, Fe^{3+}) and organic titanium complex has been studied by IR and Raman spectra, and compared with carboxymethyl cellulose (CMC-Na) and hydroxyethyl cellulose (HEC). The results show the crosslinking sites are not only the carboxymethyl groups, but the hydroxypropyl groups. The reaction mechanism is affected by the degree of substitution. Moreover, crosslinking results in a red-shift of the stretching vibration of carboxylate, a split of other frequency bands of CMHPC and CMC-Na, and in a conformational change of CMHPC. The differences between un- and crosslinked HEC appear also in FTIR. The interpretation of these results can give direct evidence of the crosslinking reaction mechanism.

INTRODUCTION

Cellulose ethers gelled with polyvalent metal ions are widely used in the petroleum industry[1-3]. However, the crosslinking mechanism is established only on a purely intuitive basis[4-5]. The intrinsic structure analysis of crosslinked cellulose ethers which can give direct evidence of the detailed reaction mechanism is particularly rare up to now. This paper reports some structural studies on the CMHPC, CMC-Na, and HEC crosslinked with chromium, aluminium and iron ions and organic titanium complex (O-Ti-C) by FTIR and Raman spectroscopy. The results give a good understanding of the stuctures and the interactions involved, especially in the case of CMHPC. It has been found that the hydroxypropyl group plays an important as well as interesting role during the crosslinking reaction.

EXPERIMENTAL

The CMHPC (sodium salt) with various DS and MS values and the CMC-Na with
DS of 0.97 were prepared by our laboratory and the HEC (MS 1.5) was
supplied by Union Carbide Corporation. The organic titanium crosslinker
was synthesized by mixing titanium tetrachloride, ethylene glycol and
lactic acid according to some ratio (Ti-EGL). After the aqueous solution
of CMHPC, CMC-Na or HEC is gelled, alcohol or acetone was added to the
hydrogel, so that the precipitate was separated from the gel. Then, these
samples were dried in a vacuum chamber at lower temperature. The dried
sample can swell to become a gel when water is added.
Powder samples were mixed with KBr and pressed into pellets for
quantitative IR analysis. All IR spectra were recorded with an Amalect
RFX-65 spectrometer.
Samples for Raman spectroscopy were previously purified and were
simultaneously treated by adding 0.1% EDTA solution while the hydrogel was
precipitated by acetone to eliminate fluorescence interference band, and
determined with Spex Raman spectrometer.

FTIR of uncrosslinked and crosslinked product.

Fig.1 shows some IR spectra of CMHPC (DS 0.91, MS 0.51) and its products
crosslinked by various metal ions. The antisymmetric stretching vibration
frequency (Vas) -CH$_2$ of uncrosslinked CMHPC appears at 2923 cm^{-1}, and the
characteristic frequency at 1385 cm^{-1} and 2976 cm^{-1} of CH$_3$ is very weak
because the MS is very low. The shape of Vas of -CH$_2$ for crosslinked
CMHPC by Cr^{+3}, Al^{+3} and Fe^{+3} has become sharp and narrow compared with
uncrosslinked CMHPC. Maybe the space for the molecule has become
relatively reduced and the vibration movement of -CH$_2$ is restricted after
crosslinking in the carboxymethyl and hydroxypropyl groups, moreover, the
initial strong induction effect of carboxyl group on -CH$_2$ and the effect
of the outside environment on -CH$_2$ are weakened because of metal ion
action. Both the above mentioned factors can change the shape of the
absorption band to become narrow.

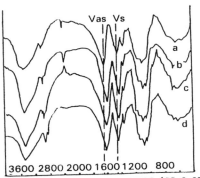

Fig.1. IR spectrum of CMHPC (DS 0.91
MS 0.51) crosslinked with various
metal ions.
a. CMHPC; b. Cr^{+3}-CMHPC;
c. Al^{+3}-CMHPC; d. Fe^{+3}-CMHPC.

The strong absorption bands 1423 cm⁻¹ and 1608 cm⁻¹ belong to respectively symmetric stretching vibration (V_s) and V_{as} of the carboxylate which exists in uncrosslinked CMHPC (sodium salt). The V_{as} bands are evidently decreased since the electron is delocalized further after crosslinking. The V_{as} bands of crosslinked products by Al^{+3}, Cr^{+3} and Fe^{+3} are decreased to 1581 cm⁻¹, 1576 cm⁻¹ and 1602 cm⁻¹, the red shift values are 27 cm⁻¹, 32 cm⁻¹ and 6 cm⁻¹ respectively. The red shift values of V_s bands appear also in the range of 8 cm⁻¹, 10 cm⁻¹ and 5 cm⁻¹ respectively. It will be seen from the above red-shift value that the linkage between Cr^{+3} and carboxyl groups is the strongest compared to the Al^{+3} or Fe^{+3} crosslinked product. The strong and wide absorption band at 1030-1200 cm⁻¹ is the V_{c-OH} characteristic frequency of the polysaccharide, which splits into several small shoulder bands since the OH group forms various hydrogen bonds, however, the V_{c-OH} of crosslinked CMHPC about 1030 cm⁻¹ has been split evidently into a sharp form, which is the V_{c-OH} of primary hydroxy. These primary hydroxys in C_6 position can be liberated themselves from hydrogen bonds associated with other groups after the metal ion is introduced into the CMHPC molecule to exhibit their characteristic bands.
CMHPC has not only the ionic carboxymethyl but also the nonionic hydroxypropyl group. It is to be noted which group is major when the CMHPC with various DS and MS value is crosslinked by metal ion.

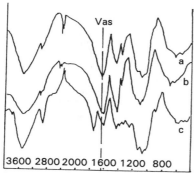

Fig.2. IR spectrum of un- and cross-linked CMC.
a. CMC; b. Cr⁺3-CMC; c. Ti-CMC.

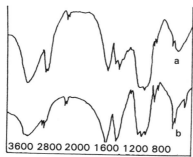

Fig.3. IR spectrum of un- and cross-linked HEC.
a. HEC; b. Cr⁺3-HEC.

First, the simple cellulose ethers with single substituting groups (e.g.
CMC-Na or HEC) are studied. Fig.2 and Fig.3 are the IR spectra of un- and
crosslinked CMC-Na and HEC respectively. The shape of $-CH_2$ Vas about 2923
cm^{-1} from crosslinked CMC-Na has become sharp as well for crosslinked
CMHPC. The absorption bands at 1620 cm^{-1} and at 1425 cm^{-1} are respectively
the Vas and Vs of carboxyl radical from CMC-Na. Fig.2 shows that the Vas
of CMC-Na crosslinked with Cr^{+3} and Ti-EGL exhibits a red shift by 20 cm^{-1}
and 18 cm^{-1} respectively. It is obvious that the major crosslinking active
group is the carboxyl group.
The absorption bands about 2929 cm^{-1} and 2881 cm^{-1} are respectively the
Vas and Vs of $-CH_2$ from uncrosslinked HEC. The Vas of crosslinked HEC,
especially, Cr^{+3}-HEC has become sharp and the absorption band about
1020-1200 cm^{-1} has further split. It may be seen that the hydroxyl group
in the hydroxypropyl is the major crosslinking active group from Fig.3.

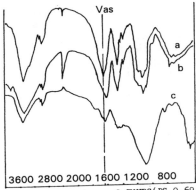

Fig.4. IR spectrum of CMHPC(DS 0.69,
MS 2.45) crosslinked with various
metal ions.
a. CMHPC; b. Cr^{+3}- CMHPC; c.Ti-CMHPC.

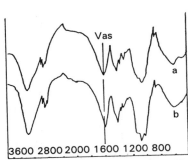

Fig.5.IR spectrum of un- and crosslinked CMHPC(DS 0.30, MS 2.38).
a. CMHPC; b. Cr^{+3}-CMHPC.

Secondly, we compare further crosslinked CMHPC having low DS and high MS
values with uncrosslinked CMHPC having the same DS and MS. Fig.4 and
Fig.5 are the IR spectra of CMHPC (MS 2.45) and CMHPC (MS 2.38)
respectively. The characteristic frequencies of $-CH_3$ in the hydroxypropyl
group appear about 1385 cm^{-1} and 2976 cm^{-1}. The $-CH_2$ Vas about 2927 cm^{-1} is
not changed after crosslinking, maybe the effect of interaction between
the hydroxypropyl group and the metal ion on the environment of $-CH$ is
not evident, however, the red shift appears in Vas and Vs of the carboxyl
group. The Vas red shift value for CMHPC (MS2.45) crosslinked with Cr^{+3}
and Ti-EGL is 40 cm^{-1} and 7 cm^{-1} respectively from Fig.4, and for CMHPC
(MS 2.38) crosslinked with Cr^{+3} is only 6 cm^{-1} because the DS value is
very low from Fig.5. However, the absorption band at 1035-1200 cm^{-1} has
been evidently changed. It is obvious that the hydroxypropyl group in
CMHPC with a high MS value is major active group for crosslinking
reaction. The structural formula of metal crosslinked CMHPC is shown as
follows:

Raman characteristic frequency for crosslinked CMHPC.

Raman shift is characteristic of vibrations of various groups. It has been pointed that the structure of cellulose and its derivative can be studied by Raman spectrum[6-8]. The initial V_{as} band of carboxyl groups from uncrosslinked CMHPC (sodium salt) about 1600 cm^{-1} in the IR spectrum is a strong absorption band, whereas the same frequency in RM is a weak depolarization band. The initial V_s band about 1407 cm^{-1} in IR is a weaker band, whereas the same frequency in RM is a stonger polarization band (Fig.6). The absorption intensity of the depolarization band has increased and one of the polarization bands has decreased, and the shift number is decreased simutaneously to 1368 cm^{-1} and 1372 cm^{-1} for the Al^{+3}-CMHPC and the Cr^{+3}-CMHPC respectively because of the effect of polarity and mass after the –COO^{-1} group interacts with the metal ion. The absorption band about 1450 cm^{-1} is due to the vibration of –CH$_2$ deformation and 1240-1370 cm^{-1} to the deformation vibration accidental frequency of –CH$_2$, C–OH. The change of these bands is not evident after crosslinking. The absorption band about 1110 cm^{-1} is due to the antisymmetric stretching vibration frequency of COC and CCO. The difference in this band between un- and crosslinked product is evident. The Raman frequency is decreased from 1110 cm^{-1} to 1108 cm^{-1} and 1088 cm^{-1} for Al^{+3}-CMHPC and Cr^{+3}-CMHPC respectively, and the absorption intensity of crosslinked product is obviously increased.

The stronger characteristic band about 900 cm^{-1} is due to the vibration frequency of the glucose ring, which relates to the conformation of macromolecular chains. After crosslinking, the metal ion that links active groups (–COO^{-1} or OH) can change initially linked hydrogen bonds to cause a macromolecular conformation change.

The mixed frequency of deformation vibration at 400-700 cm^{-1} from the CCO, COC, CCC and OH of crosslinked product has also changed.

It will be seen from these that RM linked IR can explain fully the structure of crosslinked CMHPC and give the direct evidence on the crosslinking reaction mechanism for a complicated cellulose ether crosslinked with metal ion.

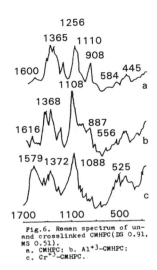

Fig.6. Raman spectrum of un-
and crosslinked CMHPC(DS 0.91,
MS 0.51).
a. CMHPC; b. Al^{+3}-CMHPC;
c. Cr^{+3}-CMHPC.

REFERENCE

[1] Gang-len Chen, Oilfield Chemistry, 1988, 5(2), 150-155. (Chinese)

[2] Chatterji, J., Soc. Pet. Eng., paper SPE9288. 1980.

[3] EU Pat. 104,009.

[4] Menjivar, J. A., Adv. Chem. Sci., 1986, 213 (water-soluble polymer), 209-226.

[5] Sandy, J.; Wiggins, M.; Venditto, J., Oil & Gas, 1986, september, 1, 52-54.

[6] Atalla, R. H.; Nagel, S. C., J. Amer. Chem. Soc., Chem. Commu., 1972, 19, 1049-1050.

[7] Atalla, R. H.; Dimik, B. E., Carbohydr. Res., 1975, 39, C1-C3.

[8] Atalla, R. H., Appl. Polym. Symposium, 1976, 28, 659-669.

50

Properties and following reactions of homo-geneously oxidized celluloses

Th. Heinze*, D. Klemm*, M. Schnabelrauch* and I. Nehls** - *Institut für Organische Chemie und Makromolekulare Chemie der Friedrich-Schiller-Universität Jena, Humboldtstrasse 10, D-0-Jena, Germany. **Institute für Polymerechemie, Kantstrasse 55, D-0-1530 Teltow, Germany.

ABSTRACT

Sodium carboxycellulose with a content of COOH groups up to 0.8 can be prepared by oxidation of cellulose with $NaNO_2/H_3PO_4$ and subsequent treatment with $NaBH_4$. The extent of oxidation increases on raising the molecular weight of the starting cellulose. There are no keto groups in the oxidized polymers. The sodium carboxycelluoses show a high tendency to form ionotropic gels. Activation and treatement of the carboxycellulose with SO_3 or HSO_3Cl leads to the corresponding sulfate esters with d.s. values up to 0.45.

INTRODUCTION

Partial oxidation of polysaccharides constitutes one of the most versatile transformations, since it provides access to various novel products and intermediates with valuable properties. The introduction of carboxylate groups into cellulose leads to polyelectrolytes with interesting properties /1/ such as ion complexation, gelation, symplexation, biological activity, e.g. heparin-

like anticoagulation activity.
There are numerous ways of oxidizing the
chain units in cellulose while the glycosidic linkages remain intact. An improved
procedure of the well-known nitrogen dioxide
oxidation method for cellulose /2/ has been
developed by PAINTER /3/ using phosphoric
acid and sodium nitrite as the oxidizing
agent. This technique yields products with
higher oxidation degree than the previous
methods do and, in addition, it offers the
advantage of lowering the polymerization
degree much less.

EXPERIMENTAL

As starting materials we used cellulose
powder, viscose staple fibre, spruce sulfite
pulp and cotton linters with d.p. values
160, 300, 600, and 1400, respectively.
Typical oxidation procedure: 5g cellulose
material was dissolved in 200 ml 85% H_3PO_4
and oxidized by addition of 15g $NaNO_2$ in
portions using a reaction time of 10h /4/.
After the destruction of the excessive
oxidizing agent with 85% HCOOH the polymer
was precipitated with ether/acetone (1:1
v/v). The conversion of the polymer in the
sodium carboxycellulose was done with NaOH
and $NaBH_4$ and precipitation with acetone.
Typical sulfation procedure: For activation
the Na-COC was dissolved in water and precipitated into DMF followed by distillative
removal of water under reduced pressure. To
a stirred suspension of 15 mmol of the
activated polymer in 80 ml dry DMF was added
30 mmol of the sulfating agent in 30 ml dry
DMF at 10°C. After stirring 3 h at 20°C the
sulfate ester was precipitated with acetone,
washed, converted into the sodium salt by 2
wt-% NaOH in ethanol/water (10:1 v/v).
The ^{13}C-n.m.r. spectra were recorded on a
Bruker MSL 400 spectrometer operating at
100.63 MHz in the pulsed Fourier transform
mode with proton decoupling. The samples
were measured at 20°C as 5-10% (w/v) solutions in D_2O using sample tubes of 10 mm and
C_6D_6 as external standard 500-2000 scans
were accumulated.

RESULTS AND DISCUSSION

While PAINTER /3/ carried out the oxidation

reaction in a large mortar, we usually
worked in stirred laboratory apparatus.

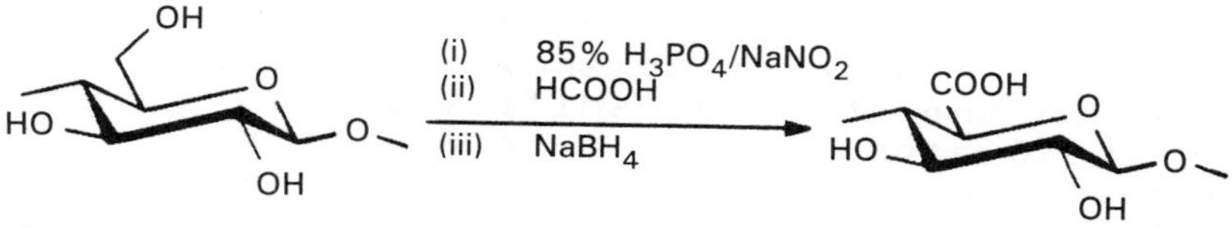

Figure 1. Reaction scheme.

The content of formed carboxy groups does
not only depend on the reaction time but
also drastically on the degree of polymerization of the starting cellulose material
(Table 1). Surprisingly, the extent of
oxidation increases with rising molecular
weight of the starting material.

Table 1. Content of formed carboxy groups
with different degrees of polymerization
(d.p.) after oxidation with $NaNO_2/H_3PO_4$
depending on the reaction time.

Starting cellulose material	d.p.	Reaction time (h)	Content of COOH-groups (%)
Cellulose powder	160	3	7.7
		5	57.9
		8	62.0
		10	63.0
Viscose staple fibre	300	3	60.1
		5	60.4
		8	65.0
		10	67.9
Spruce sulfite pulp	600	3	62.0
		5	68.9
		8	73.5
		10	75.0
Cotton linters	1400	3	35.7
		5	73.4
		8	78.0
		10	81.0

Because of the high viscosity of the solution, the liberated oxidizing agent N_2O_3
generates a foam which guarantees the contact between the cellulose and the oxidizing
agent. The foam also prevents loss of the
gaseous oxidizing agent. We have found that
the stability of the foam increases on
raising the molecular weight of the cellulose and that is why the degree of oxidation

increases in the same direction provided
comparable reaction times of at least 5 h
are considered.
The ^{13}C-n.m.r. spectra of the sodium carbo-
xycelluloses in D_2O solution show considerable changes in the chemical shift values of
the C-4,5 and 6 atom signals in comparison
with the spectra of the starting celluloses
(Figure 2).In addition, a new signal occurs
at 175.5 ppm, which can be assigned to a
carboxy group in the C-6 position of the
anhydro unit.In the case of preparing the
sodium carboxycellulose with sodium hydroxide, a further signal was found at 165.1
ppm without changes of the chemical shift
values of the other signals.

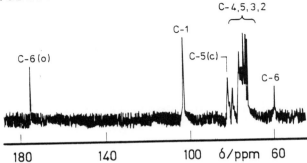

Figure 2. ^{13}C-n.m.r. spectrum of sodium
carboxycellulose (conversion into the sodium
salt by $NaBH_4$).

In our opinion the cause of this signal is
the formic acid ester groups which are
formed during the destruction of the excessive oxidizing agent N_2O_3, which is done
advantageously by means of formic acid. If
one uses sodium borohydride to prepare the
sodium salt, the carboxycellulose is free of
formic acid ester groups - that means cellulose formed selectively oxidized only in the
C-6 position. The existence of keto groups -
as assumed by PAINTER /5/ - could not be
confirmed because of the absence of n.m.r.
signals typical for CO groups.
A significant difference between carboxycellulose and carboxymethylcellulose was
also found in the gelation reaction with
calcium ions.The tendency of gelation was
determined by a gravimetric method (see
/4/).

Table 2. Calcium carboxycellulose (1%
aqueous solution, COO⁻-groups 73%) gel
portion dependence on the calcium ion concentration.

Concentration of $CaCl_2$ solution (mol/l)	Gel portion (%)
0.35	20.5
0.60	30.0
0.80	37.2
0.95	62.7
1.20	84.4
4.05	101.3

In the case of carboxycellulose an addition
of about 0.5 mol Ca^{2+} per mol carboxylate
groups leads to significant gelation (Table
2). The gelation reaches a level of about
80% - related to the initial weight - at an
addition of 1.2 mol calcium ions per mol
carboxylate groups. On the other hand a
carboxymethylcellulose with a degree of
substitution of about 1 forms only 10% gel
under comparable conditions. In our opinion
the reasons are both the carboxylate group
distribution in connection with the stiffness of the polymer backbone and the different distances between ionic groups and
polymer backbone. As mentioned above, in
carboxycellulose the carboxylate groups are
located only in C-6 position - in carboxymethylcellulose the groups are distributed
in the order C-2 > C-6 > C-3.
By dropping a carboxycellulose solution into
a cross-linking solution of calcium ions, it
is even possible to form ionotropic gels in
spherical shape /6/.
After a suitable pretreatment consisting of
the precipitation of an aqueous carboxycellulose solution in DMF and the subsequent
removal of water from the highly swollen gel
by distillation under reduced pressure, the
sulfation of the polymer with SO_3 or HSO_3Cl
yields the insoluble sulfate half esters.
Subsequent neutralisation leads to water-
soluble sodium salts.
Using 2 moles of sulfating agent per mole
modified anhydro glucose unit, sulfate-d.s.-
values between 0.35 and 0.45 are received.
Figure 3 shows the ^{13}C-n.m.r. spectrum of
carboxycellulose sulfate. The peak at 67.0
ppm can be assigned to the carbon atom 6
bearing a sulfate ester moiety. A C-2 sub-

stitution is not unambiguously detectable
because the signal which is indicating such
a substitution overlaps with that of the C-
5 atom. Therefore, we can conclude a preferred sulfation of the OH-groups at the
C-6 carbon atom. This means that the carboxy
groups at the C-6 position are not able to
avoid further substitution of the primary
OH-groups.

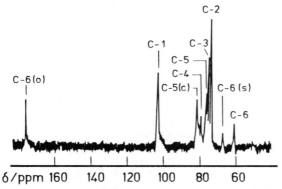

Figure 3. ^{13}C-n.m.r. spectrum of sodium
carboxycellulose sulfate.

REFERENCES

(1) See, e.g. Yalpani,M. *Polysaccharide
    Syntheses, Modification and Structure/
    Properties Relations*, Elsevier, Amster-
    dam, Oxford, New York, Tokyo 1988
    Philipp, B. *Progr. Polymer Sci 1989, 14,
    91*
(2) Yackel, E.C.; Kenyon, W. *J. Am. Chem.
    Soc. 1942 64 121*
(3) Painter, T.J. *Carbohydr. Res. 1977 55 95*
(4) Heinze, T.; Klemm,D.; Loth, F.; Nehls,I.
    Angew. Makromol. Chem. 1990 178 95
(5) Painter, T.J.; Cesaro, A.; Delben, F.;
    Paoletti, S. *Carbohydr. Res. 1985 140 61*
(6) Winkelmann, H.; Heinze, T.; Linss, W.
    *Beitr. Elektronenmikroskop. Direktabb.
    Oberfl. 1990 23 465*

51

Cellulose triacetate prepared from low-grade pulp

H. Matsumara and S. Saka - Department of Wood Science and Technology, Faculty of Agriculture, Kyoto University, Kyoto, 606-01, Japan.

ABSTRACT

As cellulose triacetate was prepared from low-grade sulfite softwood pulp (α-cellulose content; 87.5%), a considerable amount of insoluble residue was present in the acetylation solution of an acetic acid/ acetic anhydride/ sulfuric acid system. The characterization of this residue indicated that the insoluble portion retained a fiber structure of swollen form. In addition, the insoluble residue was composed of cellulose triacetate and glucomannan triacetate in aggregate with each other at the molecular level by their mutual interactions. However, it was found that glucomannan triacetate with lower molecular weight formed a less-insoluble residue and that a pretreatment of the low-grade pulps with an acetic acid and sulfuric acid mixture was effective to reduce the insoluble residues. Such a property of glucomannan acetate suggests the potential use of the low-grade pulps for cellulose acetate production.

INTRODUCTION

Cellulose acetate is one of the most important cellulose derivatives in the fiber and textile industries and is usually manufactured from cotton linters or high quality wood pulps [1]. This is because low-grade pulps contain hemicelluloses and hemicellulose acetates behave differently in the solution, resulting in the industrial problems such as filterability, turbidity and false viscosity [2]. In spite of these problems,

it is still beneficial from economical and technical viewpoints
if low-grade pulps can be used as raw materials for cellulose
acetate production. In this study, therefore, cellulose triacetate (CTA) was prepared from low-grade sulfite softwood pulp
in an acetic acid/ acetic anhydride/ sulfuric acid system. The
insoluble residue in the acetylation solution was then characterized and a remedy for reducing the amount of insoluble
residue was discussed for industrial applications.

EXPERIMENTAL

Materials and their acetylation
Cellulose triacetate was prepared from two kinds of sulfite
softwood pulps (high-grade pulp, α-cellulose content 96.0%;
low-grade pulp, 87.5%). The sample (1 part) of well-defibered
pulps was acetylated with a solution of acetic acid (160
parts), acetic anhydride (7 parts) and sulfuric acid (0.1 part)
for 3h at 40°C, and then stirred overnight at 20°C. The reaction solution was subsequently spun in a centrifuge at 7000 rpm
for 30 min to separate the soluble portion and insoluble residue. To isolate glucomannan, a low-grade pulp was extracted
under a nitrogen atmosphere with 18% NaOH aqueous solution
containing 4% H_3BO_3 to which a 0.5% $NaBH_4$ on a pulp basis was
added [3]. The obtained glucomannan was subsequently acetylated to prepare glucomannan triacetate (GTA). Cellulose triacetate I (CTA-I) and cellulose triacetate II (CTA-II) were also
prepared for studying the crystallographic nature and thermal
properties of the insoluble residue [4].

Comparative studies of acetylations
1. Acetylation after mixing; Some glucomannan (8.6wt%) was
added to the high-grade pulp (91.4wt%) so as to have the same
content of mannose (7.0%) as the low-grade pulp. The mixed
pulp with glucomannan was acetylated and the acetylated product
was fractionated into the soluble and insoluble portions.
2. Acetylation before mixing; The same amounts of glucomannan
(8.6wt%) and high-grade pulp (91.4wt%) used above were prepared
so as to have a 7.0% mannose content when mixed. Subsequently,
glucomannan and high-grade pulp were separately acetylated and
their acetylation solutions were mixed together by stirring
overnight to allow the insoluble substances to precipitate.

Evaluation methods
The degree of substitution (DS) of acetate samples was determined by a titration method [5], while neutral sugar compositions were determined by an alditol-acetate procedure [6]. To
study the crystallographic nature, the samples were annealed
[4] at 220°C for 20min under a vacuum and tablets of non-ori-
ented samples with CaF_2 were prepared to obtain X-ray diffraction patterns by a Shimadzu COMPAX S-12 type (Cu-Kα, λ=1.542A)
at 40kV and 25mA. For investigating thermal properties, the
dried samples were studied by a differential scanning calorimeter (DSC), Rigaku DSC-8230B Type, with a flow of nitrogen gas.

RESULTS AND DISCUSSION

Characterization of insoluble residue

Although the high-grade pulp fibers dissolve almost completely at the end of acetylation, some of the low-grade pulp fibers remain in the swollen state with fiber structures. In order to elucidate such insolubility, the soluble portion and insoluble residue of the low-grade pulp were separately collected. Table 1 shows the obtained DS and chemical compositions of these portions. From the results of the DS, even insoluble-fiber residue is evidently fully acetylated. Thus, the insolubility is not due to a lack of acetylation. On chemical composition, the mannose content of the insoluble residue is much greater than that of the soluble portion. Therefore, the insoluble residue must be composed of cellulose triacetate (CTA) and glucomannan triacetate (GTA).

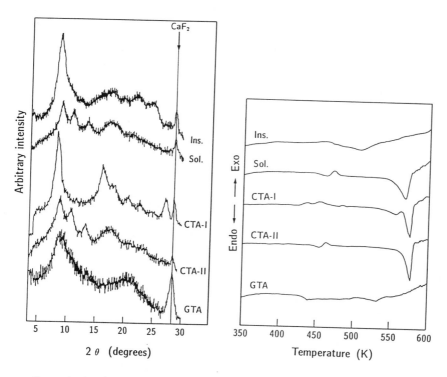

Figure 1. (left) X-ray diffraction profiles of insoluble residue and soluble portion in cellulose triacetate prepared from the low-grade pulp.

Figure 2. (right) DSC thermograms of insoluble residue and soluble portion in cellulose triacetate prepared from the low-grade pulp.

Table 1. DS and chemical compositions of soluble and insoluble portions in cellulose triacetate prepared from low-grade pulp.

Samples	Weight fractions (%)	DS	Chemical compositions (mol%)		
			Glucose	Mannose	Xylose
Low-grade pulp	100.0		90.4	7.0	2.6
Soluble	87.0	2.98	92.0	5.7	2.3
Insoluble	13.0	3.01	71.1	28.2	0.7

Such insoluble residue of low-grade pulp was studied by X-ray diffractometry (Figure 1) and differential scanning calorimetry (DSC) (Figure 2). For comparison, CTA-I, CTA-II and GTA were included as control samples. It is quite apparent from the diffraction profile and DSC thermogram, that the soluble portion (Sol.) is composed of CTA-II. However, the insoluble residue (Ins.) does not exhibit any characteristic diffraction peak (Figure 1) nor endothermic peak (Figure 2) derived from CTA-I or CTA-II. This evidence suggests that the insoluble residue is not phase-separated with CTA and GTA but molecularly aggregated by their compatible nature of these two macromolecules.

Comparative studies of acetylations

It is now apparent that the residual glucomannan in wood pulp plays an important role in the formation of insoluble-fiber residue. However, it is not known whether the insoluble residue formed is due to inherent nature of cellulose and glucomannan or due to that of CTA and GTA introduced by acetylation treatment. Therefore, comparative studies of acetylations were made and the amounts of insoluble substances were evaluated. The obtained results are shown in Table 2. Acetylation without mixing indicates a simply calculated amount of the insoluble substances, based on the data of high-grade pulp and glucomannan being 0.5% and 2.9%, respectively as separately acetylated $\{0.7(\%)=0.5(\%)X0.914+2.9(\%)X0.086\}$.

Interestingly, both acetylation systems before and after mixing resulted in much more insoluble substances formed compared with acetylation without mixing. This suggests that some mutual interactions of molecules exist between two types of polymers. If the formation of insoluble residue is due to the interaction between cellulose and glucomannan, such a considerable amount of insoluble substances was not recovered in the acetylation before mixing. Therefore, the molecular interaction must take place between cellulose acetate and glucomannan acetate.

Table 2 also shows the obtained results on chemical compositions of these insoluble substances. Interestingly, the content of GTA is less in the low-grade pulp than that in the acetylation after mixing which in turn is less than in the acetylation before mixing. This order is, in fact, parallel to

the order of the ultrastructural effects on the formation of
insoluble substances. In the low-grade pulp, glucomannan is
distributed among the cellulose microfibrils within the pulp
fibers, thus, the ultrastructural effects being large, while in
the acetylation after mixing, due to its high affinity with
cellulose, the molecule of glucomannan may attach themselves to
the cellulose at the surface of the pulp fibers with forming
slightly the ultrastructural effects. In the acetylation before
mixing, however, the fiber structure is diminished, and the
individual cellulose molecules are acetylated as mixed. Thus,
only chemical effects would be involved in the formation of the
insoluble substances without any ultrastructural effects.

Such differences in the degree of the ultrastructural effects can well explain the observed differences in the amounts
of insoluble substances between low-grade pulp and the other
two acetylation systems. Therefore, it may be concluded that
the formation of the insoluble residues would involve not only
chemical effects but also ultrastructural effects for the low-
grade pulp. The results would perhaps indicate that the amount
of CTA in the insoluble residues would be increased if the
acetylation system involves the ultrastructural effects.

Table 2. Chemical compositions of insoluble substances in comparative acetylations.

Samples[*]	Insoluble substances (wt%)	DS	Chemical compositions (mol%)			GTA:CTA
			Glucose	Mannose	Xylose	
Low-grade pulp	13.0	3.01	71.1	28.2	0.7	37:63
Acetylation after mixing	8.7	2.90	66.8	32.6	0.7	43:57
Acetylation before mixing	8.1	2.97	57.9	39.7	2.5	52:48
Acetylation without mixing	0.7	–	–	–	–	–

[*] All samples studied contain consistently 7.0% mannose.

The remedy for reducing the amount of insoluble residue

It would be effective to reduce the amount of insoluble
residue if chemical effects and/or ultrastructural effects
involved in the formation of the insoluble residue are decreased. As a first trial, the effects of the molecular
weights of glucomannan on the amounts of insoluble residues
were investigated with the acetylation after mixing. Prior to
acetylation, glucomannan in a catalyst mixture consisting of
acetic acid and sulfuric acid was treated at different temperatures to prepare glucomannan with different molecular weights.
The degraded glucomannan was then mixed with high-grade pulp
and acetic anhydride for acetylation. Table 3 shows the results obtained on the amounts of insoluble residues. It is
interesting to note that in spite of the same content of mannose for all cases, the use of glucomannan pretreated at the
higher temperatures resulted in smaller amounts of insoluble
substances formed. This indicates that glucomannan triacetate
with the lower molecular weights would form the less amounts of
the insoluble residues.

Table 3. The effects of pretreatment temperatures of glucomannan on the amounts of insoluble substances.

Pretreatments	Insoluble substances (wt%)
No treatment	8.7
20°C, 5h	7.5
40°C, 5h	5.0
60°C, 5h	3.5
80°C, 5h	2.9

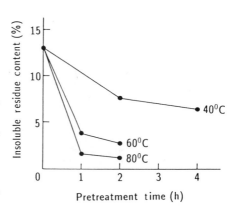

Figure 3. The effects of the pretreatments of low-grade pulp on the amounts of insoluble residues.

This finding was applied to the acetylation of the low-grade pulps. It is likely that glucomannan in wood pulps are more accessible to the catalyst mixture than crystalline cellulose microfibrils. Therefore, glucomannan in the low-grade pulps would be degraded more or less selectively during the pretreatment with the catalyst mixture. By adding acetic anhydride to the catalyst mixture, the pretreated low-grade pulps were acetylated. Figure 3 shows the obtained relation between the amounts of insoluble residues and pretreatment conditions. It is apparent that the amounts of insoluble residues can be reduced by such pretreatments in spite of the same mannose content retained in all pulps.

CONCLUDING REMARKS

It was found that glucomannan triacetate forms the insoluble residues with cellulose triacetate. However, the amounts of the insoluble residues were found to be reduced by degrading glucomannan triacetate. Since the developement of a process to reduce the insoluble residues is of great interest, such a property of glucomannan triacetate suggests the potential use of the low-grade pulp as raw materials for cellulose acetate production.

REFERENCES

1. Ichino,M. Nikkakyo Geppo 1986,39,25.
2. See,e.g. Ueda,K.;Saka,S. J.Appl.Polym.Sci.Appl.Polym.Symp. 1989,43,309.
3. Timell,T.E. TAPPI,1961,44,88.
4. Sprague,B.S.;Riley,J.L.;Noether,H.D. Text.Res.J. 1958,28,275.
5. Atsuki,K. Sen'isokagaku oyobi Kogyo,Maruzen,1956,p.418.
6. Borchardt,L.G.;Piper,C.V. TAPPI,1970,53,257.

52

Principles of the structure formation of the derivatives of microcrystalline cellulose

N.E. Kotelnikova - Institute of Macromolecular Compounds of the Russian Academy of Sciences, Bolshoy pr. 31, St. Petersburg 199004, Russia.

ABSTRACT

The preparation and properties including oxidation of chemically and structurally modified cellulose derivatives using highly crystalline cellulose-microcrystalline cellulose (MCC) are reported. In these cases the crystalline lattice of native cellulose I does not undergo structural transformation (X-ray and 13C NMR investigations). This fact indicates that the surface of MCC matrix serves as a basis of the growing layer formed as a result of the interaction between the OH-groups of cellulose and ether- and esterifying reagents. The reactions of MCC usually start on crystallographically symmetric microrelief at the location of defects and may lead to chain crosslinking.

INTRODUCTION

The present paper is a generalization of numerous data obtained by us on the properties of microcrystalline cellulose (MCC) and its reactivity in the preparation of its derivatives. MCC is a product of heterogeneous hydrolysis of native cellulose in the acid medium. Its supermolecular structure is characterized by the high crystallinity of the lattice according to X-ray data. This lattice retains the modification of cellulose I, just as native cellulose. It differs considerably from native cellulose in that it does not contain the so-called "amorphous" part. For a long time the reactivity of native cellulose was considered from the viewpoint of the accessibility of its amorphous part to reactions. The conclusion was made that the larger the amorphous part in the sample, the higher its reactivity. A detailed study of the structure and reactivity of MCC led to important new conclusions on the structure of native cellulose [I] and made it possible to formulate the main principles of structure formation of the derivatives of MCC as a highly crystalline polymer matrix. We will consider the reactions performed under heterogeneous conditions in media in which swelling and dissolution of the resulting derivatives either did not take place or was very slight.

RESULTS AND DISCUSSION

Preparation of the phosphate of MCC [2].

The method of phosphate preparation involved cellulose treatment with phosphorus pentaoxide (P_2O_5) dissolved in dimethylformamaide (DMF). When the reaction was carried out under optimum conditions, the phosphorus content of the reaction products was 6% (which corresponds to $\gamma = 37$). It was established that phosphoric acid formed by the interaction between P_2O_5 and H_2O is the phosphorylating agent. H_2O is contained in the ambient atmosphere, in air dry MCC and DMF and is also obtained by partial dehydration of elementary cellulose units with the formation of double bonds and anhydrorings with the aid of P_2O_5. The products were analyzed by IR spectroscopy, X-ray diffractometry and electron microscopy. It was shown by IR spectroscopy that MCC phosphate actually contains anhydro rings and double bonds in elementary units. An interesting feature of this process was the fact that phosphorylation is accompanied by the formation of crosslinked MCC phosphates as a result of both double bond formation between neighbouring macromolecules and the interaction between the OH-groups of neighbouring macromolecules and phosphoric acid. This was proven by determination of the solubility of MCC phosphates and by potentiometric titration. In fact the products exhibit low solubility in cadoxen but very high swelling: on keeping in cadoxen for 24h they gel. Figure 1 shows X-ray patterns of samples of MCC (1) and their phosphates with phosphorus contents of 3.7 (2) and 6.5% (3). Both phosphates exhibit the structure of cellulose I and their crystallinity indexes differ only slightly from one another and from a sample of non-phosphorylated MCC. Hence, the phosphorylation of MCC with the aid of P_2O_5 under heterogeneous conditions in DMP proceeds on the surface of the crystalline MCC matrix without any change in its structure.

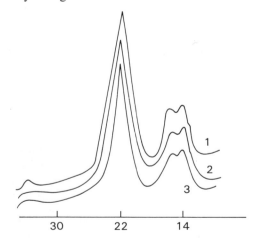

Fig. 1. X-ray diffractograms of MCC (1) and MCC phosphorus content 3.7 wt % (2) and 6.5 wt % (3).

Figure 2 (1) shows an electron micrograph of MCC phosphate (phosphorus content 2.2%). The fiber surface exhibits symmetrical formations which resemble the growing layers of esterified groups.

Preparation of the derivatives of MCC and pyromellitic acid (PMA) [3].

The reaction of MCC with pyromellitic dianhydride (PMDA) was studied. The dianhydride had previously been dissolved in DMF or dimethylsulfoxide (DMSO). PMDA readily undergoes hydrolysis. Therefore, esterification was actually performed by a mixture of PMDA, pyromellitic anhydride, and pyromellitic acid. Under optimum conditions products were obtained that contained 17% of carboxyl groups (-COOH) the quantity of which was determined by potentiometric titration. The resulting samples were analysed by the same physical methods.

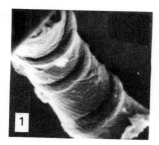

Fig. 2. Electron microphotographs of MCC phosphate (phosphorus content 2.2%) (1) and of the product of MCC reaction with PMDA (2).

Figure 3 show X-ray patterns and solid state ^{13}C NMR spectrum of the reaction product with the total content of 15% COOH-groups. Its structure is that of cellulose I with a relatively high crystallinity index −0.72. The NMR spectrum is also characteristic of cellulose I and two weak signals can be seen:~130 ppm, the signal of the benzene ring and ~170 ppm, that of the carboxyl group.

Hence, the reaction MCC with PMDA proceeds similarly to that of cellulose with other anhydrides of carboxylic acids, i.e. with the formation of an ester and an acid (scheme).

In this case the esterifying agent plays the role of the crosslinking agent for cellulose chains. In fact, the esterified products contain both bonded and free COOH-groups. Taking into account the fact that the structure of reaction products does not change as compared to that of the initial MCC, it is evident that, just as in the case of phosphorylation, the reaction proceeds on the surface of microcystalline particles and, possibly, on crystallite ends.

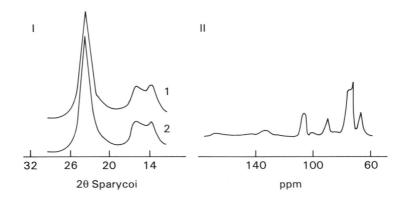

Fig. 3. X-ray diffractograms (I) of MCC (1) and of the reaction product with COOH-groups content = 15% (2) and ^{13}C NMR spectrum of (2) (II).

2 (2) shows the electron micrographs of the fiber surface of the product of MCC reaction with PMDA. The surface is separated by symmetric breaks which evidently indicate that crystallites are blocked on the end surfaces.

Preparation of mixed esters of MCC, pyromellitic and sulfuric acids [4].

The reaction of MCC was carried out after the dissolution of PMDA in DMSO and in the presence of sulfuric acid. By varying the amounts of H_2SO_4 and PMDA mixed esters of MCC containing sulfur (degree of substitution γ_s max = 84) and carboxyl groups in the amount of up to 8% were obtained. The products are insoluble in H_2O and DMSO but considerably swell in them forming a stable colloidal gel. Taking into account the fact that the content of bonded COOH groups is ~50% of their total number, just as in the preceding case, it was concluded that the products were crosslinked.

Figure 4. shows the X-ray pattern and the solid state ^{13}C NMR spectrum of the sample that contains sulfur (γ_s = 84) and COOH-groups (2.5%). Its structure is of cellulose I with the crystallinity index slightly lower as compared to that of the initial MCC.

Hence, the crystalline lattice of cellulose I does not undergo structural changes. This fact indicates that the surface of the highly crystalline MCC matrix serves as a substrate support for the growing layer formed as a result of the interaction between the OH-groups and the esterifying agents. This process is essentially similar to heteroepitaxy that characterizes the oriented crystallization of a substance on the surface of the crystalline support. Since it has not been investigated whether the structural geometric correspondence exists between the reacting substances (MCC and esterifying reagents), it is evident that the reaction temperature and the perfection and purity of the matrix exert the greatest effect on the interaction process.

It is clear taking as an example the phosphorylation of MCC and its reaction with pyromellitic dianhydride that the reaction temperature profoundly affects esterification [2, 3]. Figure 5 shows the kinetic dependence of phosphorus content on the reaction time at t = 40 ˚C (1) and 70 ˚C (2). In the second case the content of the phosphorus introduced is twice as high as in the first case.

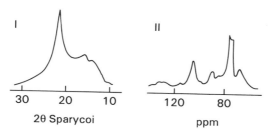

Fig. 4. *X-ray diffraction (I) and ^{13}C NMR spectrum (II) of the reaction product with COOH-groups content = 2.5% and g_s = 84.*

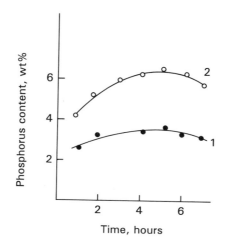

Fig. 5. *Variation of the phosphorus content in MCC phosphate versus the reaction time at 40˚C (1) and 70˚C (2).*

The reaction with pyromellitic dianhydride also shows that when the reaction temperature increases from t = 25 ˚C to 100 ˚C, the maximum content of carboxylic groups increases from 10 to 17%. (Figure 6 shows the kinetic dependences of COOH-groups content on reaction time).

As for matrix perfection, the crystalline structure of MCC is known to be much more perfect than that of native cellulose. Hence, it was not possible to carry out the above processes with native cellulose. A comparison of the rates of other reactions of MCC and native cellulose that also proceed under heterogeneous conditions show that the rate of MCC reaction is higher. We can take as an example the oxidation of MCC by sodium periodate which leads to the preparation of dialdehyde cellulose. Since the quantitative determination of aldehyde groups in oxidized products is difficult, kinetic characteristics of the reaction were obtained from the change in oxygen concentration which is determined by the polarographic method [5].

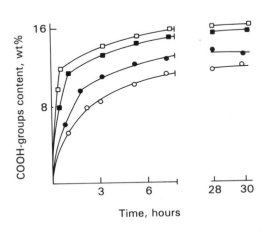

Fig. 6. Variation of the COOH-groups content versus the reaction time at 25 ˚C (1), 50 ˚C (2), 90 ˚C (3), 100 ˚C (4).

Figure 7 shows the dependence of the concentration of periodic acid on oxidation time for native cotton cellulose (2) and MCC (1). The oxidation rate constant for MCC exceeds even the rate of homogeneous oxidation of cellulose esters (AC (3) and CMC (4)) in solution.

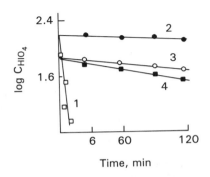

Fig. 7. Variation of the log of HI0$_4$ concentration (log C$_{HI0_4}$) versus the oxidation time.

CONCLUSION

Hence, the main relationships of the structure formation of MCC derivatives are the following: the highly crystalline structure of MCC and its ability to carry out the role of the insoluble polymer matrix for the growing esterified layer in heterogeneous reaction suggest that reaction processes start on the microrelief in defect places. Depending on the process parameters they can lead to chain crosslinking. The crystalline structure of derivatives obtained in these processes is retained in the form of cellulose modification I with a relatively high degree of crystallinity.

ACKNOWLEDGEMENTS

The author gratefully acknowledge Dr. A.A. Shashilov and Dr. A.V. Gribanov for X-ray and ^{13}C NMR research and Dr. Sh.S. Arslanov, post graduate student, for collection of material and laboratory work.

REFERENCES

[1]   Petropavlovsky, G.A. and Kotelnikova, N.E. Acta Polymerica, 1985, s. 36, N2, p. 118-123.
[2]   Petropavlovsky, G.A. and Kotelnikova, N.E. Cellul. Chem. Technol., 1985, Vol. 19, N6, p. 591-600.
[3]   Petropavlovsky, G.A. and Kotelnikova, N.E., Arslanov, Sh.S. J. Appl. Chem. (Russ.), 1989, Vol. 62, N12, p. 2803-2808.
[4]   Arslanov, Sh.S., Petropavlovsky, G.A., Kotelnikova N.E. J. Appl. Chem. (Russ.), 1990, Vol. 63, N1, p. 177-183.
[5]   Petropavlovsky, G.A., Chernova, Z.D., Kotelnikova N.E. J. Appl. Chem. (Russ.), 1977, Vol. 50, N6, p. 1348-1352.

53

Preparation of cellulose esters with aromatic carboxylic acids

Y. Shimizu, A. Nakayama and J. Hayashi - Department of Applied Chemistry, Faculty of Engineering, Hokkaido University, North 13 West 8, Kitaku, Sapporo, Hokkaido, 060 Japan.

ABSTRACT

Esterification of cellulose with aromatic carboxylic (substituted benzoic) acids in the system of pyridine (Py) containing 4-toluenesulfonyl chloride (TsCl) was investigated. Cellulose could be readily acylated with a substituted benzoic acid, such as nitro-, chloro-, methyl- and methoxy-benzoic acid, by the use of the Py/TsCl/acid system, and cellulose esters with a high degree of substitution (DS) were obtained.(Eq.1) The difference between 2-, 3- and 4-isomers of the acid did not influence the reaction significantly.

$$\text{Cell-OH} + \underset{X}{\overset{COOH}{\bigcirc}} \xrightarrow{\text{Py/TsCl}} \text{Cell-O-C} \underset{X}{\overset{O}{\bigcirc}} \qquad \text{(Eq.1)}$$

X: $-NO_2$, $-Cl$, CH_3, $-OCH_3$

With an amino-substituted benzoic (4-aminobenzoic) acid, the corresponding cellulose ester could not be obtained, because acylation of the $-NH_2$ group of the other 4-aminobenzoic acid took place in preference to the $-OH$ group of cellulose. When the $-NH_2$ group was protected with a methyl group, however, the corresponding ester could be prepared by this method.

INTRODUCTION

It is well known that organic esters of cellulose can be prepared by the use of carboxylic acid anhydride with an acid catalyst, carboxylic acid chloride in the presence of a base and carboxylic acid in combination with trifluoroacetic anhydride [1,2]. In any method, either the anhydride or the chloride, i.e. an activated carboxylic acid is necessary. A method for the preparation of cellulose ester with carboxylic acid, in which neither the anhydride nor the chloride is used, will lead to not only a decrease in the reaction steps but the preparation of a cellulose ester of a specific carboxylic acid of which the derivatives is difficult to obtain.

We have reported a method for the preparation of cellulose acetate with acetic acid (AcOH) by the Py/TsCl/AcOH system [3] and showed that this method can be generally applicable to the esterification with higher aliphatic and benzoic acids [4]. In this paper we describe the esterification with substituted benzoic acids by similar systems. Recently, a variety of cellulose triesters are utilized for such functional materials as chromatographic optical resolution [5]. This method seems to have the possibility to prepare a new cellulose ester directly with the specific acid.

EXPERIMENT

Acetate grade pulp (α-cellulose 96.1%) was used as a cellulose sample. After grinding to less than 40 mesh, the pulp was immersed in distilled water at room temperature for 1 hr to increase activity. Then the sample was squeezed on a glass filter, rinsed completely with Py to remove the water and then squeezed again.

The pretreated sample (1g) was introduced into a solution of 7g of TsCl in 30g of Py. Subsequently an equimolar amount of an aromatic carboxylic acid with TsCl was added into the solution. After reacting for 2-20 hr at 50 °C, the reaction mixture was poured into an excess amount of ethanol. The product was collected in a glass filter, washed with ethanol and finally Soxhlet-extracted with ethanol for 6 hr, then dried at room temperature. Elemental and infrared (IR) analyses were conducted on the product.

RESULTS AND DISCUSSION

Substituted benzoic acids used in this work and acidity constants of their isomers are listed in table 1.

Figure 1 shows the rate of esterification with isomers of nitrobenzoic acid at 50 °C. Equimolar amounts of TsCl and the acid, each 2 moles to OH group of cellulose, was used for the reaction because of the most rapid reaction at the amounts on the acetylation with AcOH [3,4]. DS of the products was determined by elemental analysis. As can be seen from the figure, the reaction proceeded readily and DS of the products reached almost 3 within 2 hr resulting in dissolution of the product into the reaction mixture.

Although there is a large difference in the acidity constant between the 2-isomer and 3-, 4-isomers, the difference of reactivity between them was not so significant.

Table 1   Acidity constant Ka of substituted benzoic acids.

Substituent	$Ka \times 10^{-5}$		
	2 –	3 –	4 –
$-NO_2$	6 7 0	3 2	3 6
$-Cl$	1 2 0	1 5. 1	1 0. 3
$-CH_3$	1 2. 4	5. 4	4. 2
$-OCH_3$	8. 2	8. 2	3. 3

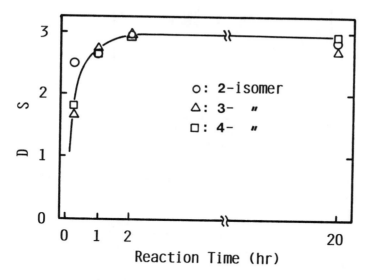

Fig.1 The rate of esterification with Py/TsCl/nitrobenzoic acid

Figure 2 shows the IR spectra of original pulp and the products obtained with the nitrobenzoic acids after 20 hr at 50 °C. The products had common absorptions in strong ester C=O (about 1750 cm⁻¹) and C-O (about 1260 cm⁻¹) stretching, and the absorption based on OH group of cellulose (about 3400 cm⁻¹) almost disappeared. Further, absorptions of C-H (3000-3100 cm⁻¹) and C=C (1400-1600 cm⁻¹) based on the aromatic ring and absorptions characteristic of the substituent NO_2 (1540 and 1350 cm^{-1}) were observed in the spectra. These results indicate that the products are cellulose esters of the corresponding nitrobenzoic acids.

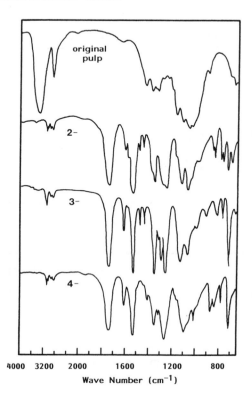

Fig.2   IR spectra of original pulp and the products prepared with nitrobenzoic acids for 20 hr at 50 'C.

IR spectra of the products prepared with 2-isomers of chloro-, methyl- and methoxy-benzoic acids are shown in the figure 3. The spectra of the products with the other isomers of these acids also showed similar characteristics to that of the 2-isomer. These spectra have absorptions based on the ester and the aromatic ring similarly to those of nitrobenzoic acids esters, indicating cellulose esters of these acids.

DS of the esters obtained in this work are summarized in table 2. From these results, it was found that the esters with high DS could be prepared independently of the substituents and isomers of a substituted benzoic acid by the Py/TsCl/acid system.

Esterification with p-aminobenzoic acid by the system, however, was unsuccessful because the acylation of the NH_2 group took place in preference to the OH group of cellulose (Eq.2). When a

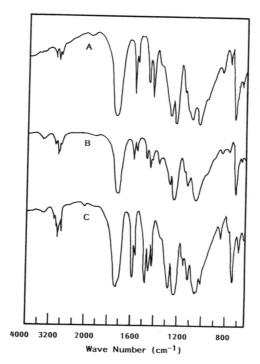

Fig.3   IR spectra of the products prepared with 2-isomers of (A)
        chloro-, (B) methyl- and (C) methoxy-benzoic acids for 20 hr
        at 50 ΄C.

Table 2   DS of the products obtained after reaction for 20 hr
          at 50 °C.

Substituent	D S		
	2 –	3 –	4 –
$-NO_2$	2. 8 5	2. 7 0	2. 9 5
$-Cl$	2. 9 0	2. 9 0	2. 8 0
$-CH_3$	2. 4 0	2. 6 0	2. 5 0
$-OCH_3$	2. 6 0	2. 7 0	2. 5 0

amino-substituted benzoic acid of which the NH₂ group was protected,
for example p-(N,N-dimethylamino)benzoic acid, was used as the acid,

the corresponding cellulose ester could be prepared (Eq.3).

(Eq.2)

R: CH$_3$-   (Eq.3)

CONCLUSION

The Py/TsCl/acid system is a useful method for the preparation of cellulose esters with a variety of substituted benzoic acids.

REFERENCES

[1] E.J.Bourne, M.Stacey, J.C.Tatlow and J.M.Tedde, J. Chem. Soc., 2976-2979 (1949)

[2] K.S.Barclay, E.J.Bourne, M.Stacey and M.Webb, ibid., 1501-1505 (1954)

[3] Y.Shimizu and J.Hayashi, SEN-I GAKKAISHI, 44, 451-456 (1988)

[4] Y.Shimizu and J.Hayashi, Cellulose Chem. Technol., 23, 661-670 (1989)

[5] T.Shibata, I.Okamoto and K.Ishii, J. Liq. Chromatogr., 9, 313-340 (1986)

54

Formation of polyelectrolyte complex from chitosan wool keratose

Won Ho Park and Wan Shik Ha - Department of Textile Engineering, College of Engineering, Seoul National University, Seoul 151-742, Korea.

ABSTRACT

Chitosan, a cationic polysaccharide, was heterogeneously depolymerized with hydrogen peroxide to control its molecular weight. The chitosan with different molecular weights was mixed in an acidic aqueous solution with the α–keratose obtained by oxidizing the wool keratin, and the formation behavior of their polyelectrolyte complex(PEC) was observed with a turbidimetric measurement. The effects of sodium chloride, urea, and guanidine hydrochloride on the PEC formation were also examined on the basis of the change of electrostatic and hydrogen-bonding forces probably responsible for the chitosan-α–keratose complex formation reaction. In addition, the effect of the chitosan on the secondary structure of the keratose in the complex is discussed using circular dichroic spectroscopy and differential scanning calorimetry .

INTRODUCTION

The mixing of oppositely charged polyelectrolyte solutions leads to the formation of a complex whose composition is sensitive to such reaction conditions as the composition of the reaction mixture, the order of mixing, pH, and the polyelectrolyte concentration. Up to now many researches in this field have been reported by lots of investigators. However, only a few papers have dealt with the polyelectrolyte interaction between charged polysaccharide and protein in spite of its importance relating to a molecular interaction of biopolymer in biological systems. These studies have suggested that the interactions are principally ionic but they can be dependent on the geometry and structure of the polymer component as well as on the number of the ionic groups. Hence, a thorough understanding of polyelectrolyte complex (PEC) formation should lead to the control of complex formation and facilitate the industrial applications of natural origin to macromolecular electrolytes. While chitosan, a cationic polysaccharide, was dealt with as a polyelectrolyte component for PEC formation in several reports[1-3], no one has investigated the complex formation between the chitosan and the α-keratose. This paper describes the novel results on the PEC formation of α-keratose with the chitosan, especially the effect of molecular weight of chitosan on the PEC formation. The conformational change of α-keratose will also be reported.

EXPERIMENTAL

Materials

Chitosan(Sigma, lot : 89F0383, $\overline{M}v = 5.91 \times 10^5$, degree of deacetylation : 80%) was purified by precipitating its acidic solution with a sodium hydroxide solution. After the liquor ratio had been adjusted to 1:100 at pH10-11, the purified chitosan was depolymerized while being stirred at 70°C for 1hr by adding different concentrations of aqueous hydrogen peroxide solution. The molecular weight of the chitosan samples was determined using the equation $[\eta] = 1.81 \times 10^{-3}\overline{M}v^{0.93}$ from the viscosity measurements in 0.1M acetic acid solution containing 0.2M sodium chloride at 25°C [4]. The oxidation of scoured Lincoln wool fiber with performic acid[5] was allowed to proceed for 24hr at 0°C and the samples were freeze-dried[6]. The α-keratose was fractionated by using the conventional precipitation procedure[6,7] from the oxidized wool keratin. Table 1 lists the information on $\overline{M}w$, the weight average molecular weight, $\overline{M}o$, the average molecular weight of amino acid residue, and the content of ionizable groups of α-keratose.

Table 1. Ionizable group contents of α-keratose

Polymer	$\overline{M}w$	$\overline{M}o$	Ionizable group	
			Anionic (mole %)	Cationic (mole %)
α-keratose	~50,000	114.9	Cysteic acid 7.2	Lysine 3.6
			Aspartic acid 10.0	Histidine 1.0
			Glutamic acid 18.8	Arginine 8.8

PEC

The chitosan and α-keratose were dissolved in 0.1N hydrochloric acid and sodium hydroxide solutions respectively. The pH of both solutions was adjusted to 5.2 with a sodium hydroxide or hydrochloric acid solution. 0.1% (wt.) chitosan solution was added in drops to a certain amount of corresponding α-keratose solution, and the complex formation was followed by a turbidimetry after standing for 10 min. A mixing ratio (R) was defined as $R = C_{pc}/(C_{pc} + C_{pa})$, where C_{pc} and C_{pa} represent the total number of cationic groups in the chitosan and anionic groups in the α-keratose respectively.

Separation of PEC

After standing for 30 min the insoluble PEC was separated by centrifugation, washed with a deionized water and methanol, and vacuum dried at 60°C.

Measurement

The turbidity produced was calculated from the absorbance data at 540 nm on a spectrophotometer (Macbeth Color Eye 3000, Kollmorgen Co., U.S.A.). The pH measurement was done with a pH meter HI 8418 (Hanna Instruments, Italy). The circular dichroism (CD) measurement was carried out with a JASCO J-20c CD/ORD spectropolarimeter equipped with a quartz cell of path length 10 mm. The DSC curve was obtained on a Perkin-Elmer DSC 7 differential scanning calorimeter. The sulfur analysis of the PEC and the amino acid analysis of α-keratose were performed at the Korea Institute of Science and Technology and the Korea Basic Science Center respectively.

RESULTS AND DISCUSSION

Turbidity was measured as an index of insoluble complex formation. As shown in Fig.1, the turbidity observed was plotted against the mixing ratio for α-keratose-chitosan systems at pH 5.2. Maximum turbidity appeared at the almost same mixing ratio (0.2) regardless of the molecular weight difference of chitosan samples. Although the turbidity was a good indicator for complex formation, it was not directly related to the amount of complex formed under some experimental conditions. This occurred when extensive complex formation resulted in sedimentation and lowered the turbidity of the mixture. The difference of turbidity values in Fig.1 showed that the PEC formation between the chitosan and α-keratose was dependent on the molecular weight of the chitosan.

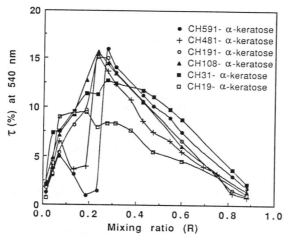

Fig. 1. Turbidity curves plotted against the mixing ratio(R) for α-keratose-chitosan systems at pH 5.2. CH, chitosan; 591, 481, 191, etc, molecular weight of chitosan(x10^{-3}).

The complex precipitated near the R_{max} value was isolated and the composition ratio was determined from sulfur content. As shown in Table 2, the composition ratio observed agreed with the R_{max} value observed. In the stoichiometric reaction of chitosan with α-keratose the following equation should hold:

$$C_{SO_3H} + \alpha C_{COOH} = \beta C_{NH_2},$$

Table 2. R_{max} values and composition ratios for polyelectrolytes complex formed at R_{max} in the reaction between α-keratose and chitosan with different molecular weights

Polyelectrolyte complex	pH	S(%)	R_{max}	Composition ratio	
				Obs'd	Calc'd
CH481-α-keratose	5.2	1.80	~0.2	0.17	0.48
CH191-α-keratose	5.2	1.81	~0.2	0.17	0.48
CH108-α-keratose	5.2	1.78	~0.2	0.18	0.48
CH31-α-keratose	5.2	1.70	~0.2	0.22	0.48
CH19-α-keratose	5.2	1.67	~0.2	0.23	0.48

where α and β represent the degree of dissociation for COOH groups of α-keratose and NH_2 groups of chitosan at a given pH value respectively. C_{SO_3H}, C_{COOH}, and C_{NH_2} are numbers of ionizable groups, SO_3H, COOH, and NH_2 respectively. Since $C_{SO_3H} = 0.25C_{COOH}$ (see Table 1), the composition(R_s) of the complex formed by the stoichiometric reaction was obtained from the following equation :

$$ R_s = \frac{C_{NH_2}}{C_{SO_3H} + C_{COOH} + C_{NH_2}} = \frac{C_{NH_2}}{1.25C_{COOH} + C_{NH_2}} $$

$$ = \frac{\alpha + 0.25}{\alpha + 1.25\beta + 0.25} $$

Hence the composition ratio(R_s) could be calculated from the degree of deacetylation for the chitosan and the number of dissociated groups. With respect to the degree of dissociation, we obtained α and β from the potentiometric titration data. As shown in Table 2, the R_{max} values (0.20) observed at pH 5.2 were much lower than those calculated (0.48). This discrepancy seems to be ascribable to the chain rigidity (conformational hindrance) and/or the intramolecular ionic linkage owing to the coexistence of oppositely charged NH_3^+ groups in the α-keratose molecules. Thus it was noted that the complex formation required a larger amount of α-keratose than that calculated.

The effect of sodium chloride, urea, and guanidine hydrochloride on the formation of α-keratose-chitosan complexes was examined by measuring the transmittance of the reaction mixture. Fig.2 shows that the addition of sodium chloride and guanidine hydrochloride increases the transmittance of PEC solution, while urea, the hydrogen bond-breaking agent, does not influence that of PEC solution. The inhibition of α-keratose-chitosan complex formation by sodium chloride and guanidine hydrochloride suggests that an electrostatic interaction may be the primary mechanism of the reaction involved in the PEC formation. Furthermore, it was suggested that the hydrogen bond between the α-keratose and the chitosan had only a minor effect.

When an ionized α-keratose is reacted with the oppositely charged chitosan, the conformational change of the α-keratose may be accompanied by the complex formation depending on circumstances. The effects of complex formation on the conformational changes of α-keratose were investigated by comparing CD spectra. Fig.3 exhibits the CD spectra of chitosan and α-keratose at the same concentrations.

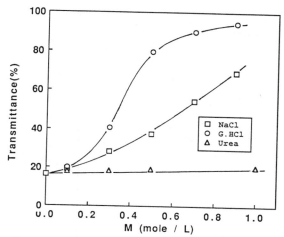

Fig. 2. Effect of inhibitors on the formation of polyelectrolyte complexes at R=0.27.

The CD curve of α-keratose suggested the conformation containing some of α-helix. From Fig.4 which shows the CD spectra of chitosan-α-keratose mixture with the varied mixing ratio, it was found that the characteristic bands for the α-helix did not change with the variation of mixing ratio except for a specific mixing one. At the mixing ratio near R_{max}, the α-helical conformation of α-keratose seemed to be destroyed by the complex formation with the chitosan, showing the spectrum shift to a longer wavelength. Fig.5 shows the DSC thermograms of α-keratose, chitosan, and insoluble α-keratose-chitosan PECs obtained at R=0.23. From these it was noted that the unfolding peak of α-helix in the α-keratose at 228°C disappeared. This finding agrees with the results obtained by the CD mesasurement.

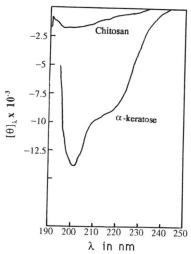

Fig. 3. CD spectra of α-keratose and chitosan in a concentration of 0.003g / dL.

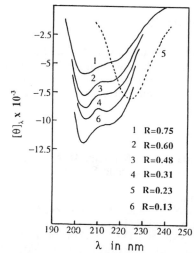

Fig. 4. CD spectra of α-keratose and CH191 mixture at various mixing ratios.

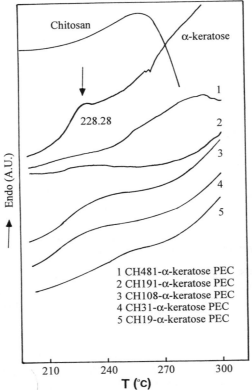

Fig.5. DSC thermograms at a heating rate of 20°C/min
for various PECs formed at R = 0.23.

CONCLUSIONS

It is concluded, through a series of studies on the formation of PEC between the α-keratose and the chitosan with the different molecular weights, that the formation reaction of PEC shows a maximum yield near the mixing ratio of 0.2, and the α-helical conformation of α-keratose is affected by the chitosan in the specific mixing ratio region.

REFERENCES

1. H. Fukuda and Y. Kikuchi, *Makromol. Chem.*, **180**, 1631 (1979).
2. V. Chavasit, C. Kienzle-Sterzer, and J.A. Torres, *Polym. Bull.*, **19**, 223 (1988).
3. K.Y. Kim, O.S. Min, and H.S. Chung, *Polymer (Korea)*, **12**, 234 (1988).
4. G.A.F. Roberts and J.G. Domszy, *Int. J. Biol. Macromol.*, **4**, 374 (1982).
5. G. Toennis and R. Homiller, *J. Am. Chem. Soc.*, **64**, 3054 (1942).
6. R.S. Asquith and D.C. Parkinson, *Text. Res. J.*, **36**, 1064 (1966).
7. M.C. Corfield, A. Robson, and B. Skinner, *Biochem. J.*, **68**, 348 (1958).

55

Synthesis and thermal analysis of aromatic polyethers derived from degradation products of lignin

S. Hirose, T. Hatakeyama* and H. Hatakeyama - Industrial Products Research Institute, 1-1-4 Higashi, Tsukuba, Ibaraki 305, Japan. *Research Institute for Polymers and Textiles, 1-1-4 Higashi, Tsukuba, Ibaraki 305, Japan.

ABSTRACT

Aromatic polyethers were synthesized by the polycondensation of 4,4'-difluoro-3,3'-dinitrobenzophenone with potassium salts of the following hydroquinone derivatives: 2,5-dimethoxyhydroquinone, 2,5-dimethylhydroquinone and hydroquinone. Average molecular weights of the obtained polyethers were in the order of 10^4. Thermal properties of the polyethers were studied by thermogravimetry and differential scanning calorimetry. The polyethers started to decompose at ca. 250 $^{\circ}$C and their glass transition temperatures were in the range of 140 $^{\circ}$C to 189 $^{\circ}$C. The liquid crystalline phase of 4-chlorophenol solutions of the polyethers was observed using a polarizing microscope.

1. INTRODUCTION

The degradation of lignin produces phenols having characteristic chemical structures such as p-hydroxyphenyl, guaiacyl and syringyl groups [1]. Synthetic linear polymers have been extensively synthesized on the basis of molecular design in our laboratory, and they have been studied with reference to the relationship between chemical structures and thermal and mechanical properties of polymers [2-7]. Recently, we have reported that aromatic copolyesters having syringyl groups showed thermotropic liquid crystalline properties [4,5] and that aromatic polyethers having 4-hydroxyphenyl structures had a high degree of thermal stablility [6,7].

having a rod-like structure form a liquid crystalline phase in concentrated solution and in the molten state [8]. However, liquid crystalline properties of aromatic polyethers have scarcely been reported, since the rotation of aromatic rings which are attached to the bonds, occurs more freely compared with those of amide and ester groups. Accordingly, it may be possible for aromatic polyethers to become liquid crystalline, if the rotation of aromatic groups could be controlled by the steric hindrance which due to the existence of substituent on aromatic rings. Therefore, if the above assumption is correct, the characteristic chemical structure of lignin-related phenols such as guaiacyl and syringyl structures can be effectively utilized to obtain liquid crystalline aromatic polyethers.

In the present study, in order to obtain aromatic polyethers having o-substituted aromatic rings of ether bonds, polyethers were synthesized from 3,3'-dinitro-4,4'-difluorobenzophenone and hydroquinone derivatives such as methoxyhydroquinone, methyl hydroquinone and hydroquinone. Thermal properties of the obtained polyethers were studied by thermogravimetry (TG) and differential scanning calorimetry (DSC). Furthermore, the liquid crystalline phase was observed by polarizing microscopy.

2. EXPERIMENTAL

2.1 Materials

Dimethoxy-1,4-hydroquinone was prepared from 1,2,4-trimethoxybenzene [9]. 2,5-Dimethyl-1,4-hydroquinone was prepared from 2,5-dimethylphenol [10]. 3,3'-dinitro-4,4'-difluorobenzophenone was synthesized by nitration of 4,4'-difluorobenzophenone. Polymerization was carried out by solution polycondensation of 3,3'dinitro-4,4'difluorobenzophenone with a hydroquinone derivative in the presence of potassium carbonate and 18-crown-6, under an atmosphere of nitrogen.

2.2 Measurements

Average molecular weights of the obtained polyethers were determined by gel permeation chromatography in a mixed solvent of N,N-dimethylformamide (DMF) and tetrahydrofuran (THF) (1:1, volume). Monodisperse polystyrenes were used as standard materials. TG measurements were carried out using a Seiko TG 220 thermogravimeter at heating rates of 3, 5, 10 and 20 $^{\circ}$C/min. DSC measurements were carried out using a Seiko DSC 200 calorimeter at a heating rate of 10 $^{\circ}$C/min. The liquid crystalline phase of the solutions of the obtained polyethers was observed using a Leitz Orthoplan Pol polarizing light microscope at 20 $^{\circ}$C.

3. RESULTS AND DISCUSSION

3.1 Synthesis

The synthetic route of polyethers is indicated in Scheme 1. The results of polymerization are summarized in Table 1. In order

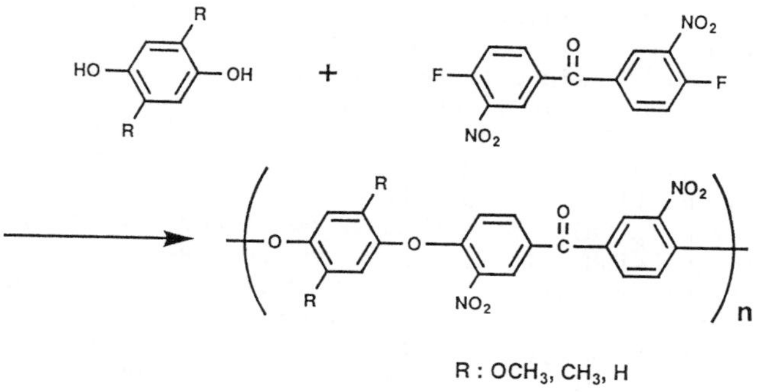

R : OCH₃, CH₃, H

Scheme 1. Synthetic route of polyethers.

to   obtain   polyethers   having   high   molecular   weights,
polymerization conditions were varied (No 1 - 4). Dimethyl
sulfoxide (DMSO) was found to be an appropriate solvent for
polymerization, since polymerization took place in solution
during the reaction. It was also found that the molecular weight
of obtained polyether increased with the increase in temperature
up to 60 $^{\circ}$C (No 4). Therefore, both polyethers having methyl and
methoxyl groups were synthesized by the polymerization at 60 $^{\circ}$C
in DMSO.

Table 1. Results of Polymerization.

No.	R	Solvent	Temp. $^{\circ}$C	Time h	Result	M_n $\times 10^4$
1	H	CH_2Cl_2	20	2	ppt.	--
2	"	NMP	20	6	"	--
3	"	DMSO	20	24	soln.	1
4	"	"	60	6	"	6
5	CH_3	"	20	24	"	--
6	"	"	60	6	"	2
7	OCH_3	"	60	6	"	2*

NMP: N-methyl-2-pyrrolidone, DMSO: dimethyl sulfoxide
* η inh = 0.76 dL/g, 0.25 g/dL 4-chlorophenol soln., 60 $^{\circ}$C.

3.2 Thermal properties

A typical halo pattern was observed in X-ray diffractograms of
polyethers   having   hydrogen atoms and   methyl   groups   as
substituents, R (R= H and CH_3, in Scheme 1), suggesting that the
above two polyethers are amorphous. The X-ray diffractogram of
polyether (R=OCH_3) showed peaks at ca. 10 and 20 degrees,
indicating that this polyether is crystalline.

Phase transition of the above polyethers was studied by DSC. Measurements were carried out in the temperature range from room temperature to 240 $^{\circ}$C, since thermal decomposition of polyethers was observed at 250 $^{\circ}$C in TG measurements. Fig. 1 shows DSC curves of polyethers. Glass transition was observed in DSC curves of polyethers (R=H and R=CH$_3$) and glass transition temperatures (T$_g$'s) were 189 and 144 $^{\circ}$C, respectively. The existence of methyl groups reduces the T$_g$ value of the above polyether. It is thought that there are two kinds of effects which are caused by methyl groups. One effect is related to the restriction of rotation of phenylene groups. This restriction may cause an increase in the T$_g$ value of the above polyether. The other effect is related to the reduction in the interaction among polyether chains. This reduction may a cause the decrease in T$_g$ values. The obtained results clearly indicate that T$_g$ values were reduced after the introduction of methyl groups in phenylene groups of polyether. Therefore, it can be said that methyl groups have a greater influence on the intermolecular interaction than on the intramolecular interaction in this polyether system.

Thermal degradation of polyethers having R=H, CH$_3$ and OCH$_3$ was studied by TG. Fig. 2 shows the TG curves of polyethers. As shown in Fig. 2, starting temperatures of thermal decomposition (T$_{di}$'s) of polyethers are almost the same, regardless of the kind of substituent R in the polyethers. However, the decomposition rate of polyether (R=H) is smaller than that of the polyethers (R=CH$_3$ and R=OCH$_3$). The activation energy (E) of thermal

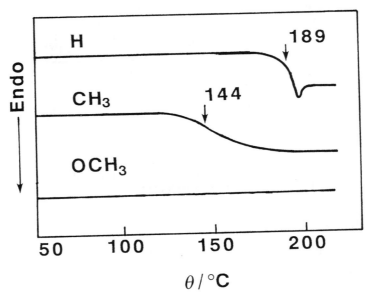

Fig. 1. DSC curves of polyethers. Heating rate: 10 $^{\circ}$C/min.

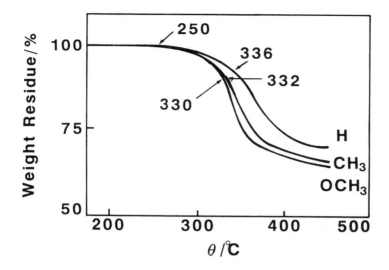

Fig. 2. TG curves of polyethers. Heating
rate: 10 °C/min.

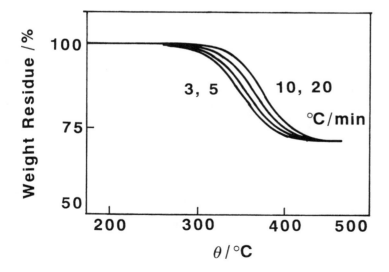

Fig. 3. TG curves of polyether (R=H).
Heating rate: 3, 5, 10 and 20 °C/min.

decomposition was also calculated by the integral method [10]. Fig. 3 shows TG curves of polyether (R=H) which were obtained by measurements at heating rates (ϕ's) of 3, 5, 10 and 20 ^{O}C. The temperatures (T's), at which the weight loss becomes constant, were obtained from TG curves shown in Fig. 3. Fig. 4 shows the plots of log ϕ versus 1/T. The calculated value of E from the gradient of the straight lines was 176 kJ/mol. E's for polyethers (R=CH$_3$ and R=OCH$_3$) were also calculated and their values were 151 and 146 kJ/mol, respectively. These results suggest that the existence of methyl and methoxyl groups decrease the thermal stability of polyethers. However, we have reported that the T_{di} of aromatic copolyesters having methoxyl groups was ca. 360 ^{O}C [4]. T_{di} of the obtained polyethers in this study was ca. 250 ^{O}C. Therefore, it can be said that the existence of nitro groups in polyethers is mainly attributed to the initiation of thermal decomposition of polyethers.

3.3 Liquid crystalline properties of polyethers

The liquid crystalline properties of the obtained polyethers were studied by polarizing light microscopy. A liquid crystalline phase was not observed when polyether samples were heated until 240 ^{O}C. However, the typical pattern of nematic liquid crystalline phase was observed at room temperature for 40 wt% solutions of polyethers in 4-chlorophenol. This indicates that higher-order structure of the polyethers can be controlled by the steric hindrance of aromatic rings having substituents.

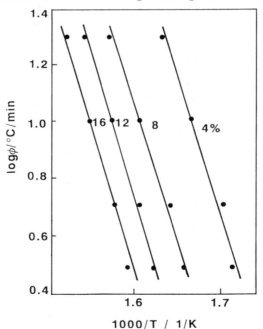

Fig 4. Relationship between log ϕ and 1/T for polyether (R=H).

REFERENCES

[1] Goldstein, I. S. J. Appl. Polym. Sci. **1975**, 28, 259.
[2] Hatakeyama, H.; Hirose, S.; Nakamura, K.; Hatakeyama, T. in Glasser, W. G.; Sarkanen, S. (Eds.) Lignin: Properties and Matrerials, ACS Symposium Series 397, American Chemical Society, Washington DC., 1989, 205.
[3] Hirose, S.; Hatakeyama, H.; Nakamura, K.; Hatakeyama, T. in Scheurch, C. (Ed) Cellulose: Chemistry and Technology, Wiley-Interscience, New York, 1989, 1133.
[4] Hirose, S.; Hatakeyama, H.; Hatakeyama, T. in Kennedy, J. F. ; Phillips, G. O.; Williams, P. A. (Eds.) Wood Processing and Utilization, Ellis Horwood Ltd., Chichester, 1989, 181.
[5] Hirose, S.; Hatakeyama, T.; Hatakeyama, H. Sen'i Gakkaishi, 1991, 47, 388.
[6] Hirose, S.; Nakamura, K.; Hatakeyama, T.; Hatakeyama, H. Sen'i Gakkaishi, 1987, 43, 595.
[7] Hirose, S.; Yoshida, H.; Hatakeyama, T.; Hatakeyama, H. in Glasser, W. G.; Hatakeyama, H. (Eds.) Viscoelasticity of Biomaterials, ACS Symposium Series 498, 1992 in press.
[8] Blumstein, A. (Ed.) Polymeric Liquid Crystals, Plenum Press, New York, 1985
[9] Wehrli, P. A.; Pigott, F. Org. Synth. 1972, 52, 83.
[10] Davidge, H.; Davies, A. G.; Kenyon, J.; Mason, R. F. J. Chem. Soc. 1958, 4569.

PART 3

56

Super absorbent polymers from graft copolymers of acrylonitrile onto water-soluble cellulose derivatives

M. Sakamoto, H. Hasuike, M. Itoh, Y. Ishizuka, K. Yoshida, H. Watamoto and K. Furuhata - Department of Organic and Polymeric Materials, Faculty of Engineering, Tokyo Institute of Technology, Meguro-ku, Tokyo, 152, Japan.

ABSTRACT

Graft copolymerization of acrylonitrile onto water-soluble cellulose derivatives, that is, methyl-, hydroxyethyl-, and carboxymethylcelluloses, was studied with ammonium cerium(IV) nitrate as an initiator. Graft copolymers were saponified, neutralized, and cast into films at ambient temperature. Most of the films showed high water absorbencies. In some cases, heat treatment was required to obtain water-indispersible absorbent films. Some properties of the absorbent films were studied. When the absorbent film was placed in a very dilute cupric sulfate solution, the film swelled at the initial stage, then, shrank spontaneously. Cupric ions could enter into the film to form crosslinks only after it became swollen.

INTRODUCTION

Super absorbent polymers, which can absorb hundreds of times their weight of water in a few minutes were first discovered in the early '70s by a group of Northern Regional Research Laboratories in the course of application study of graft copolymers of starch [1, 2]. When starch-g-polyacrylonitrile (S-g-PAN) was saponified with aqueous alkali, nitrile groups were converted into amide and carboxylate groups. Films made from the resultant starch-g-(polyacrylamide-co-polyacrylate) (S-g-PAN-S) by casting from its solution at ambient temperature became insoluble in water and absorbed large quantities of water, to yield highly swollen gel sheets, which could retain water even under pressure.

The super absorbency is considered to come from the very lightly crosslinked polyelectrolyte structure. Water can enter into the polymer network by the osmotic pressure until the elasticity of the network prevents the polymer chains from extending further. Later work [3] indicated that S-*g*-PAN-S was already crosslinked. Coupling of growing polymer radicals during graft copolymerization gave crosslinks. Additional crosslinking took place during hydrolysis because hydroxide and alkoxide ions induced polymerization of nitrile groups. It is now considered that the mixture of S-*g*-PAN-S and water is not a real solution but a dispersion of swollen microgels. When the S-*g*-PAN-S films were heated, their water absorbencies were decreased. Additional crosslinks may be introduced during heating. There remains a question: why does the swollen gel film not disintegrate into original microgels when placed in water? This question has not been answered yet.

A number of super absorbent polymers, either starch-based materials or totally synthetic polymers, are now in commercial production [4]. Most of the commercial super absorbent polymers are crosslinked intentionally by after-treatment with crosslinking agents or by copolymerization with polyfunctional monomers. Thus, the super absorbent polymers derived from S-*g*-PAN are unique compared with others, because the crosslinks are introduced unintentionally in this particular case.

In this article, we will briefly describe our work on the graft copolymerization of acrylonitrile (AN) onto water-soluble cellulose derivatives (WSC) and some properties of the super absorbent polymers made from the graft copolymers, WSC-*g*-PAN.

MATERIALS AND METHODS

One type of carboxymethylcellulose sodium salt (CMC), two types (HEC$_S$ and HEC$_L$) of hydroxyethylcellulose (HEC), and two types (MC$_S$ and MC$_L$) of methylcellulose (MC) were used as starting materials. Degrees of substitution per anhydroglucose unit (DS) of CMC, MC$_S$, and MC$_L$ were 0.69, 1.7 and 1.7, respectively. Degrees of molar substitution per anhydroglucose unit (MS) of HEC$_S$ and HEC$_L$ were 2.0 and 1.8. Degree of polymerization of CMC was 430. Viscosities of 2% aqueous solutions for HEC$_S$, HEC$_L$, MC$_S$ and MC$_L$ were 302, 7620, 16, and 520 cP, respectively.

The graft copolymerization was carried out with ammonium cerium(IV) nitrate (ACN) as an initiator under nitrogen. No attempt was made to remove homopolymer (PAN) except for CMC-*g*-PAN. *g*-PAN was separated after complete acid hydrolysis of WSC-*g*-PAN, and the average molecular weight (M$_{BR}$) was determined by viscometry. The % PAN, that is, weight % of *g*-PAN in the total graft copolymer was calculated from nitrogen content of WSC-*g*-PAN determined by Kjeldahl method. Experimental code number n for graft copolymerization is used to differentiate the graft copolymer samples, in a way such as CMC-*g*-PAN-4, when necessary.

WSC-*g*-PAN was saponified typically at 100 °C for 3 h with 0.1 N NaOH and the solution was neutralized with acetic acid to pH 7.0. Films were cast from a water solution/dispersion of the product, WSC-*g*-PAN-S, at ambient temperature. Some films were heat-treated at elevated temperatures. The sample code for the heat-treated film is WSC-*g*-PAN-S-H or WSC-*g*-PAN-S-H(T,t) where T is the heat treatment temperature in °C and t is the heating time in h.

Water absorbency was determined by a filtration method with a sieve and expressed as gram water per gram dry film used. No calibration was made for partly dissolved film, if any.

PREPARATION

Graft copolymerization of CMC was carried out at 30 °C for 2.5 h at pH 2.0. The experimental parameters were varied as follows:

Concentration of CMC	16 - 28 g/L	(2 levels)
Concentration of ACN	1.54 - 46.2 mmol/L	(4 levels)
Amount of AN	16 - 32 g/(L aq. solution)	(4 levels)

The yield was found to be dependent on [ACN]/[CMC]: the yield was low ($\leq$ 30 %) when the molar ratio was below 0.17 and high ($\geq$ 70 %) when the ratio was above 0.17. This is because ceric ions and carboxyl groups in CMC form complexes even at low pH, and therefore an excess amount of ceric ions is necessary for graft copolymerization to proceed effectively.The absorbencies of the films obtained from CMC-*g*-PAN by saponification followed by heat treatment at 95 °C for 1 h were 220 - 440 g/g.

Graft copolymerization of HEC was carried out at 35 °C for 3 h at pH 1.0.The experimental parameters were varied as follows:

Concentration of HEC	2 - 10 g/L	(2 levels)
Concentration of ACN	0.6 - 30.4 mmol/L	(11 levels)
Amount of AN	354 - 705 g/(L aq. solution)	(2 levels)

Except for one case in which [ACN]/[HEC] was the lowest (= 0.01 mol/mol), yields of HEC-*g*-PAN exceeded 70 % and the % PAN ranged between 73 and 94 %. The films made from the saponified products, HEC-*g*-PAN-S, showed high absorbencies (500 - 1400 g/g) without heat treatment. Heat treatment of the absorbent films reduced the absorbency and the films made from HEC-*g*-PAN of lower % PAN were more influenced by the heat treatment temperature.

Graft copolymerization of MC was carried out at 40 °C for 3 h at pH 1.0. The experimental parameters were varied as follows:

Concentration of MC	1.0 - 26.7 g/L	(5 levels)
Concentration of ACN	2.0 - 15.0 mmol/L	(5 levels)
Amount of AN	10.6- 159 g/(L aq. solution)	(2 levels)

When the graft copolymerization onto MC$_S$ was carried out with [AN]/[MC$_S$] below 10 mol/mol, MC$_S$-*g*-PAN samples obtained were with % PAN equal to or below 70. Films made from them after saponification were found soluble in water. Some of them were still soluble after heat treatment at 130 °C for 2 h. Other MC$_S$-*g*-PAN samples with % PAN higher than 80 % gave insoluble

---

*Molar concentrations of WSC are based on their masses of the polymer repeating units.

films when saponified and cast into films. Samples with M_{BR} below 200,000 gave low absorbency and the absorbency increased by the heat treatment. Heat treatment at higher temperatures, however, led to a decrease in absorbency.

Most of the samples, as cast or heated, were dissolved in hot water at 80 °C. An easy method was not available for determining if the mixture was a true solution or a dispersion of swollen microgels. It is possible that the interaction between microgels in the film is weak in hot water.

For the graft copolymerization of MC_L, high $[AN]/[MC_L]$ ratios were used. M_{BR} of MC_L-g-PAN were above 200,000, except one case. The absorbency of the saponified samples after heating at 120 °C for 2 h was between 300 and 2800 g/g.

Partial experimental data of graft copolymerization onto various WSC described above are given in Table I.

Table I   Graft Copolymerization of AN onto Various WSC

WSC	Exp. code	[WSC]	[AN]/ [WSC]	[ACN]/ [WSC]	Yield	M_{BR}	% PAN	Absorbency
	n	mmol/L	mol/mol	mol/mol	%	×10⁵		g/g
CMC	1	73	4	0.11	22	1.2	45	269[a,b]
CMC	4	73	8	0.21	85	2.7	63	330[a,b]
HEC_S	1	40	167	0.05	74	3.9	79	1420[c]
HEC_S	3	40	167	0.46	84	1.7	75	840[c]
HEC_L	1	8	804	0.22	74	8.1	94	730[c]
HEC_L	10	42	161	0.13	86	3.2	73	520[c]
HEC_L	14	42	322	0.73	91	2.5	85	1100[c]
MC_S	2	22	9	0.37	74	0.4	70	17[d]
MC_S	3	22	9	0.46	73	0.3	70	Dissolved[d]
MC_S	6	22	46	0.23	86	2.5	86	1008[e]
MC_L	9	22	47	0.37	90	2.7	89	995[f]
MC_L	11	22	139	0.23	83	13.9	97	2064[f]

[a] Saponified with 0.1 N NaOH at 80 °C for 1 h. Other graft copolymer samples were saponified with 0.1 N NaOH at 100 °C for 3 h. [b] Heated at 95 °C for 1 h. [c] Without heat treatment. [d] Heated at 130 °C for 2 h. [e] Heated at 100 °C for 2 h. [f] Heated at 120 °C for 2 h.

PROPERTIES

The super absorbent polymers swell in water and collapse in organic solvents such as ethanol. A swollen gel film of CMC-g-PAN-4-S-H (140,1) was placed alternatively in ethanol and in water and the change in length of the film was measured. Figure 1 shows that shrinkage-expansion cycles can be repeated by alternative change of the external solvent between ethanol and water. The rate of shrinkage was lower than that of expansion. Shrinkage proceeded from the surface to the inner part. The shrunk layer formed acts as a barrier for diffusion of solvent molecules.

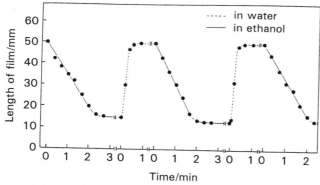

Figure 1. Shrinkage-expansion cycles of swollen CMC-*g*-PAN-4-S-H (140,1).

Phase transition of ionic gels between the swollen state and the collapsed state was discovered by Tanaka and the theory of the transition has been developed [5, 6]. Small changes in an external parameter, such as temperature, pH and composition of solvent in a mixed solent system may induce a phase transition. There are a few papers [7] reporting phase transition of super absorbent polymers. Another example is given in this article.

The absorbency of HEC_L-*g*-PAN-10 was studied in water-acetone and water-methanol systems at room temperature. Gel transition took place clearly at 50 volume % of acetone and 70 volume % of methanol for the water-acetone and the water-methanol systems, respectively.

One of the important factors which determine the absorbency of crosslinked polyelectrolytes is the osmotic pressure. When carboxylate-bearing anionic absorbent polymers are placed in salt solutions, the absorbency is decreased significantly. Furthermore, when divalent metal cations are in the solutions, they form crosslinks with carboxylate ions of the absorbent polymers, hence, the absorbency drops further.

The rate of shrinkage of swollen gels of MC_L-*g*-PAN-11-S-H (120, 2) in 200 μmol/L cupric sulfate was studied at different temperatures (10, 20, and 50 °C). The rate was much higher at higher temperatures. The behaviour of the dry absorbent polymer was compared in dilute solutions containing monovalent and divalent cations at 20 °C. Figure 2 shows that the dry absorbent film swells gradually in a dilute solution of sodium chloride and reaches an equilibrium state in 1 h. The absorbency at the equilibrium in the solution is lower than that in pure water (2064 g/g). The behaviour of the absorbent film in a dilute solution of cupric sulfate is very different: the absorbent film swells rapidly in the initial stage, then, the swollen gel starts to shrink spontaneously. The final equilibrium absorbency is very low.

The swelling behaviour of MC_L-*g*-PAN-9-S-H(120,2) at different temperatures was studied in 200 μmol/L cupric sulfate. Both swelling in the initial stage and shrinkage in the later stage proceeded faster at higher temperatures. The maximum absorbency, however, was not dependent on the temperature. The swelling of MC$_L$-g-PAN-9-S-H(120, 2) was studied at 20 °C at different concentrations of aqueous cupric sulfate. The time to reach the highest absorbency was shorter in a solution of higher concentration (400 μmol/L) than in a lower concentration (100 μmol/L) and the maximum absorbency was lower in the solution of higher concentration.

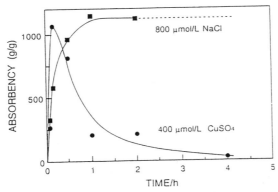

Figure 2. Swelling behaviour of MC$_L$-g-PAN-11-S-H (120, 2) in dilute solutions of sodium chloride and cupric sulfate.

These observations suggest that when a dry absorbent polymer is placed in a dilute salt solution, only water molecules can enter into polymer networks at the initial stage. To prove this, another experiment was made with MC$_L$-g-PAN-11-S-H(120, 2) in 40 μmol/L cupric sulfate at 10 °C in which the change in copper concentration of the external solution was followed by atomic absorption spectrometry. At 10 min after the immersion, the film absorbed 670 times weight of water, while the copper concentration of the external solution increased slightly. This increase was due to the concentration of the solution by preferential absorption of water to the film. At 30 min after the immersion, the absorbency reached a maximum value and the copper concentration of the external solution became lower than the original concentration, indicating that cupric ions were now present in the gel. Thereafter, the gel started to shrink while it absorbed cupric ions further.

REFERENCES

1. Fanta,G.F.;Weaver,M.O.;Doane,M.O. *Chemtech.*, **1974**, *4*, 675.
2. Weaver, M. O.; Bagley, E. B.; Fanta, G. F.; Doane, W. M. *J. Appl. Polym. Sci., Appl. Polym. Symp. Ed.*, **1974**, *25*, 97.
3. Fanta, G.F.; Burr, R.C.; Doane, W.M. *ACS Symp. Ser.*, **1982**, *187*, 195.
4. Masuda, F. *Ko-kyuusuisei Porima*, Kyouritsu Shuppan, Tokyo, **1987**.
5. Tanaka, T. *Polymer,* **1979**, *20*, 1404.
6. Tanaka, T. *NATO ASI Ser., Ser. E*, **1987**, *133*, 237.
7. e.g., see Miyata, N.; Sakata, I. in Kennedy, J.F.; Phillips, G.O.; Wedlock, D.J.; Williams, P.A. (Eds.) *Wood and Cellulosics*, Ellis Horwood, Chichester, **1987**, p.491.

57

Hydroxyethylcellulose graft copolymers as super water-absorbents

Namiko Miyata* and Isao Sakata** - *Nakamura Gakuen College, 5-7-1, Befu, Jonan-ku, Fukuoka, 814-01 Japan. **Kyushu University, 6-10-1, Hakozaki, Higashi-ku, Fukuoka, 812, Japan.

INTRODUCTION

Super water-absorbents can absorb a lot of water, namely 1000g water/g absorbent, and they are now widely used in many fields, as sanitary materials, agricultural and gardening agents, industrial dehydration agents, drying preventives and so on.

We have investigated the synthesis and the properties of the super water-absorbents of graft copolymers based on hydroxyethylcellulose(HEC), other polysaccharides such as mannan, alginic acid, starch, agar, sodium alginate[1,2], and lignin[3]. In this study the effects of the degree of grafting and the cross linking density, namely, the amount of cross linking agent on the morphology of the graft copolymers of HEC containing partially hydrolyzed polyacrylamide(HEC-P-Hyd-PAM) and on the texture of the hydrogels of HEC-P-Hyd-PAM swollen in water were examined by using a scanning electron microscope(SEM), a CRYO SEM and an optical microscope(OM). Also HEC-P-Hyd-PAM on which copper ions(Cu++) were absorbed were examined by energy dispersive X-ray spectrometers(EDS). The relations between the morphology of the HEC-P-Hyd-PAM, the texture of the hydrogels, and the viscoelasticity, the water absorbency and the observed results by EDS were examined.

EXPERIMENTAL

Materials: HEC-P-Hyd-PAM as super water absorbents were synthesized as reported in the previous papers[1, 2].

Observation by Microscopy and EDS: The morphology of HEC-P-Hyd-PAM and the texture of the hydrogels were observed by using JSM-300 and JSM-5400 CRYO scanning electron microscopes with accelerating voltages of 15.0kV and 3.0kV, respectively. The gels dyed with Methylene Blue(MB) were observed by an Olympus optical microscope. After equilibrium of absorption of copper ions(Cu^{++}) on HEC-P-Hyd-PAM was reached, HEC-P-Hyd-PAM samples were washed with methanol and were dried in a vacuum at 40°C. Then the samples were examined by EDS.

Viscoelasticity and Water Absorbency: Measurements of viscoelasticity of the hydrogels of HEC-P-Hyd-PAM and the water absorbency were reported in the previous paper[2].

RESULTS AND DISCUSSION

Fig. 1 shows that the morphology of HEC-P-Hyd-PAM changed from a granular structure to a globular structure and a further small bundle-like structure as the degree of grafting increased.

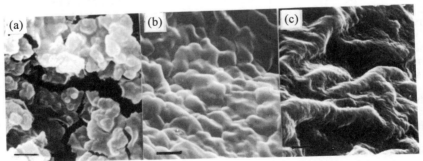

Fig. 1 The effects of the degree of grafting on the
        morphology of HEC-P-Hyd-PAM observed by SEM :
        Grafting and hydl.; (a)161.7%, 59.0%(b)486.8%,
        57.1%(c)1816.7%, 59.2%:Crosslinking agent;(a),
        (b) and (c) 0.3% . Bars represent 1 μm.

Fig. 2 shows the effects of the cross linking density, namely, the amount of cross linking agent on the morphology of HEC-P-Hyd-PAM observed by SEM. The granular structure can be seen when using a relatively small amount of cross linking agent, for

example,   0.1% and 0.3%,   whereas when using a large
amount, i.e. 2.5%,   one can see irregular projec-
tions in shape on the surface of HEC-P-Hyd-PAM.

Fig. 2 The effects of the amount of a cross linking
       agent on the morphology of HEC-P-Hyd-PAM ob-
       served by SEM: Amount of   crosslinking agent,
       grafting and hydl.,;(a)0.1%, 464.6%, 59.4%
       (b)0.3%, 161.7%, 59.0%, (c)2.5%, 2105.0%, 68.0%
       Bars represent 1 μm.

Fig. 3 shows the   SEM photographs of the hydrogels
of HEC-P-Hyd-PAM swollen in water by using CRYO SEM
by which we can observe hydrogels in a frozen state,
and also Fig.4 shows the photomicrographs of the
hydrogels dyed by MB.   The hydrogels form looser
networks as the water absorbency of HEC-P-Hyd-PAM
increases.

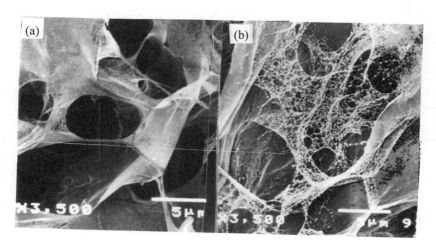

Fig. 3 The CRYO SEM photographs of the hydrogels of
       HEC-P-Hyd-PAM(grafting:727.5%,hydl.:57.7%):
       Absorbency;(a)521g/g, (b)1551g/g. Bars repre-
       sent 5 μm.

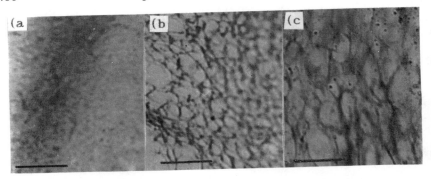

Fig. 4   Photomicrographs of the hydrogels of HEC-P-
         Hyd-PAM(grafting:727.5%,hydl.:57.7%) dyed by
         M.B.:Absorbency;(a)12.9g/g, (b)250g/g, (c)
         1551g/g. Bars present 100 μm.

Fig. 5 presents microphotographs showing the effects
of the amount of cross linking agent on the tex-
tures of hydrogels of HEC-P-Hyd-PAM dyed by MB.   The
hydrogels form closer networks as the cross linking
density, namely, the amount of cross linking agent
increases.   It is also interesting that the hydrogel
looks like   films by observing with CRYO SEM when
using large amounts of cross linking agent on the
graft copolymerization of PAM onto HEC.

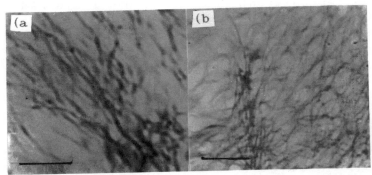

Fig.5 The effects of the amount of   cross linking
      agent on the textures of the hydrogels of HEC-
      P-Hyd-PAM observed by OM:Grafting, hydl., amo-
      unt of   crosslinking agent and absorbency;(a)
      464.3%, 55.7%, 0.1%,1125g/g, (b)727.5%, 57.7 %,
      0.3%, 1000g/g. Bars represent 100 μm.

Fig. 6 shows the photographs of EDS analysis of HEC-
P-Hyd-PAM on which copper ions(Cu++) were absorbed.
EDS analysis indicates that the branch polymers
i.e., P-Hyd-PAM of HEC graft copolymers were grafted

relatively uniformly and densely when the degree of
grafting and the amount of cross linking agent were
large. On the other hand, the branch polymers were
grafted relatively randomly and loosely, that is to
say, the graft distribution of the branch polymers
was random in the case of the HEC graft copolymers
of low degree of grafting and with a small amount
of cross linking agent.

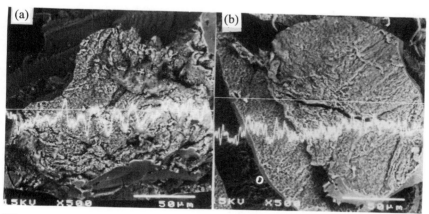

Fig. 6 Photographs of EDS analysis of HEC-P-Hyd-PAM
       on which copper ions (Cu++) were absorbed:
       Grating, hydl. and crosslinking agent ; (a)
       141. 4%, 49. 8%, 0. 3% (b)590. 0%, 62. 1%, 1. 5%.

Fig. 7 shows storage modulus (E'), loss modulus (E")
and loss tangent (tan δ) of the hydrogels of HEC-P-
Hyd-PAM and commercial super water-absorbents, and
the amount of cross linking agent on grafting vs.
water absorbency. E' of the hydrogels of HEC-P-Hyd-
PAM and commercial water-absorbents decreased and
the tan δ increased with an increase of water absorbency.   E' of the hydrogels of HEC-P-Hyd-PAM increased with an increase of the amount of
crosslinking agent.   These results may be reflected
by the difference in the structure of the hydrogels;
that is to say, the hydrogels are in looser or
denser networks structures.

The conclusions are as follows; the surface of HEC-
P-Hyd-PAM and its hydrogel were composed of
granular and network structures, respectively. The
morphologies of HEC-P-Hyd-PAM and its hydrogel
changed from granular and network structures to
bundle-like and closer network structures like
films, respectively, and also the branch polymers
i.e., P-Hyd-PAM of HEC graft copolymers were grafted
relatively uniformly and densely as the degree of
grafting and the amount of cross linking agent are
large; E' of the hydrogels of HEC-P-Hyd-PAM

decreased and the tan δ increased with an increase of water absorbency. E' of the hydrogels increased with an increase of the amount of crosslinking agent. These results may be reflected by the difference in the structure of the hydrogels; that is to say, the hydrogels are in looser or denser networks structures.

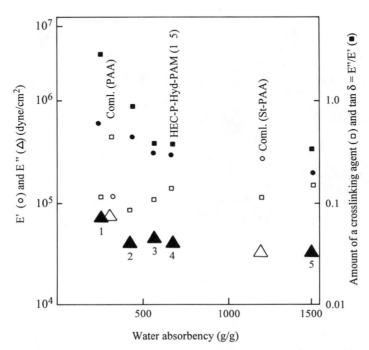

Fig. 7 Storage modulus(E'), loss modulus(E") and loss tangent(tan δ) of the hydrogels of HEC-P-Hyd-PAM and commercial water-absorbents, and the amount of cross linking agent on grafting vs. water absorbency.

REFERENCES

1. N. Miyata and I. Sakata, "Wood and Cellulosics", J. F. Kennedy, G. O. Phillips, D. J. Wedlock and P. A. Williams Eds., Ellis Horwood, Chichester, U. K., p. 491(1987);Sen'i Gakkaishi(Fiber), Vol. 47, pp. 95-101(1991)
2. N. Miyata and I. Sakata, Sen'i Gakkaishi(Fiber), Vol. 47, pp. 428-433(1991)
3. N. Miyata and I. Sakata, Sen'i Gakkaishi(Fiber), Vol. 46, pp. 356-359(1990)

58

High-performance dielectric polymers from cellulose derivatives

T. Sato, M. Minoda and T. Miyamoto - Institute for Chemical Research, Kyoto University, Uji Kyoto 611, Japan.

INTRODUCTION

Polymers with high–dielectric constants find practical use as binding polymers in electroluminescence cells and capacitors. The properties required for these polymers are listed in Table 1. In order to obtain polymers with these properties they must have a large permanent dipole. One method for preparing such polymers is to introduce substituents with large dipole moments, such as cyano groups, to polyhydroxy polymers. Cyanoethylated polysaccharides such as cyanoethylated derivatives of cellulose [1], amylose and pullulan [2] have actually been commercialized as binding polymers in organic dispersion types of electroluminescent cells. However, the properties of these polymers are not entirely satisfactory, especially, in terms of their dielectric constants. Organic dispersion electroluminescent devices are used as a luminescent source, i. e. a light source for liquid crystal displays. The brightness of the luminescence is directly proportional to the dielectric constant of the binding polymers. Therefore, if polymers having a higher–dielectric constant can be developed, further miniaturization of liquid crystal displays is possible.

Table 1. Properties required for binding polymers in organic dispersion electroluminescent devices

---

1. High dielectric constant and low dielectric loss tangent.
2. Thermoplasticity and excellent adhesiveness.
3. Good malleability and high optical transparency.
4. Low moisture regain.

Figure 1 shows the chemical structure of dihydroxypropyl cellulose (DHPC). DHPC is a recently developed polyhydroxy cellulose derivative [3]. If the dihydroxypropylation reaction can be controlled, derivatives which have more than three hydroxyl groups per glucose unit can be obtained. It is of practical interest to investigate this polymer as a potential starting material for the preparation of dielectric polymers. In the present work, we have attempted to prepare at first DHPC and then dielectric polymers from DHPC by subsequent chemical modifications.

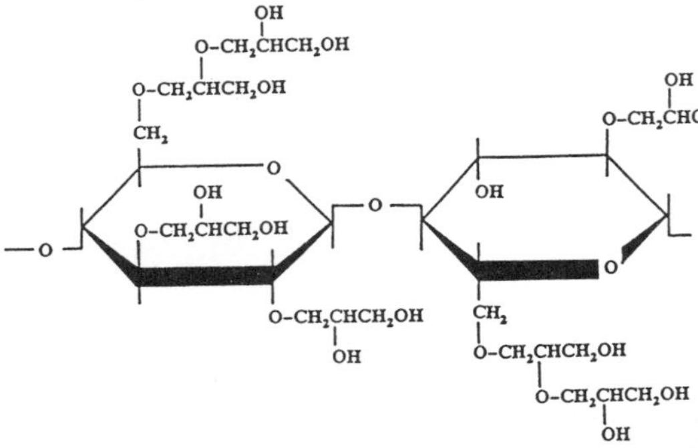

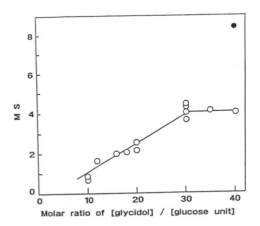

(MS=3.5, DS=2.5)

Figure 1. Chemical structure of O–(2,3–dihydroxypropyl) cellulose (DHPC).

RESULTS and DISCUSSION

Preparation and Characterization of DHPC

It is well known that the properties of cellulose derivatives strongly depend not only on the degree of substitution(DS) but also on the specificity of substitution on the glucose units and the distribution along the cellulose chain [4]. In the case of DHPC, the MS value, i. e. the number of dihydroxypropyl units per glucose unit, is also important. The control of the substituent distribution provides a means of generating a wide variety of unique and useful polymers. The development of non-aqueous solvent systems has enabled such a functionalization of cellulose. Here we attempted to prepare two different types of DHPC samples using homogeneous and heterogeneous reaction conditions and examined the difference in properties between homogeneous and heterogeneous samples.

At first, homogeneous DHPC samples were prepared using a Lithium chloride / Dimethyl-acetamide solvent system [5]. Finely powdered

Figure 2. Relationship between the MS value of DHPC samples and the amount of glycidol added.

sodium hydroxide was used as catalyst. The molar ratio of sodium hydroxide to glucose unit was fixed at 4, and glycidol was added dropwise to the cellulose solution. The reactions were carried out at 50 °C for 19h. Figure 2 shows the results. The MS values were estimated from the ^{13}C–NMR spectra of the fully acetylated derivatives [5]. It can be seen that the MS value increases with an increasing amount of glycidol. However, the products having an MS value higher than 4 could not be obtained under the reaction conditions employed.   This is due to the homopolymerization of glycidol. In order to obtain samples with an MS value higher than 4, it was necessary to add glycidol stepwise. This is very important to avoid the homopolymerization of glycidol. A sample having an MS value of about 8, shown by the closed circle, was obtained by a four–portion–wise addition of glycidol. Each addition was made dropwise.

Table 2 shows the solubility of the DHPC samples, compared with that of commercial hydroxypropyl cellulose (HPC) and hydroxyethyl cellulose (HEC). The number attached to the sample code indicates an approximate MS value of the sample. DHPC prepared by homogeneous reaction is alkali–soluble and does not form a gel even in heated aqueous solutions.

Table 2. Solubility of DHPC, HPC, and HEC in various solvents

Solvent	Solubility[a]					
	DH–1	DH–2	DH–4	DH–8	HPC[b]	HEC[c]
Water	o	o	o	o	o	o
Aq. alkali	o	o	o	o	g	g
Dil. acid	o	o	o	o	o	o
Methanol	x	x	x	x	o	Δ
DMSO	o	o	o	o	o	x
Chloroform	x	x	x	x	o	x
THF	x	x	x	x	o	x

a) o, soluble; x, insoluble; Δ, swelling; g, gelation.
b) O–(2–Hydroxypropyl)cellulose (MS = 4.1, DS = 2.4).
c) O–(2–Hydroxyethyl)cellulose (MS = 1.9, DS = 1.1).

As already mentioned, the preparation of DHPC by the conventional alkali cellulose process has been already reported in the literature [3]. It is well known that the alkali cellulose process results in a non–uniform distribution of substituents along the chain. In the present study, however, we attempted to prepare DHPC with a more sequentially nonhomogeneous distribution of substituents, that is, block–copolymer–like DHPC.

Recently, Sikkema succeeded in the preparation of block–copolymer–like carboxymethyl cellulose [6]. In the light of his results, heterogeneous DHPC samples were prepared using the similar reaction conditions. At first, a sufficient amount of alkali was added to destroy the crystallinity of cellulose, but before the addition of glycidol the solution temperature was kept low (0 °C) and the alkalinity of the solution decreased, so as to promote the recrystallization of cellulose. This procedure results in the formation of microcrystals. It is expected that segments of the cellulose backbone are protected from the subsequent chemical modifications. The products thus obtained are considered to consist of highly substituted and poorly substituted cellulose sequences. Acetic acid was used for the partial neutralization. The reactions were carried out under almost the same conditions as used for the homogeneous reactions.

Figure 3 shows the relationships between the DS and MS values of the samples. Although the difference between homogeneous and heterogeneous products is not so large, it is obvious that the substituent distribution of the products prepared under heterogeneous reaction conditions is more sequentially nonhomogeneous than the products derived from the homogeneous reaction. The results also show that the products having a DS value higher than 1.5 cannot be obtained using the heterogeneous reaction employed. This is due to considerable graft copolymerization of glycidol which takes place during the reaction.

The sequentially nonhomogeneous distribution in the heterogeneous DHPC was evident from the difference in crystallinity between the homogeneous and heterogeneous products (Figure 4). From a comparison of these X-ray diffraction spectra, it is clear that the crystallinity of the heterogeneous sample is higher than that of the homogeneous sample. Next, we examined the difference in properties between the homogeneous and heterogeneous products.

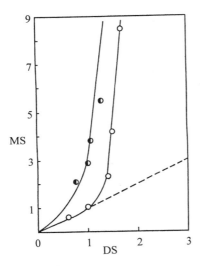

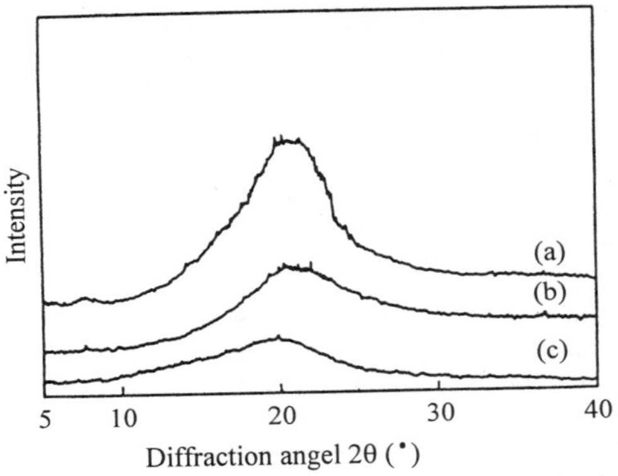

Figure 4. X-ray diffraction patterns of DHPC samples and amorphous cellulose.
(a) Heterogeneous DHPC (HDH–3).
(b) Homogeneous DHPC (DH–3).
(c) Amorphous cellulose.

Figure 3. Relationship between the DS and MS values of the DHPC samples prepared.
○ , Homogeneous reaction system.
◑ , Heterogeneous reaction system.

Table 3 shows the difference in water solubility. It can be seen that the lower limit MS value for water-soluble samples prepared by the homogeneous reaction is about 0.6. This value is much lower than that of the heterogeneous samples. It is obvious that such a difference is due to the difference in the distribution of substituents along the cellulose chain. Another difference can be seen in the solution viscosities. Table 4 shows the viscosity of DHPC aqueous solutions. The samples examined have almost the same MS and molecular weight. It should be noted that the viscosity of heterogeneous DHPC is very high, about twice as large as that of homogeneous DHPC.

Table 3. Water solubility of homogeneous
and heterogeneous DHPC samples prepared.

Sample code	MS[a]	Solubility[b]
Homogeneous		
DH–1	0.6	O
DH–2	2.3	O
DH–3	3.1	O
DH–4	4.2	O
DH–8	8.5	O
Heterogeneous		
HDH–1	1.5	X
HDH–2A	2.3	Δ
HDH–2B	2.5	Δ
HDH–3	2.9	O
HDH–4	3.8	O

a)Calculated from ^{13}C–NMR of acetylated DHPC.
b) For 2%(w/v) aqueous solution.
  O; soluble, X; insoluble, Δ; swelling.

Table 4. Comparison of viscosity between
homogeneous and heterogeneous DHPC samples.

Sample code	Mn[a]	MS	Viscosity[b] (cp)
DH–3	18000	3.1	40
HDH–4	17000	3.8	75

a) Estimated by a poly(ethylene oxide)–
  calibrated GPC analysis.
b) Measured for 3 wt.% aqueous solution
  by rotary viscosity meter.

Dielectric Properties of Chemically Modified DHPC Derivatives

Polar polymers may be roughly classified into three categories: (1) polymers with permanent dipoles parallel to the main chain; (2) polymers with permanent dipoles perpendicular to the main chain; and (3) polymers with flexible, polar side groups [7]. The polymers prepared in the present study are considered to belong to the third category.

Table 5 shows the dielectric constants and dielectric loss tangents (tan δ) of chemically modified DHPC derivatives. Dielectric measurements were carried out at room temperature at 1 KHz. Cyanoethylation was achieved by a Michael condensation with acrylonitrile using sodium hydroxide as a catalyst. To avoid side reactions, the concentration of sodium hydroxide was kept low [8]. Benzoylation was achieved by the acid chloride–pyridine procedure. In any case, these reactions proceeded quantitatively. Complete cyanoethylation and esterification were confirmed by IR spectroscopy.

Table 5. Dielectric constants(ε') and dielectric loss tangents (tan δ) of DHPC derivatives

Sample code	Substituent	ε'[a]	tan δ[a]
DH–1–CN	–OCH$_2$CH$_2$CN	18	0.01
DH–2–CN	–OCH$_2$CH$_2$CN	27	0.07
DH–4–CN	–OCH$_2$CH$_2$CN	31	0.04
DH–8–CN	–OCH$_2$CH$_2$CN	26	0.13
DH–4–CB	–OC(=O)–⟨O⟩–CN	5	0.02
DH–4–DN	–OC(=O)–⟨O⟩(NO$_2$)(NO$_2$)	7	0.09

a) Measured at 1 KHz and 27 °C.

It can be seen that the dielectric constants of the cyanoethylated DHPC derivatives are very high, compared with those of other high–dielectric polymers previously reported [1,2]. However, 3,5–dinitro– and 4–cyano–benzoylated DHPC derivatives exhibited lower dielectric constants than expected. We later learned that the carbonyl groups also have a relatively high dipole moment with a direction opposite to the dipole moment of the polar groups introduced.

Last to be mentioned is the relationship between the dielectric constant and the phase structure of the DHPC derivatives. Cellulose can form thermotropic liquid crystalline phases when substituted with appropriate substituents. In the case of homogeneous DHPC samples, we found that fully cyanoethylated samples having a moderate MS value can also form thermotropic cholesteric liquid crystalline phases over a wide temperature range [5]. Table 6 shows the relationship between the phase structure and dielectric constant of polymers prepared from homogeneous and heterogeneous DHPC samples. The first two homogeneous samples having low MS values were in the solid state at room temperature, while DH–4–CN was in the liquid crystalline state and DH–8–CN was in the isotropic state. From this table, it may be concluded that the dielectric constant of polymers is almost independent of their phase structure. In the case of heterogeneous DHPC, all of the cyanoethylated samples were in the solid state, independent of their MS value. This is the most significant difference between the homogeneous and heterogeneous DHPC samples. The results also show that the dielectric constant of cyanoethylated DHPC is independent of the substituent distribution along the chain. This is probably due to the fact that homogeneous and heterogeneous samples with the same MS value have almost the same content of cyano groups by weight.

Table 6. Relationship between phase structure and dielectric constant(ε') of sample polymers prepared.

Sample code	Dielectric constant (ε')[a]	Phase structure[b]
DH–1–CN	18	rubbery solid
DH–2–CN	27	rubbery solid
DH–4–CN	31	liquid crystalline (cholesteric)
DH–8–CN	26	isotropic melt
HDH–2–CN	28	rubbery solid
HDH–3–CN	27	rubbery solid
HDH–5–CN	30	rubbery solid

a) Measured at 1 KHz and 27 °C.
b) At room temperature.

REFERENCES

1   S. Tasaka, N. Inagaki, S. Miyata, T. Chiba, *J. Soc. Fib. Sci. Thechnol. Jpn.* **44**, 546 (1988).
2   Y. Onda, H. Mutoh, H. Suzuki, JP. 79/93,557 (1979); *Chem. Abstr.* **94**, 193997c (1981).
3   Y.–X. Zhang, J. C. Chen, D. Patil, G. B. Butler, T. E. Hogen–esch, *J. Macromol. Sci., Chem.* **A25(8)**, 955 (1988).
4   S. Takahashi, T. Fujimoto, T. Miyamoto, H. Inagaki, *J. Polym. Sci., Polym. Chem. Ed.* **25**, 987 (1987).
5   T. Sato, Y. Tsujii, M. Minoda, Y. Kita, T. Miyamoto, *Makromol. Chem.* in press.
6   D. J. Sikkema, H. Jannsen, *Macromolecules* **22**, 364 (1989).
7   W. H. Stockmayer, *IUPAC Symposium on Macromolecular Chemistry, Tokyo–Kyoto* (1966).
8   W. H. Daly, A. Munir, *J. Polym. Sci., Polym. Chem. Ed.* **22**, 975 (1984).

59

Immobilization of hyaluronic acid onto cellulose

M. Shimada, S. Takigami, Y. Nakamura, N. Kobayashi*, P.A. Williams** and G.O. Phillips** - Department of Chemistry, Faculty of Technology, Gunma University, Kiryu, Gunma 376, Japan. *School of Medicine, Gunma University, Maebashi, Gunma 371, Japan. **Research Division, The North East Wales Institute, Deeside, Clwyd CH5 4BR, UK.

ABSTRACT

Hyaluronic acid was immobilized onto cellulose using cyanuric chloride as a crosslinking reagent. The amount of hyaluronic acid immobilized increased by carrying out the reaction simultaneously with carbamoyl- and carboxy-ethylation. The biocompatibility of the samples evaluated by complement activation and activation of blood clotting cascade was good.

INTRODUCTION

Hyaluronic acid occurs naturally in tissues of animals. The vitreous humour of the eyes and synovial fluid of the knees are rich in hyaluronic acid. Cellulose shows fairly good biocompatibility when used in haemodialysis membranes. In other words, there is little adverse reaction between the human body and cellulose. However, it causes some problems to patients when used for a long time. As hyaluronic acid is a biomolecule found in many tissues of the human body, it is expected to increase the biocompatibility of materials introduced into it.

EXPERIMENTAL

Carbamoyl- and Carboxy-ethylation of Cellulose

Cotton cellulose was carbamoyl- and carboxy-ethylated with acrylamide at 30 °C for 24 h under alkaline conditions.

Immobilization Reaction

The immobilization reaction of hyaluronic acid onto cellulose was carried out under alkaline conditions using cyanuric chloride as a crosslinking reagent.

In some reactions, the carbamoyl- and carboxy-ethylation was performed simultaneously with the immobilization reaction.

The degree of reaction was evaluated by per cent of weight increase according to the following equation:

$$\text{Weight increase (\%)} = \frac{\text{Weight increase by reaction (g)}}{\text{Weight of sample before reaction (g)}} \times 100$$

Complement Activation

A sample was immersed in complement solution at 37 °C for 60 min. An aliquot of the solution was withdrawn and transferred to another test tube. Sensitized sheep red blood cells were added to the complement solution and incubated at 37 °C for 60 min. The mixture was centrifuged at 2000 rpm for 10 min. The optical density of the supernatant was measured at 541 nm and the value was compared with that of complement.

Activation of Blood Coagulation Cascade

APTT (Activated Thromboplastin Time) was measured using commercial APTT reagent which contains rabbit cephalin and ellagic acid (Sigma APTT reagent). After incubating the reagent at 37 ° C for 1 min, plasma which was removed from a test tube with a sample was added and incubated for 3 min. The warm calcium chloride solution was added and the timer of the coagulometer was started to give the time required to form a clot.

RESULTS AND DISCUSSION

Immobilization Reaction

The mechanism of immobilization of hyaluronic acid onto cellulose is shown below.

| Cellulose | Cyanuric chloride | Hyaluronic acid | | Cellulose with hyaluronic acid |

The percent of weight increase by immobilization reaction onto cellulose is shown in Figure 1. It was very small in all reactions. It appears that a low concentration of sodium hydroxide is sufficient. The role of sodium hydroxide is to neutralize the acid formed. If all hyaluronic acid reacted with cellulose, the weight increase would be 20 %. As cellulose has a rigid crystal structure, it may be difficult for hyaluronic acid to react with it. Therefore, the immobilization reaction onto carbamoyl- and carboxy-ethylated cellulose was

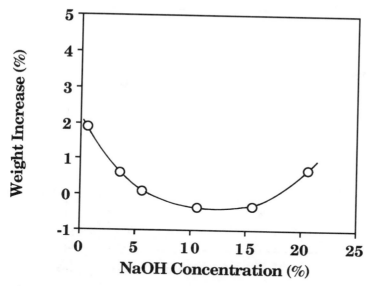

Figure 1.   Immobilization of hyaluronic acid onto cellulose.   The
concentrations of hyaluronic acid and acrylamide were 0.2 % and 0%,
respectively.

performed simultaneously with the derivatizing reaction in which cellulose has
a less rigid structure.   As shown in Figure 2, an increased amount of
hyaluronic acid was immobilized.   Here, the weight increase by hyaluronic
acid was estimated as follows:

$$\begin{pmatrix} \text{Weight} \\ \text{increase by} \\ \text{immobilization} \\ \text{of hyaluronic} \\ \text{acid} \end{pmatrix} = \begin{pmatrix} \text{Weight increase by} \\ \text{immobilization reaction} \\ \text{carried out simultaneously} \\ \text{with carbamoyl- and} \\ \text{carboxy-ethylation} \end{pmatrix} - \begin{pmatrix} \text{Weight} \\ \text{increase by} \\ \text{carbamoyl- and} \\ \text{carboxy-} \\ \text{ethylation} \end{pmatrix}$$

Complement Activation
       All samples including cotton, and cellulose samples with and without
hyaluronic acid   did not activate complement, as shown in Table I.   This
indicates that the samples are biocompatible.   If the samples had activated
complement, the OD_{sample} / $OD_{complement}$ ratio would have been more than
two.

Activation of Blood Coagulation Cascade
       Effect of samples on clotting time of plasma is shown in Figure 3.   The
clotting time of plasma itself increased with incubation time at 37 °C.   This
phenomenon is explained by the aging of the plasma.   Clotting time of plasma
with cotton cellulose decreased slightly with short incubation time and
increased slightly with long incubation time.   Other samples with and without

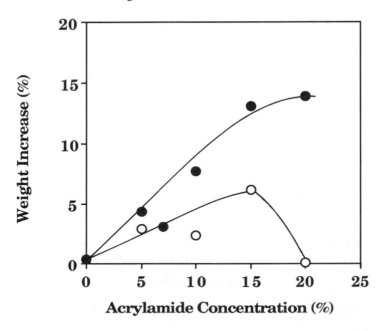

Figure 2. Effects of swelling on immobilization reaction of hyaluronic acid. Immobilization reaction was carried out simultaneously with carbamoyl- and carboxy-ethylation. The concentrations of sodium hydroxide were 5.6 % (○) and 15.6 % (●).

Table I. Complement activation

Sample*	Weight increase by hyaluronic acid (%)	$OD_{sample}/OD_{complement}$
Cotton	-	1.11
Cotton	2.3	1.09
CE(L)	-	1.08
CE(L)	2.8	1.04
CE(H)	-	1.07
CE(H)	13.9	1.06

* See legends in Figure 3

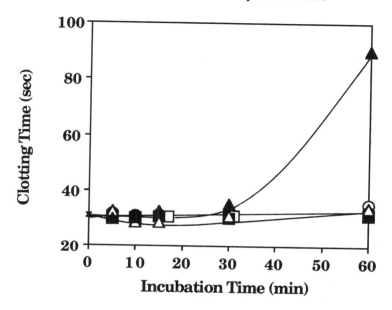

Figure 3.   Effects of samples on clotting time of plasma.   The plasma was incubated with a sample at 37 °C.   Plasma (×),   Cotton cellulose (○),   Cotton cellulose with 2.3 % hyaluronic acid   (●), Carbamoyl- and carboxy-ethylated cellulose at low concentration of sodium hydroxide "CE(L)"(□), CE(L) with 2.8% hyaluronic acid (■), Carbamoyl- and carboxy-ethylated cellulose at high concentration of sodium hydroxide "CE(H)" (△), and CE(H) with 13.9 % hyaluronic acid (▲).

hyaluronic acid except the carbamoyl- and carboxy-ethylated cellulose with 13.9 % hyaluronic acid showed almost a similar result to that obtained with cotton cellulose.   Clotting time of the plasma contacted with the sample with 13.9 % hyaluronic acid decreased slightly and then increased very much when the sample was immersed in plasma for 60 min.

    Changing the sample/plasma ratio, the sample was immersed in plasma for 60 min and the clotting time of the plasma was measured and the results are shown in Figure 4.   With increasing amount of sample, the clotting time first decreased and then increased progressively.   The results show that the carbamoyl- and carboxy-ethylated celluloses with 13.9 % hyaluronic acid have properties that increase clotting time.

    To elucidate the effect of hyaluronic acid on clotting time of plasma, the concentration of hyaluronic acid was changed and clotting time observed was

as shown in Figure 4.   Though the clotting time increased with increasing
hyaluronic acid concentration, it is possibly due to a polyanion effect which is
observed in some charged biomolecules.

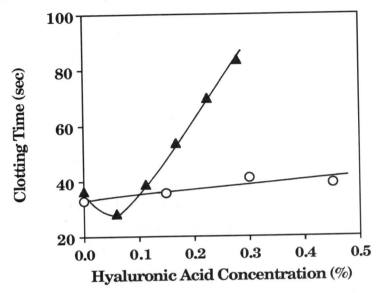

Figure 4.   Effect of hyaluronic acid concentration on clotting time
of plasma.   CE(H) with 13.9 % hyaluronic acid was immersed in
plasma (O) and hyaluronic acid was mixed with plasma (▲).

When we look at the data on cellulose with hyaluronic acid, we look at the
effect of the sample on plasma.   One possible explanation for the phenomenon
could be that one or more of the plasma proteins which take part in clotting are
removed by binding on the surface of the sample.   As more hyaluronic acid is
present, more of the protein or proteins bind and the clotting time is effected
progressively.   This process will take some time and an incubation period of
longer than 30 min will be required for the binding to show effect.   A coating of
protein onto the hyaluronic acid would make the surface biocompatible.

CTA hemodialysis membrane design for β_2-microglobulin removal

M. Ohno, M. Suzuki, M. Miyagi, T. Yagi, H. Sakurai and T. Ukai - TOYOBO Research Institute, 2-1-1 Katata, Ohtsu-shi Shiga, Japan 520-02.

ABSTRACT

Hemodialysis is one of the successful organ substitution treatments as far as saving life. However, long-term hemodialysis patients tend to suffer amyloid-related complications since conventional membranes can not remove β_2-microglobulin. In this report, CTA(cellulose triacetate) hemodialysis membrane is described for its design to remove β_2-microglobulin by diffusion without albumin leakage, and for its crimped configuration to remove small molecules sufficiently.

β_2-microglobulin and albumin are so close in size that it is hard to remove only β_2-microglobulin by simply increasing membrane pore size, which results in a serious albumin leakage. Based on "Pore Theory", solute transfer through the membrane is determined in two steps. First step is steric hindrance at a pore entrance. Second step is friction in the pore between solute and pore wall. The idea for the prevention of albumin leakages is to increase only steric hindrance by partial pore blockage. For this purpose, protein adsorption is positively employed by selecting material and by controlling spinning conditions.

In addition to membrane performance, a module design is also important especially for small molecules since a dialysate side resistance is high enough to consider. In order to make dialysate flow uniformly in a module, crimped configuration is applied for CTA membrane, which keeps adequate space among fibers.

REMOVAL TARGET

Molecular range to be removed has been extended from small molecular uremic toxins such as urea, creatinine, and uric acid to low molecular proteins since β_2-microglobulin was identified[1),2)] as a pathogenic substance of hemodialysis related amyloidosis. Meanwhile, β_2-microglobulin is so close to albumin in size as shown in Table 1 that it's quite hard to separate these two proteins only by membrane pore sieving. The membrane is designed to achieve the target "sufficient β_2-microglobulin removal without albumin leakage".

Table 1 Physical Property

		β_2-microglobulin	Albumin
Molecular Weight	[Dalton]	11,600	67,000
Diffusion Coefficient	[cm²/sec]	13.7×10^{-7}	6.3×10^{-7}
Solute Radius	[Å]	15.6	35.1
Concentration	[g/l]	0.06	50

DESIGN BACKGROUND

Based on "Pore Theory"[3),4),5)], solute transfer is determined by diffusion and convection. Though a pressure difference ΔP and a concentration difference ΔC can be controlled, solute permeability and sieving coefficient are determined automatically by the function of a ratio of solute radius to pore radius.

Solute Transfer Equations in "Pore Theory"

$$\text{Solute Transfer} = \begin{bmatrix} \text{Solute Transfer} \\ \text{by Diffusion} \end{bmatrix} + \begin{bmatrix} \text{Solute Transfer} \\ \text{by Convection} \end{bmatrix}$$

$$\begin{bmatrix} \text{Membrane} \\ \text{Solute Permeability} \end{bmatrix} \times \Delta C \qquad \begin{bmatrix} \text{Membrane} \\ \text{Sieving Coefficient} \end{bmatrix} \times Lp \times \Delta P \times C$$

ΔC : Concentration Difference
Lp : Water Permeability
ΔP : Pressure Difference
C : Blood Concentration
D : Solute Diffusion Coefficient
Ak : Ratio of Pore Area to Membrane Surface Area
Δd : Membrane Thickness

$$\text{Solute Permeability} = \underbrace{\begin{array}{c} \text{Steric Hindrance} \\ \text{at pore entrance} \end{array}}_{(1-q)^2} \times \underbrace{\begin{array}{c} \text{Friction between} \\ \text{solute and pore wall} \end{array}}_{\frac{(1-2.1q+2.1q^3+1.7q^5+0.73q^6)}{(1-0.76q^5)}} \times D \times Ak / \Delta d$$

contants

$$\text{Sieving Coefficient} = \underbrace{\begin{array}{c} \text{Steric Hindrance} \\ \text{at pore entrance} \end{array}}_{2(1-q)^2-(1-q)^4} \times \underbrace{\begin{array}{c} \text{Friction between} \\ \text{solute and pore wall} \end{array}}_{\frac{(1-(2/3)q^2+2.0q^5)}{(1-0.76q^5)}}$$

q : functon of the ratio of solute radius to pore radius

MEMBRANE DESIGN

EFFECT OF PORE SIZE

Fig.1 shows that the relationship between pore size and sieving coefficient. Though conventional membrane with the pore radius 30Å will not leak albumin, β_2-microglobulin removal is not sufficient. In order to remove more than 60 percent of β_2-microglobulin, more than 50Å of pore radius is required. Which, however, will result in a serious albumin leakage.

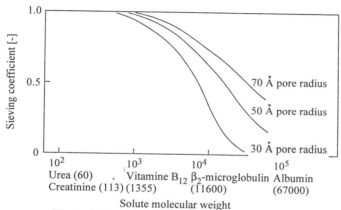

Fig.1   Effect of Pore Size on Solute Transfer

BREAKTHROUGH IDEA

According to the equations, steric hindrance and friction in the pore will change simultaneously as long as pore dimension is constant. A breakthrough idea for the prevention of albumin leakage is to increase steric hindrance independently by partial pore blockage at the entrance, as illustrated in Fig.2 since larger molecular albumin will be hindered more strictly than β_2-microglobulin.

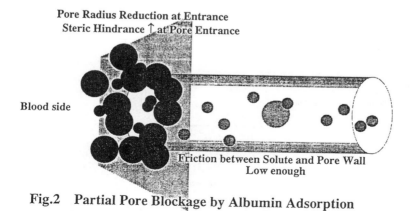

Fig.2   Partial Pore Blockage by Albumin Adsorption

Fig.3 shows how solute sieving coefficient will change with a partial pore blockage - 20, 30, and 35Å pore radius reduction, when 80 Å is the original pore radius. Fig.3 suggests if the partial pore blockage can be controlled, it is possible to
achieve sufficient β_2-microglobulin removal without albumin leakage. For this
purpose, albumin adsorption is positively employed.

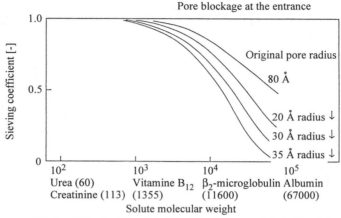

Fig.3   Effect of Partial Pore Blockage on Solute Transfer

MATERIAL SELECTION & SPINNING

Protein adsorption properties of membranes were examined by liquid chromatog-
raphies as shown in Fig.4. Membranes ($0.1m^2$) were incubated 2 hrs at 37°C in
100ml of citrated bovine plasma. Free proteins on the membranes were washed
out by 300ml PBS three times. Adsorbed proteins were separated by 1wt% sur-
factant(Triton X100)/20ml PBS with 2 hr ultrasonic vibration. TOSO column
SW3000 was used with PBS as eluent. Proteins were detected at UV 280 nm,
and concentrations were also analyzed by the Lowry-Folecateau method[6].

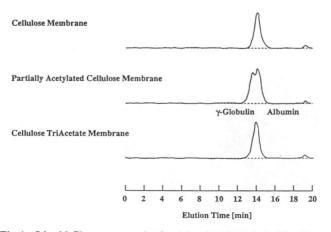

Fig.4   Liquid Chromatography for Adsorbed Protein in Membranes

The kind of adsorbed proteins differs from albumin to globulin, depending on the zeta potentials of the membrane surface, which are influenced by hydroxyl groups and acetyl groups in cellulosic membranes. Cellulose triacetate was selected since only albumin would be adsorbed and hydroxyl groups are replaced with acetyl groups - these are regarded as good blood compatible features. Fig.5 shows the albumin adsorption properties of three types of cellulose triacetate membranes. Draft ratio is the ratio of final winding velocity to initial spinning velocity, which is directly related to the microphase separation property in axial direction, while membrane thickness is related in radial direction.

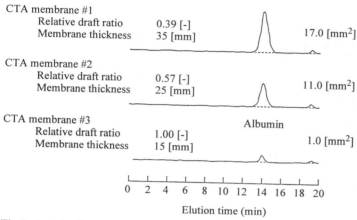

Fig.5   Liquid Chromatography for Adsorbed Protein in CTA membranes

Separation profile of designed CTA membrane is shown in Fig.6, compared with conventional membrane. By selecting material and spinning conditions, CTA membrane can remove β_2-microglobulin sufficiently without albumin leakage.

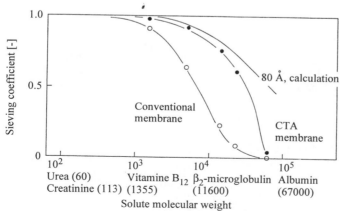

Fig.6   Separation Profile of CTA Hemodialysis Membrane

CRIMPED HOLLOW FIBER CONFIGURATION

The overall mass transfer resistance of hemodialysis module is expressed by three resistances - membrane resistance, boundary layer resistances in blood side and in dialysate side. Membrane resistance is automatically determined and blood side resistance is also constant as far as hollow fiber inner diameter is fixed. However, dialysate side resistance will change module to module, depending on hollow fiber packing conditions. In order to pack hollow fibers uniformly, crimped configuration[7] is applied as shown in Fig.7, which keeps an adequate space between fibers. Therefore, dialysate flows uniformly in a module, resulting in a lower dialysate side resistance. Fig.7 shows the effect of crimped configuration on clearance, compared with straight hollow fiber.

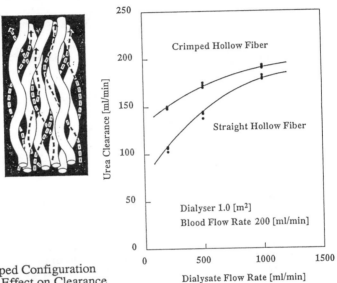

Fig.7   Crimped Configuration
          & its Effect on Clearance

CONCLUSION

Sufficient β_2-microglobulin removal is achieved by increasing the pore size from 30 Å to 80 Å in radius. At the same time, albumin leakage is prevented by albumin thin layer adsorption in initial blood contact, which is indirectly regulated by selecting CTA material and by controlling the spinning conditions.
High dialysis efficiency as a module performance is achieved by crimped hollow fiber configuration.

REFERENCES

1. F. Gejyo, et al. : Jpn. J. Clinical Dialysis 2,61(1986)
2. F. Gejyo, T. Yamada, et al. : Biochem. Biophys. Res. Commu. 129,701(1985)
3. A. Verniory, R. Du Bios, et al. : J. Gen. Physiol. 62,489(1973)
4. K. Sakai, S. Takesawa, et al. : Jpn. J. Chem. Eng. 20,351(1987)
5. E. M. Renkin : J. Gen. Pysiol. 38,225(1954)
6. O. Lowry, et al. : J. Biol. Chem. 193,265(1951)
7. M. Ohno, M. Miyagi, T. Ukai : Membrane (Jpn.) 13,6(1988)

61

Structure and permeation characteristics of cellulosic and related membranes

T. Uragami - Chemical Branch, Faculty of Engineering, Kansai University, Suita, Osaka 564, Japan.

ABSTRACT

Permeation and separation characteristics of aqueous acetic acid (CH_3COOH), dimethyl sulfoxide (DMSO) and ethanol (EtOH) solutions through cellulose triacetate (CTA), chitosan and cellulose nitrate (CN) membranes were studied by evapomeation (EV) and temperature difference controlling evapomeation (TDEV). Permselectivity and separation mechanism for these aqueous organic liquid mixtures through the above membranes by the EV and TDEV method are discussed.

INTRODUCTION

We have already reported that chitosan and its derivative membranes were excellent water permselective membranes for aqueous alcoholic solutions by pervaporation and EV (1),(2). In this paper, based on these results, cellulosic and related membranes such as CTA, CN and chitosan membrane were applied to investigate permeation and separation characteristics for aqueous CH_3COOH, EtOH and DMSO solutions by EV and TDEV, which give higher membrane performance for the permeation and separation of aqueous organic liquid mixtures. Also, permeation and separation mechanisms for aqueous organic liquid mixtures through these cellulosic and related membranes are discussed in detail.

EXPERIMENTAL

CTA (degree of acetylation, 64.1 %), produced by Aldrich Chemical Company, Inc., CN (degree of nitration, 11.1 %), produced by Daicel Chemical Industries, Ltd., and chitosan (degree of deacetylation, 99.1 %), produced by Bioscience Laboratory of Katokichi Co. Ltd., were employed as membrane material. All reagents used in this study were supplied by commercial sources.

CTA membranes were prepared by pouring the casting solution, consisted of CTA (2 g) and methylene chloride (98 g), onto a glass plate and allowing the casting solvent to evaporate at 25 $^{\circ}$C until a dried CTA membrane was obtained. Preparations of chitosan and CN membranes were described in earlier papers (1), (3).

The apparatus for EV (1), (4) and TDEV (5), (6) experiments have been reported in earlier papers. Compositions of the feed mixture and permeate for aqueous EtOH and DMSO solutions were measured by means of gas chromatography (Shimadzu GC-9A). Compositions of the feed, adsorbed in the membrane and permeate for aqueous CH_3COOH solutions were determined by ion chromatography (Shimadzu HIC-6A).

PERMEATION CHARACTERISTICS BY EV

The permeation and separation characteristics for aqueous CH_3COOH solutions through the CTA membrane by EV is shown in Figure 1. These results support the assumption that water molecules predominantly permeate through relatively hydrophobic CTA membranes. In order to reveal these results, the composition adsorbed in the CTA membrane was measured and shown in Figure 2. These results show that the CTA membrane selectively adsorbed CH_3COOH into the membrane from aqueous CH_3COOH solution. In spite of the CH_3COOH in the feed being preferentially sorbed by the CTA membrane, water was predominantly permeated through this membrane from aqueous CH_3COOH solution. This result is interpreted as follows. Namely, the CH_3COOH molecules are preferentially incorporated into the CTA membrane but it is very difficult for the CH_3COOH molecules to diffuse through the CTA membrane, because the affinity between the CH_3COOH molecules and the CTA membrane is very strong and also the molecular size of CH_3COOH is rather large. On the other hand, it is difficult for the water molecules to dissolve into the CTA membrane.

However, once the water molecules are incorporated
into the membrane, they can easily diffuse through
the CTA membrane, because the interaction between the
water molecules and the CTA membrane is weak and also
the molecular size of the water is smaller than that
of CH_3COOH.   Consequently, the water molecules are
predominantly permeated through the relatively
hydrophobic CTA membrane.

The permeation and separation characteristics for
aqueous DMSO solutions through the chitosan membrane
by EV and the DMSO concentration adsorbed into the

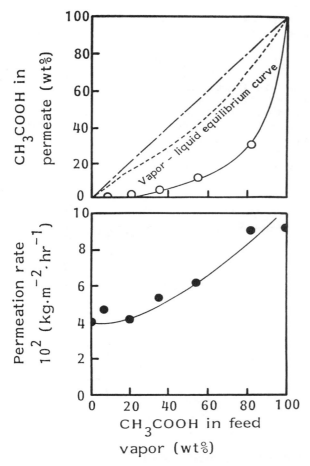

Figure 1.   Effect of the acetic acid concentration in
the feed vapor on the permeation rate and acetic acid
concentration   in the permeate through the   cellulose
triacetate membrane by evapomeation.

chitosan membrane were investigated. From these results, it was found that the chitosan membrane permeated predominantly water from aqueous DMSO solutions and adsorbed preferentially water into the membrane. High permselectivity for water through the chitosan membrane from aqueous DMSO solutions is attributed to both high solubility of water molecules into the chitosan membrane and high diffusivity of water molecules in the chitosan membrane, because the chitosan membrane is highly hydrophilic and the molecular size of water is smaller than that of DMSO.

PERMEATION CHARACTERISTICS BY TDEV

Figure 3 shows the effect of temperature of the membrane surroundings on the permeation rate and the concentration of CH_3COOH in the permeate for an aqueous solution of 10 wt% CH_3COOH through the CTA membrane by TDEV. In this figure, the temperature of the feed solution was kept constant at 40°C. When the temperature of the membrane surroundings was dropped, the permeation rate decreased but the permselectivity for water increased. This decrease of the permeation rate is due to a decrease of motion of the permeating molecules with a reduction of the temperature of the membrane surroundings.

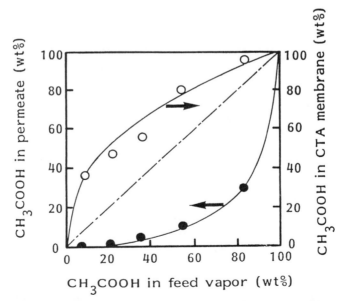

Figure 2. Relation among the acetic acid concentrations in the feed vapor, in the permeate, and in the cellulose triacetate membrane by evapomeation at 40 °C.

The permeation and separation characteristics for the permeation of an aqueous DMSO solution through the chitosan membrane by TDEV were studied by keeping constant at 40 °C the temperature of the feed solution and changing the temperature of the membrane surroundings. Both the permeation rate and the permselectivity for water increased with a reduction of the temperature of the membrane surroundings.

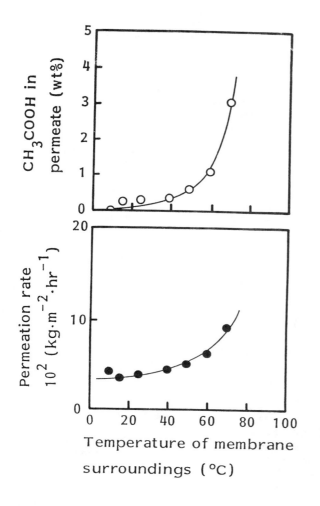

Figure 3. Effect of temperature of the membrane surroundings on the permeation rate and acetic acid concentration in the permeate for an aqueous solution of 10 wt% aectic caid through the cellulose triacetate membrane by TDEV. Feed temperature: 40 °C.

PERMEATION AND SEPARATION MECHANISM IN TDEV

These improvements of permselectivity for water from aqueous solutions of CH_3COOH or DMSO through the CTA or chitosan membrane is explained as follows. When the CH_3COOH, DMSO and water molecules vaporized from the feed mixture are coming close the membrane surroundings, the CH_3COOH or DMSO vapor is condensed much easier than the water vapor and tends to be liquefied as the temperature of the membrane surroundings becomes lower. This condensation of CH_3COOH or DMSO molecules is responsible for the increase of permselectivity for water through the CTA or chitosan membrane. In order to reveal such a tentative mechanism for the separation of aqueous organic liquid mixtures through polymer membranes by TDEV, CN membranes were selected because CN has a stronger affinity for EtOH than water. The CN membrane selectively permeated EtOH from an aqueous EtOH solution by TDEV. Also, the permselectivity for EtOH increased on dropping the temperature of the membrane surroundings. This increase of permselectivity for EtOH can be explained as follows. Namely, when the EtOH and water molecules vaporized from the feed mixture are coming close to the membrane surroundings, the water vapor is condensed much easier than the EtOH vapor and tends to be liquefied as the temperature of the membrane surroundings was dropped. This aggregation of water molecules is responsible for the increase of permselectivity for EtOH through the CN membrane.

ACKNOWLEDGMENT

This work was financed by grants from Izumi Science and Technology Foundation.

REFERENCES

1. Uragami, T.; Saito, M.; Takigawa, K. Makromol. Chem. Rapid Commun. 1988, 9, 361.
2. Uragami, T.; Takigawa, K. Polymer, 1990, 31, 668.
3. Tamura, M.; Uragami,T.; Osumi, Y.; Sugihara, M. Angew. Makromol. Chem. 1981, 95, 35.
4. Uragami, T.; Saito, M. Sepa. Sci. Technol. 1989, 24, 541.
5. Uragami, T.; Morikawa, T. Makromol. Chem. Rapid Commun. 1989, 10, 287.
6. Uragami, T.; Shinomiya, H. Makromol. Chem. 1990, 192, 2293.

62

Cellulose hollow fiber membranes for blood purification

D. Paul, H.-J. Gensrich and G. Malsch - Institute for Polymer Chemistry, KantstraBe 55, O-1530 Teltow, Germany.

ABSTRACT

Porous membranes are the most important part of extracorporeal systems of blood purification to separate toxic from nontoxic substances.
For this purpose the possibilities of the viscose process to prepare cellulose hollow fiber membranes was studied. By this process the morphology and, therefore, the separation properties of membranes can be altered over a wide range by changing different parameters in the spinning process.
In this way it is possible to produce membranes for dialysis and ultrafiltration with pore diameters from 1.8 to 15 nm. By chemical modification of the membrane surface the level of human complement activation may be decreased.

MANUFACTURING OF HOLLOW FIBER MEMBRANES

Porous membranes are the most important part of extracorporeal systems of blood purification to separate toxic from nontoxic substances.
Today the medical-technology industry produces for this purpose many types of dialyzers with hollow fiber membranes of different polymers. The favoured material is cellulose. Two processes are of importance for the production of cellulosic hollow fiber membranes from

linters:
- melt spinning of cellulose diacetate and saponification to cellulose /1/
- dry/wet spinning of a solution of cellulose in cuprammonium and coagulation in sodium hydroxide /2/.

Our aim was to study the potential of the viscose process to prepare cellulose hollow fiber membranes from pulp by wet spinning. Important steps of this process are the dissolution of cellulose xanthate in sodium hydroxide and the extrusion of the polymer solution into different coagulation and precipitation baths. The hollow fiber membranes are formed by using a special nozzle. Such a spinneret has a circular orifice for the polymer spinning solution to form the membrane wall and in the centre an aperture for the lumen filling media (gas or fluid) which is pressed into the formed membrane. Hollow fiber membrane can be spun with fatty acid esters or air in the lumen. The direction of viscose spinning process is from the bottom to the top (Fig. 1).

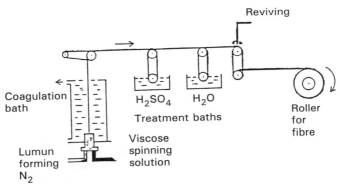

Fig.1 Cellulose viscose hollow fiber spinning system

Passing a precipitation bath, a stretching bath, a system of washing baths and the addition of plasticizer (glycerin) the membrane is wound up after drying. The process includes many parameters which have an effect on the coagulation and, therefore, on the morphology of the membrane /3,4/. At the interface of the polymer solution and coagulation bath, a very thin primary membrane is formed. This skin covers a network structure of tubules and/or connected pore systems. The porous structure situated more centrally in the wall and well visible and detectable by electron microscopy can be understood as a support. The internal surface of the membrane wall at the lumen is also covered by a skin layer.

Generally,
- strong coagulation causes fibers with higher pore volume, mild coagulation has the contrary effect
- stress deforms the pore system and decreases the pore volume
- drying by evaporation of water decreases the pore volume /5/.

In the formation process of hollow fiber membranes for blood purification we have to obtain a morphology suitable for dialysis or ultrafiltration.

SEPARATION PROPERTIES

The variation of the cellulose concentration in the spinning solution and the composition of the coagulation bath did not make evident a systematic influence on the dialytic properties of hollow fiber membranes.
On the contrary for ultrafiltration there exists a correlation between both quantities flux and rejection: Increasing cellulose concentration and decreasing sulfuric acid concentration (mild coagulation) led to high rejection and low flux.
From the data of ultrafiltration, which depend on the sieve mechanism, the rejection has to be calculated in dependence on the molecular weight and corresponding hydrodynamic molecular diameter of solute /6/. A molecule that is too large to fit through the pores of the membrane is absolutely rejected. Plotting in logarithmic systems (Fig. 2) a linear correlation of rejection with the molecular weight is obtained from the obtained straight lines.

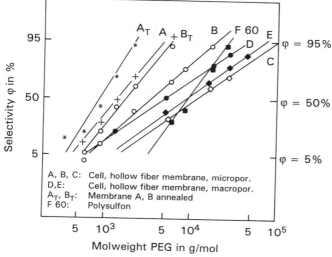

Fig.2   Selectivity of different cellulose membranes

It was proved for hollow fiber membranes manufactured by viscose process the variation of the morphology of the membrane wall with different pore diameters conducts to different separation properties. Investigations by using test permeators permit a comparison permeability of hollow fiber membranes with other membranes under the same technical conditions (Fig. 3). Hereby it will be come clear, that hollow fiber membranes from cellulose according to their morphology are suitable especially for blood purification by dialysis.

Material \ Parameter	Diffusive Permeability $P_v[cm/min]\ 10^3$	Hydraulic Permeability UFR $[ml/h\ m^2\ kPa]$
PMMA	30	10
EVAL	36	10
PSu F 6	41	41
AN 69 S Filtral	38	325
CA CDAK	37	156
Cell Cupro	48	43
Cell Viskose	43	150

Fig.3   Permeability of dialysis membranes

BLOOD COMPATIBILITY

However, during dialysis using membranes made from unsubstituted cellulose, the free hydroxyl groups on the membrane surface are believed to activate the complement system in the blood flowing through the dialyzer /7/. Consequently research in this scientifically and medically stimulating field has led to development of modified cellulose /8,9/ membranes, with the favoured aim of a reduced complement activation potential. In this order, some selected chemical cellulose modification routes have been developed (Fig.4).
The derivatisations are represented by an urethanation (pathway A) and combined two-step reactions of either acrylation/sulfation (pathway B) or chloroacetylation-/sulfonation (pathway C), whereas the cellulose graft copolymerization is characterized by a polyacylation reaction (pathway D) using reactive acyl functionalized copolymers.

INFLUENCE OF CHEMICAL MODIFICATION ON SELECTED PARAMETERS

Type of Modification	Introduced Groups	Contact Angle Air/Water (°)	Interfacial Free Energy (mN/m)	C3a des Arg (% Cell.)	Electrophoretic Mobility (% Cell.)
Cellulose (Cell.)		12,1±0,4		100	100
A Urethanation	Masking OH Time (min)				
	7	15,1		84	97
	10	26,3		83	95
	20	42,6		52	19
B Acrylation/ Sulfation	~CH·CH$_2$	43,1	8,3	83	13
	~CH$_2$-CH$_2$-$\fbox{OSO$_3^\ominus$}$	21,8	0,2	24	7
C Chloroacetylation/ Sulfonation	~CH$_2$Cl	48,9			
	~CH$_2$-$\fbox{SO$_3^\ominus$}$	19,8	0,1	25	12
D Polyacylation	$\fbox{~COO$^\ominus$}$				
P (AACl-CO-AA) I C$_{AA}$: 30 Mol%		49,9	5,8	22	0
P (AACl-CO-AA) II C$_{AA}$: 70 Mol%		36,4	2,3	59	0

Fig.4   Pathways of cellulose membrane surface modification

In case of urethanation (pathway A) the underwater contact angle is raised continuously with increasing reaction time, indicating a decrease of the surface hydrophilicity. This change in polarity is induced by the crosslinking which reduces the number of polar hydroxyl groups onto the surface. From the lowered C3a level it can be established that only by altering the content of hydroxyl groups without functionalization, an improved blood compatibility can be approachable. Using the two-step derivatisations for cellulose modifying the alteration of hydroxyl groups is combined with introducing ionic groups onto the surface by sulfation (B) or sulfonation (C). In both cases the former reactions of acrylation (B) and chloroacetalytion (C) were accompanied by loss of hydrophilicity, which is recovered in the second step of functionalization.
Both pathway B and C are characterized by an effective reduction in the C3a levels of up to 24 and 25 %, respectively.
The polyacylation graft modifications of cellulose by pathway D with a series of poly(acryloyl-chloride-co-acrylic acid) copolymers/P (AACl-co-AA) of defined composition were used for introducing carboxylic functionalities onto the surface.

The best level of C3a (22 % of cellulose) was obtained with P(AACl-co-AA), characterized by a molar copolymer composition of 70 % acryloyl chloride and 30 % acrylic acid.
It can be summarized that in model cellulose modifications in addition to a simple alteration of hydroxyl groups by a cross-linking reaction, the introduction of specific ionic groups is very effective for a further decrease in complement activation. The availability of hydroxyl groups to approaching complement factors is reduced most effectively when their steric hindrance is combined with ionic functionalization.
The wide variability of the separation and biomedical properties of cellulose hollow fiber membranes made by viscose or other processes for blood purification is an essential presupposition for their optimal application in dialysis or filtration. Surely a further improvement is possible and can be reached earlier when the mechanism of membrane formation as well as the interaction between cellulose and blood are understood better.

REFERENCES

1   UK Pat. Appl. GB 2 065 546
2   DE-PS 736 321
3   Klare, H.: Über den Lösungszustand von Viscose und den Bildungsmechanismus von Celluloseregeneratfäden, Lenzinger Berichte (1965), 19, 21 - 29
4   Gröbe, V.; Bartsch, D.; Bossin, E. et al: Hohlfäden nach dem Viskoseverfahren für Dialyse und Ultrafiltration, Acta Polym. (1979), 30, 343-347
5   Bartsch, D.; Müller, K.: Möglichkeiten zur Blutentgiftung mit Chemiefaserstoffen, Formeln, Faserstoffe, Fertigwaren (1982), No. 3, 23 - 25
6   Gensrich, H.-J.; Scholz, C.; Holtz, M. et al: Hohlmembranen nach dem Viskoseverfahren - vergleichende Charakterisierung, Wiss. Beiträge IHS Köthen (1988) H. 4, 88 - 107
7   Chenoweth, D. E., Complement activation during hemodialysis : clinical observations, proposed mechanism and theoretical implications. Artificial Organs, 8 (1984) 281-7.
8   Paul, D., Malsch, G., Bossin, E., Wiese, F., Thomaneck, U., Brown, G.S., Werner, H. & Falkenhagen, D., Chemical modification of cellulosic membranes and their blood compatibility. Artificial Organs, 14 (1990) 122-5.
9   DE-OS 3 814 326

63

Selecting alpha cellulose filter papers for use in dry chemistry test strips

M. Rapkin and R.D. Kremer* - Boehringer Mannheim Corporation. *Schleicher and Schuell, Incorporated.

INTRODUCTION:

The selection of such papers depends on essentially six factors, some or all of which may be of importance to their performance as a test strip media. They are:

1. Purity of fiber prior to processing.
2. Uniformity of sheet formation.
3. Basis weight/sheet caliper relation.
4. Absorptivity.
5. Retention.
6. Additives.

The importance of these factors, with regard to test strip performance, are as follows:

PERFORMANCE FACTORS:

PURITY OF FIBER PRIOR TO PROCESSING

Alpha cellulose is the cellulose fraction of highest quality and purity. It is that fraction of the raw fibrous material which remains after treatment with 17.5% sodium hydroxide, at 20°C, for 45 minutes.

The fiber sources from which this alpha cellulose is obtained are as diverse as wood chips or cotton linters, materials which vary considerably in total cellulose content and non-cellulosic impurities. Cotton linters contain a cellulose content of 85% to 90% and only minor percentages of other materials, such as 3% lignin. On the other hand wood chips, from a typical coniferous wood, such as white spruce, have a cellulose content of only 60% and a lignin content of 28%.

Raw materials of a higher non-cellulosic content require a more vigorous and concentrated chemical pulping and bleaching action to remove these impurities. As such, there is a potential for higher residuals of such impurities and process chemicals in the cellulosic pulps produced from raw materials with high levels of non-cellulosic content. Also, the more vigorous chemical action required for their removal can affect the remaining cellulose itself, imparting a degree of ion exchange activity to the resulting pulp. For these reasons a high purity raw material, such as cotton linters requiring only relatively mild chemical action, is preferred for the fabrication of test strip media.

UNIFORMITY OF SHEET FORMATION

This is determined by the evenness of the fiber distribution in the sheet.

Sheets with good formation exhibit little variation in light transmission when examined over a light box. However, when one observes a cloudy quality in such light transmission, this means that fiber masses and clumps have formed in the pulp, prior to its having been "set" by the dewatering action of table rolls, foils and vacuum suction boxes on the forming section of the paper machine. As these clumps and masses of fiber increase in size, frequency and density the light transmission becomes extremely variable and the sheet is referred to as having a "wild" formation.

With regard to test strip media, the affect of such variation in sheet formation is that it results in variations in the values of almost all the other sheet properties, i.e., absorption, flow, capillary migration/wicking speed, retention, strength, etc.

Determinations of good or bad sheet formation are still a matter of visual observation. As such, a consensus of what is suitable and obtainable sheet formation should be determined by the producer and user, on the basis of actual sheet samples, to be used as comparative standards, to reduce variation in individual visual estimates to a minimum.

Poor formation results from such factors as fiber length (the longer the fiber the greater the tendency to form localized fiber masses or clumps), high pulp stock consistency (the concentration of fibers in a given volume of pulp), rapidity of deposition of the pulp on the wire/fabric web of the formation section and the rapidity of movement and rate of dewatering on the web.

BASIS WEIGHT/SHEET CALIPER RELATION

This provides for a measure of the apparent sheet density or void volume that is available for the imbibition of fluid material into the sheet.

Just as it is necessary to have uniform sheet formation, it is also necessary to have a uniform sheet density/void volume for the consistent uptake of fluid impregnants, reagents, samples, etc. in test strip media.

It is important to consider this not only from the producer/paper makers responsibility to provide such uniformity, but also with regard to the user/converter, who may apply compressive forces in the form of pressure or vacuum to assist in the processing of such media into finished test strips.

Applicable TAPPI test methods are:

- Grammage (basis weight) of Paper and Paper Board (weight per unit area) – T 410 OM-88.
- Thickness (caliper) of Paper, Paper Board, and Combined Board – T411 OM-84

ABSORBENCY

This is the measure of the amount of time required for the media to take up a specified volume of liquid, the area wetted in a specified time, or the rate of migration or capillary rise of a liquid, in a vertically suspended strip of the media, in contact with the liquid at its base.

These are critical processing and consumer time factors for the user/converter of test strip media. The amount of application time required for a particular reagent to be fully absorbed by a test strip media has definite process economic considerations; as do the area wetted and the amount absorbed.

Such considerations also apply to the speed of migration of liquids in the media particularly if chromatographic applications are of importance. Of particular importance is the rapidity of sample or reagent migration from one area of application to another, in terms of test time to the end user of the converted finished product. In this latter context it is important to remember that fluid flow is faster in the machine direction of a paper than in the cross machine direction, as the fibers tend to orient themselves longitudinally in the direction of wire movement, i.e. in the machine direction. For this same reason spot/drop application of sample or reagent will tend to form elliptical areas of application, a tendency which can be reduced or eliminated by application of vacuum, on the reverse side of the media, at the time of application.

Differences in the absorption characteristics of different fluids, i.e. aqueous reagents, solutions containing organic polymers, saliva, blood serum and whole blood not only differ considerably from one another, but also exhibit an appreciable range of difference within their own class. For example, the absorptive characteristics of whole blood from newborns, containing a hematocrit (relative volume occupied by cells and plasma in blood) of 55% differs significantly from that of adult blood with a hematocrit of 48%.

Application TAPPI test methods for absorptive behavior and wetting are:

- Water Absorbency of Bibulous Papers – T432 OM-87
- Machine Direction of Paper and Paperboard – T409 OM-88
- Surface Wettability of Paper (Angle of Contact Method) – T 458 OM-84.

RETENTION

As absorption is related to available void space, so retention is related to the subdivision of that space into units of appropriate size, commensurate with what one desires to retain, along with the desired rate of flow or migration through the media.

There is a trade off here. With increasing segmentation of the void space into smaller units to increase retention, slower rates of fluid flow and migration through the media are also encountered.

To increase the retention of a cellulosic filter paper, two courses of action can be taken. These are:

   a. Increase the degree of refining of the stock. In a refiner, the cellulosic fiber is subjected to two types of processing action as it passes between the narrow clearance of the stationary and rotating metal bars of the refiner.

       The first of these processing actions is the brushing and rupturing of the outer cellular wall of the fiber which in turn releases and disperses many very small diameter microfibers or fibrils contained within the cell. This action produces many more points of fiber to fiber contact, resulting in increased paper strength and subdivision of the void space into smaller and smaller units of volume.

       The second processing action is that of the cutting of the parent fiber and fibrils into smaller and smaller segments, which tends to reduce fiber strength but also results in smaller units of void volume by acting as a form of cellulosic filler, filling up void space. During sheet formation on the paper machine where water is removed by the mechanical action of drainage through the machine wire, many or most of these smaller cellulose fragments or "fines" are removed from the wire side of the formed sheet. As such, the wire side will tend to have a more open texture than the opposite felt side and may also exhibit a different affinity for reagents, such as indicator dyes, resulting in a difference in color intensity, shade and hue on the two sides of the sheet.

   b. The second course of action which may be taken is that of compression or calendering, to reduce the thickness/caliper of the sheet and thus the total void volume, compressing each subdivision of void volume into a smaller unit of space. This reduces the absorptive capacity of the sheet but retains its strength, which may be lost through further refining action.

       Retention is important to the user/converter who may be applying fluid suspensions of finely divided solids such as resins or reagents and needs to confine them to a small area of application, or concentrate them near the surface of the media on which they are applied. These concentrated, thin surface layers of particulate application can in turn provide for an enhanced intensity of visualization of test results.

       Should the test involve the precipitation of a fine particulate reaction product, that product can be retained at the site of sample application, while the bulk of the applied fluid sample, and potentially obscuring soluble secondary reaction products, are wicked away into areas of the sheet media adjacent to the reaction site.

       Applicable testing method for retention of fine precipitates, may be found in TAPPI Useful Test Methods (UM), i.e.

Testing Analytical Filter Papers, Useful Method 572.

Interaction of Paper Properties on Absorbency Performance.

Changes in these paper chacteristics:			result in	these changes in filtration/absorbency characteristics:		
Basis Weight	Caliper	Extent of Defilibration[1]		Filtration Speed/Air Permeability	Retention	Absorbency
Unchanged	Increases	Unchanged	⇒	Increases	Decreases	Increases
Unchanged	Decreases	Unchanged	⇒	Decreases	Increases	Decreases
Increases	Unchanged	Unchanged	⇒	Decreases	Increases	Decreases
Increases	Increases	Unchanged	⇒	Decreases[2]	Increases	Increases
Unchanged	Unchanged	Increases	⇒	Decreases	Increases	Unchanged
Increases	Decreases	Increases	⇒	Decreases to Minimum Extent	Increases to Maximum Extent	Decreases

[1]Paper fibers are composed of cells which contain a mass of much finer paper fibrils enclosed within the cell walls. As the paper fibers are defilibrated, the cell walls are ruptured, releasing this mass of finer aper fibrils. This effect produces a much finer sieving medium, increasing retention, and decreasing filtration rate/air permeability, without any change in paper density.

[2]Paper is considered to be a depth filter medium. Therefore, when both basis weight and thickness are increased, more is gained in retention than is lost in filtration speed.

ADDITIVES

With regard to high quality alpha cellulose filter papers, used for test strip media, there is generally only one type of additive that may be added to the sheet in the paper making process, and that would be a wet strength agent. Other additives may be used once compatibility with the selected reagent system has been established.

A wet strength agent may be required by the user/converter to maintain the sheet's structural integrity during processing stages when the sheet is wetted, particularly when it is being processed as a continuous strip moving under tension.

The most common type of wet strength additive currently employed is melamine/formaldehyde resin. To obtain sufficient wet strength with this resin fairly significant levels of addition, i.e. 1.0% to 1.5% of resin based on dry fiber weight must be employed. In addition, the resin continues to exhibit a very significant level of crosslinking/reactive activity over a period of several months, an activity which could interfere with reagents and reactions of interest to the user/converter of the paper into test strips.

Recently at S&S comparative trials were made with some of the polyamide resins versus melamine/formaldehyde. An example of the wet strength development of a polyamide resin treated sheet, at an addition level of 0.5%, is compared with that of three melamine/formaldehyde sheets at addition levels of 0.5%, 1.0% and 1.5%. Wet strength levels were determined in terms of wet mullen bursting strength, and the tests were conducted over a 32 week period, with time intervals following sheet production of 1, 2, 4, 8, 16 and 32 weeks.

As can be seen from Fig. 1, the polyamide resin, at a loading of 0.5%, develops a wet mullen burst strength equivalent to that eventually attained by the melamine/formaldehyde at a loading of 1.0%. Furthermore, it attains this level in a much shorter time, about 8 weeks as compared to the 32 weeks for the melamine/formaldehyde, and levels off exhibiting little if and further strength

Fig. 1

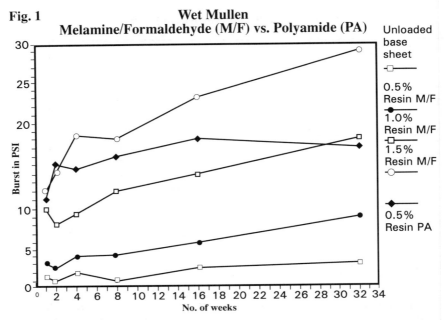

Wet Mullen
Melamine/Formaldehyde (M/F) vs. Polyamide (PA)

Unloaded
base
sheet
-□-

0.5%
Resin M/F
-●-
1.0%
Resin M/F
-□-
1.5%
Resin M/F
-○-

0.5%
Resin PA
-◆-

Burst in PSI

No. of weeks

addition. This means that for test strip media that may require additional levels of wet strength, a polyamide resin could supply this at a lower level of addition and exhibit less chemical activity in the sheet, at a reasonable time interval, following production.

CONCLUSION:

Attention to each of these individual factors and the setting of realistic target and performance range values, by mutual interaction between the producer of the paper and the converter/ value added processor of the final product, are what is required to produce that most vital of factors, i.e. *consistency of performance.*

Too often papers are selected on the basis of economic considerations alone, usually from among the relatively inexpensive general filtration, blotting or wicking grades. Here performance specifications are relatively few and performance ranges fairly wide. Assumptions may be made by the converter/value added processor, based on their own independently derived performance data, quite possibly on a single production lot of the paper, that such performance criteria are of importance to the producer of the paper and are controlled by him. This is not always true, nor in the case of the general filtration, blotting, or wicking grades is it a valid assumption.

When selecting a paper for test strip use, concentrate on its consistency of performance characteristics and communicate your needs to the paper supplier, to find if these characteristics are of equal importance to him and at what target levels and performance ranges the product is being produced. With the producer, make a mutual and realistic assessment of what can be achieved and the criteria by which it may be validated. Then consider economics, for the worst economic choice is the product which will not perform according to expectations.

64

A bio-degradable material composed of polysaccharides

J. Hosokawa, M. Nishiyama, K. Yoshihara and T. Kubo - Government Industrial Research Institute Shikoku, 2-3-3 Hananomiya-cho, Takamatsu 761, Japan.

ABSTRACT

A composite material formation from chitosan and fine cellulosic aq. suspension was successfully completed by drying, and resulted in the formation of non-thermoplastic moldings including films having good wet strength, gas shielding property and bio-degradability. The oxygen gas shielding property of the composite film deteriorated under an atmosphere with high humidity. The combination of the two materials was essential to form the composite moldings which are insoluble in water. The amino groups of chitosan and carbonyl groups of cellulosic material played an important role in the formation of the composite moldings. The composite film had a maximum wet strength (600 kg/cm^2) at 10-20% chitosan on cellulosic material, and was softened by the addition of glycerol (75% for cellulosic material). The bio-degradability of the moldings could be controlled by adjusting the conditions of molding formation, e.g., the temperature at formation stage. Adjusting the carbonyl group content in cellulosic materials was also effective in controlling bio-degradability of the moldings.

INTRODUCTION

The authors have been interested in effective utilization of cellulosic materials without any chemical modification or dissolving, because chemical treatments raise the cost of cellulosic material.

Fine cellulosic fiber or powder can be produced from bleached pulp by use of various pulverizing machines, or occasionally obtained as waste pulp sludge from some special

paper mills.   On the other hand, chitosan is   one of a few
natural cationic polysaccharides which can be derived from
crustaceans or various fungi.   In the case of chitosan sole
film formation, the involvement of an alkali treatment is
required in order to prepare a film which is insoluble in
water.   Sheet film from fine cellulosic powder or fiber
shows no water resistance.

The authors have found recently that a combination of
chitosan and fine cellulosic powder or fiber forms useful
moldings (film, sponge and nonwoven fabrics) and results in
the formation of various kinds of strong and water-resistant
composite films by only cast-drying the material without any
complicated treatment like alkali treatment [1,2].   The
composite moldings can be expected to be bio-degradable
and, further, there is   a possibility of controlling the
period of   degradation.   Therefore, the preparation
conditions of this composite moldings and its properties
including bio-degradation are discussed in this report.

EXPERIMENTAL

Materials. Commercial-grade chitosans made from prawn
shell and   crab shell were used to prepare composite
moldings.   The degree of deacetylation was 99.8% for
chitosan from prawn, and ca.85% for that from crab.   Fine
cellulosic fiber was obtained   as MICRO FIBRIL CELLULOSE:
MFC-100 from Daicel Chemical Industries Ltd., Japan.   Fine
cellulosic powders were obtained as bleached pulp sludges or
pulp powders from pulp mills, and some parts of them were
refined in   our   laboratory.   The cellulosic materials,
therefore, contain hemicellulose and trace amounts of
lignin.

Formation of moldings.
The film formation was
conducted in our lab.
according to **Scheme 1.**
In this procedure, the
cellulosic materials are
not dissolved in the
solution, but exist as
aq. suspensions. The
sponge formation was
conducted in our
collaborating company:
Nishikawa Rubber Co. LTD,
Japan, and the method is
now under a patent
application.

Bio-degradability.
The bio-degradation was
checked in two ways. One
was by using a chitosan
degrading bacterium;
another was by using
cellulase. A bacterium
which is degrades
chitosan and is
ubiquitous in many
regions in Japan was
isolated from soil, and

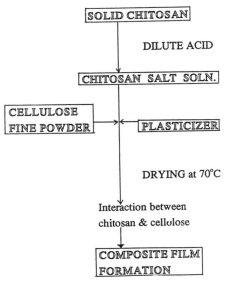

SOLID CHITOSAN

DILUTE ACID

CHITOSAN SALT SOLN.

CELLULOSE
FINE POWDER   PLASTICIZER

DRYING at 70°C

Interaction between
chitosan & cellulose

COMPOSITE FILM
FORMATION

Scheme 1 Flow of Composite Film Formation.

used as an indicator of bio-degradation of the composite film. A cellulase, Meicelase (Meiji Seika Kaisya, LTD. Tokyo, Japan), was also used at 0.08% in 0.1M acetic acid buffer pH 5.0 as another indicator. The culture medium for the bacterium comprised 0.2% potassium phosphate, 0.2% chitosan-hydrochloride solution, 0.05% magnesium sulfate, and trace amounts of other minerals. The pH of the culture medium was 6.0-6.2. Three test pieces of composite film (7mm x 7mm) and 6-8mesh glass beads (0.3g) were added into test tubes together with 5ml of the medium, and then inoculated with the bacterium after sterilization. The test tubes were shaken for three weeks in a reciprocal shaking incubator (300rpm, 2cm reciprocal distance) at 28°C. The bio-degradability of the composite film was expressed as the period within which the test pieces were degraded to fine fragments by shaking with the bacterium.

RESULT AND DISCUSSION

Effect of chitosan content on film strength. Composite films were prepared in the range 0 - 40 w/w% chitosan on MFC-100. The dry strength of a composite film increases with the chitosan content. For a chitosan content of 10% or more, the film is sufficiently strong (ca. 1000 kg/cm^2). On the other hand, the wet-strength has shown a maximum point (600 kg/cm^2) for a chitosan content of 10-20%. A film composed of MFC-100 alone has no wet tensile strength, suggesting that chitosan plays a role in binding cellulose, as reported for pulp sheet [3].

A film composed of chitosan alone has been conventionally produced by drying the aqueous solution of chitosan and acetic acid salt, followed by removing acetic acid from the film with an alkali-treatment [4]. In the case without an alkali-treatment, the chitosan film dissolves easily in water. The reduction of the wet tensile strength of a composite film above 20% chitosan content can be explained by the fact that chitosan without an alkali-treatment tends to be dissolved in water.

Cellulosic materials from bleached pulp generally contains trace amounts of carboxyl and carbonyl groups. MFC-100 contained 25.8 mmol/kg of carboxyl and 33.3 mmol/kg of carbonyl groups. Such functional groups would bond with chitosan and, thus, a composite film insoluble in water is produced, as mentioned later.

The composite film is strong with respect to tensile strength, but slightly rigid. Therefore, glycerol as a softening agent was added at 0, 50, 75, and 100% on the cellulosic material with the aim of increasing the flexibility of the composite film. Though the composite films containing 100% glycerol were too soft and weak, composite films of 75% glycerol had relevant flexibility as wrapping film. The tensile strength of a composite film containing 75% glycerol is lower than those without glycerol, but almost the same as those of general plastic films. The tensile strength per cross sectional area of test pieces decreases with an increase in the glycerol content. Though the elongation of the composite film increased with glycerol content, the value of elongation, 10-20%, was not high in comparison with general plastic films. In both cases of 0% and 75% glycerol, the maximum of the wet tensile strength occurred at 10-30% chitosan content on cellulose.

Composite films were also prepared from cellulosic powders. The appearance of the films was semi-transparent and similar to that from MFC-100, but their strengths were below 80% of that. The strength of the composite films seemed to correspond to the aspect ratio of fiber of the cellulosic materials used. The difference between chitosans from prawn shell and crab shell on the strength of composite film was not so discernible.

Other properties of the composite film. From the data mentioned above, the composite film was found to be formed attractively under the following conditions: 20% chitosan, 75% glycerol on cellulosic material, and a drying temperature of 70°C. Therefore, the various properties of the composite film formed under these conditions are discussed.

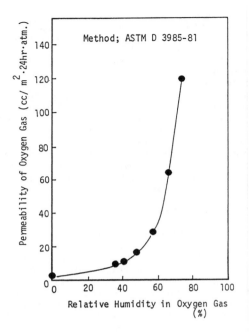

Fig.1 Relationship between oxygen gas permeability of composite film and humidity of the oxygen gas.

The film (80 micrometers thick) showed a low oxygen gas permeability of 2-8 ml/m^2 *24hrs*atm (ASTM 1434-66T). This value is much lower than that of polyethylene, but comparable to that of nylon and PET film which are known as oxygen-shielding plastic films. However, the oxygen gas permeability of the composite film increased with increasing humidity in oxygen gas as shown in **Figure 1**. At 60% humidity its permeability attained to 30ml/m^2*24hrs*atm. The phenomenon presumably relates to the fact that the film showed a high water-vapor transmission rate of 6500 g /m^2*24hrs according to the Japanese Industrial Standard Z-0208.

Effect of heat treatment on film properties. The effect of the temperature in heat treatment is discussed regarding the strength and water-absorption of the composite film. The tensile strength did not increase with a rise in temperature of the heat treatment. Films formed at high temperatures absorb less water than those formed at low temperatures. This fact suggests that the bonding points in the film increase with increasing temperature, and that the film obtained has difficulty in swelling.

Effect of carbonyl and carboxyl groups of cellulosic material on film properties. The effects of carbonyl and carboxyl groups of cellulosic fiber on the water-absorption of the composite films are shown in the upper part of **Figures 2**.

It is well known that carbonyl groups form Schiff-base compounds with amino groups of chitosan, and carboxyl groups form ionic bonds with the amino groups of chitosan. The authors therefore had expected that an increase in the number of carbonyl or carboxyl groups in cellulosic material would result in an increase in the number of bridging bonds in composite films. The degree of swelling of the composite film decreased with an increase in carbonyl groups, not carboxyl groups as shown in **Figure 2**, suggesting that only carbonyl groups contribute to the bridging bonds between chitosan and cellulosic material.

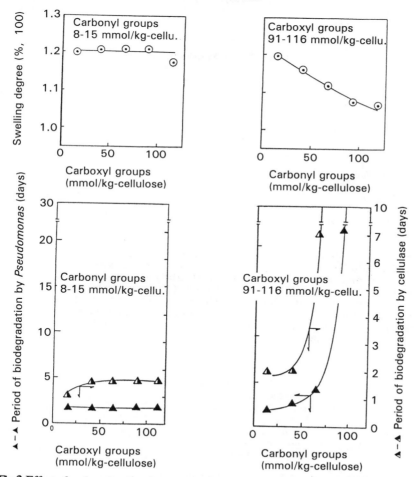

Fig.2 Effect of carboxyl and carbonyl groups of cellulosic material on swelling degree and bio-degradability of composite film obtained.

Bio-degradation of composite films. The increase in temperature of heat treatment prolonged the degradation periods by the attack of bacterium and cellulase on the composite film[1]. This suggests that the enhancement of temperature in the heat treatment increased the number of bridging points of the composite film.

The bio-degradability of various composite films made from cellulosic materials having carbonyl or carboxyl group is shown in the lower part of **Figures 2**. The bio-degradability of the composite film was little influenced by the carboxyl group content in cellulosic material. On the other hand, carbonyl group affected the bio-degradability, and the composite film made from carbonyl-rich cellulosic material can prolong the bio-degradation period. This phenomenon corresponds to the situation of swelling degree of the composite films. This fact shows that bridging points initiated from carbonyl groups of cellulosic material would prevent bio-degradation.

TABLE I   PHYSICAL PROPERTIES OF MISASA ECOFOAM

	Open Cell	Closed Cell	Film
Apparent Density (g/cm3)	0.02-0.05	0.1-0.3	1.3
Tensile Strength (kgf/cm2)	0.5-10	5-20	700-1000
Elongation (%)	15-40	10-20	ca.10
Hardness (Degree, Asker F)	5-70	60-80	--
Water Absorption (%)	1000-4000	50-200	--
Period of Bio-degradation	1/3	1/2	1

Properties of sponge moldings. Various types of sponge named "MISASA ECOFOAM" were prepared from chitosan, fine cellulosic fiber and glycerol in our collaborating company. The physical properties of their samples [5] were shown in **Table 1**. The strengths of the sponges were not so high in comparison with that of polystyrene. These sponge samples are, therefore, now being tested for medical application like wound dressing materials etc.

REFERENCES

1. Hosokawa,J.; Nishiyama,M.; Yoshihara,K.; Kubo,T.
    Ind. Eng. Chem. Res.1990, 29, 800-805.
2. Hosokawa,J.;   Nishiyama,M.;   Yoshihara,K.;   Kubo,T.;
    Terabe,A. Ind. Eng. Chem. Res.1991, 30, 788-792.
3. Allan,G.G.; Fox,J.R. et al. Trans. BPBIF Symp. Fiber-
    Water Interact. Papermaking (Oxford),1977, 2, 765.
4. Miya,M.; Iwamoto,R. et al. Kobunsi Ronbunsyu 1985,42,
    139.
5. See, Sample Pamphlet of Nishikawa Rubber Co. LTD.

65

Reconciliation of conformational analysis and applied results in the separation of phenol formaldehyde oligomers on native cellulose

A. Pizzi and G. De Sousa - Department of Chemistry, University of the Witwatersrand, Johannesburg, South Africa.

ABSTRACT

The minimum energy interactions of the conformations of the three isomers of dihydroxydiphenylmethane averaged each over a number of 24 sites on an elementary model of crystalline Cellulose II, composed of 20 anhydroglucose residues distributed over five chains, were calculated by a semiempirical and a force field molecular mechanics methods. The percentage differences in the averaged minimum energies were found to correspond to the percentage differences in paper chromatography Rf's of the three dihydroxydiphenylmethane isomers. Of interest was the apparent correspondence of the two sets of results independently of the polarity of the solvent employed, indicating the example chosen might be one of the simpler cases possible. Although the number of sites used must be considered a physical approximation, it appeared to already give sufficient correspondence between numerical and experimental results. The indications from the obtained correspondence of results appears to be that separation of non—enantiomeric isomers on achiral cellulose substrates could also be modelled by computational methods.

INTRODUCTION

Recent investigations by Alvira et al [1—3] on the resolution of enantiomers from racemic mixtures of alanine on an amorphous cellulose substrate have shown that separation processes on cellulose can be modelled by calculation. It would be of interest if this approach could be taken for the separation of non—enantiomeric isomers on an achiral cellulose substrate. Dihydroxydiphenylmethane is the first oligomer formed during the acid—catalyzed reaction of phenol with formaldehyde [4]. Three isomers of dihydroxydiphenylmethane are known, namely the *ortho—ortho*, *ortho—para* and *para—para* [4]. Their interaction with cellulose is of interest in modelling mechanisms of specific adhesion [5] of phenol— formaldehyde oligomers on cellulose substrates. The adsorption/specific adhesion theory proposes that an adherend and a substrate will adhere because of the interatomic and intermolecular forces which are established between the atoms and molecules in the surface of the adherend and of the substrate [5]. For phenol—formaldehyde oligomers on cellulose substrates such forces can be limited to van der Waals, H—bond and electrostatic interactions; and ionic, covalent, metallic and donor—acceptor bonds can be disregarded [5].

Chemically processed cellulose appears as a mixture of crystalline Cellulose II and amorphous cellulose [6]. The percentage of crystalline Cellulose II in this mixture is high, often over 90% [6]. In crystalline cellulose the monomers are positioned on an achiral lattice, the parameters of which depend on the cellulose species. In crystalline Cellulose II they form linear alternating sheet—antiparallel chains.

The aim of the present paper is to report calculations on the relative separation achievable on crystalline Cellulose II of the three isomer forms of dihydroxydiphenylmethane and to compare this with experimental paper chromatography results obtained for the same system.

COMPUTATIONAL METHODS

Two computational programmes have been used for the calculation of the secondary forces interactions between the three isomers of dihydroxy—diphenylmethane and crystalline cellulose. Firstly a non—force field programme necessary to refine at first the complex systems involved, was used. This was originally developed as programme SZEN01 and 02 at the University of Rome, Italy, by the Liquori research group on conformational analysis of proteins and polypeptides. It can accommodate 30 bond rotational angles (30 $\Phi°$s and 30 $\Psi°$s) maximum. It has been extensively modified for the calculation of conformational energies, allowed conformations, and atom coordinates of polymeric carbohydrates. It has now several more capabilities beyond the original programme. Its algorithms have already been extensively reported [7,8].

Secondly, the use of the above computational system was supplemented and checked by the use of a known force—field programme, MM2 [9]. We preferred to use this force field as a consequence of the favourable comparative study of Gundertofte *et al.* [10].

MODEL

Three phenyl—formaldehyde (PF) dimers coupled *ortho—ortho, para—para and ortho—para*, hence all the possible dihydroxydiphenylmethane isomers obtainable by the reaction of phenol with formaldehyde [4,11], the tridimensional structure and conformations of minimum total energy have of which have already been reported [12] were used for this study. The structure of a five chains, four glucoses residues each, of a schematic elementary crystallite, already reported [13], was used as the substrate.

The number of degrees of freedom for such calculations is considerable and the following technique was devised to facilitate computation. Eight positions on the surface of the cellulose crystallite were taken. These are shown in Fig. 1. Positions A, B, C and D, correspond to the geometrical central points of each of the four surfaces of the crystal. Positions AB, BC, CD and AD correspond to the central points of each long edge of the crystallite. The detailed model has already been reported [8].

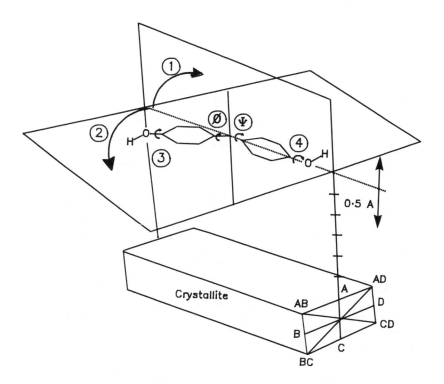

Figure. 1   Schematic representation of rotational angles. Rotation planes and direction of approach of a dihydroxydiphenyl— methane isomer to cellulose crystallite (position A). Note the position of the other sites (B,C,D,AB,BC,CD,AD) indicated on the end of the schematic crystallite.

The procedure was repeated using the computational programme BONDS (constrained force—field) [7,8] and MM2 [19]. The average results obtained are shown in Table 1.

Table 1 Energy values of minimum total energy for the interaction of *para–para*, *ortho–para* and *ortho–ortho* phenol–formaldehyde dimers with cellulose crystallite surfaces.

	VdW	H–bond	Electrostatic	Total Energy	E_{tot}Cellul. isomer interaction	E_{tot} within isomer
			(kcal/mol)			
Para–para						
TOTAL AVERAGE	46.4	–4.4	0.7	42.857	–10.203	53.060
Ortho–para						
TOTAL AVERAGE	46.1	–5.8	0.7	41.017	–12.085	53.102
Ortho–ortho						
TOTAL AVERAGE	47.2	–8.1	0.8	39.904	–12.329	52.233

Original Atomic Coordinates
The initial atomic coordinates of crystalline Cellulose II were derived from the refined x–ray crystal structure of Kolpak, Weih and Blackwell [14] refined by Pizzi and Eaton [13]. The valence bond angles at the glycosidic oxygen atom between the two glucopyranose rings was 115.8°, from the x–ray work of Ham and Williams [15]. The coordinates for the three dihydroxydiphenylmethanes were derived from the x–ray structure of Whittaker [16] refined by computational methods by Smit *et al* [12]. In the initial approach with a non–force–field computational method the atomic coordinates of the five chains, each of 4 anhydroglucose residues, as elementary model of crystalline cellulose II were fixed. In the dihydroxydiphenylmethanes the systematic rotations about bonds to the methylene bridge were first carried out with increments of rotation of 20° followed by refinement of $\Phi°$ and $\Psi°$ in 2° increments. The angles of rotation $\Phi°$ and $\Psi°$ were the dihedral angles between C(4)–C(5′) and C(7)–C(4) and between C(4)–C(5) and C(7)–C(4′) respectively. The initial conformation $(\Phi°, \Psi°) = (0,0)$ was defined as the conformation of minimum energy of the dihydroxydiphenylmethanes alone, defined by Smit *et al.* [18]. The rotation is considered to be positive when viewing along the C(4′)–C(7) [or C(4)–C(7)] towards the methylene bridge carbon atom the rotation is performed anticlockwise.

Preparation of phenol–formaldehyde oligomers
The phenol–formaldehyde condensation from which the three dihydroxydiphenylmethane isomers were isolated was as follows: to 94.0g phenol were added 55.0g of 97% para– formaldehyde powder, 40g of a 80/20 W/W water/methanol at 40°C for 10 minutes. 5ml 33% NaOH solution were then added followed 10 minutes later by another 5ml of 33%

NaOH solution. The mixture was maintained at 40°C for another 10 minutes and then the temperature increased up to 94°C over a period of 30 minutes. Two other amounts of 5ml each of 33% NaOH were added during the period of temperature increase. At 94°C the reaction refluxed and the reflux was maintained for another 30 minutes. The resin was then cooled in an ice bath and stored.

Preparative and thin layer chromatography

A 9.6% phenol—formaldehyde solution in benzene was applied to the 20 × 20 cm silica gel plate (PLC 2.0mm thick and TLC 0.5mm thick plates were used) in a narrow strip using a 0.50ml syringe. 0.30ml of solution was applied each time. The solvent was allowed to evaporate, and the plate was developed in a 10 × 30 × 25 cm chromatography tank in benzene:THF (9:1 v/v). The separated bands were located by UV light and were marked off. Four bands were then scraped off the plate separately into a vial. Each band was extracted separately using AR acetone. The silica gel was then filtered off from the extract, and the acetone was evaporated under reduced pressure at 48°C to give the product. Each product was weighed. Each product was run on 2 × 6cm TLC silica gel plates in benzene:THF (9:1 v/v) to verify purity.

Identification of Compounds by C^{13} NMR and H^1 NMR spectra

50—100mg of each of the separated products was required in for identification. Products were dissolved in acetone—d_6. The spectra were recorded on a Bruker AC 200. The same two solutions of products 1 and 2 used for C–13 NMR were used to record proton NMR spectra on a Varian EM–360A (60MHz) spectrometer (Fig.2).

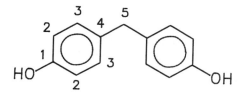

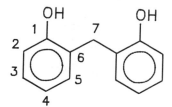 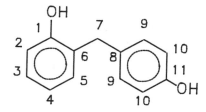

Figure 2   The three isomers of dihydroxydiphenylmethane with the numbering of their carbon atoms alloted to C^{13} NMR shifts.

Paper Chromatography

10% solutions of the separated products were made up in benzene. These solutions were used to spot 60 × 150mm strips of chromatography paper and run them in combinations of two solvents (in varying proportions). The following solvents were used: acetone, benzene, ethyl acetate, and methanol.

NUMERICAL AND EXPERIMENTAL RESULTS

Correlation of computational and experimental results

The calculated energies of interaction between each of the three dihydroxydiphenylmethane isomers and an elementary crystal of cellulose II, indicate that the average minimum total energy of the interactions calculated were, for:

para–para isomer	−10.203 kcal/mol
ortho–para isomer	−12.085 kcal/mol
ortho–ortho isomer	−12.329 kcal/mol

This corresponds to a 18.5% difference in energy between the *para–para* and *ortho–para* dimers, 20.4% between *para–para* and *ortho–ortho*, and 2.2% between *ortho–para* and *ortho–ortho*.

Table 1 shows the energy values and conformations of minimum total energy of the interactions between each phenol–formaldehyde dimer isomer and a cellulose II crystallite.

Product 1 had already been identified as the *para–para* PF dimer, and product 2 as a mixture of the *ortho–para* and *ortho–ortho* PF dimers, when paper chromatography was performed.

Paper chromatography of product 1 resulted in a Rf value of 0.65, and product 2 gave Rf 0.55 (Table 2). This corresponds to a 18.2% difference in the energies of interaction between products 1 and 2 with the cellulose paper; ie. between the *para–para* isomer, and the average of the *ortho–para* and *ortho–ortho* isomers. This value compares favourably with the values of the minimum conformations averages obtained by computational methods. The computed expected difference of 2% between the Rf of *ortho–para* and *ortho–ortho* isomers cannot be detected by paper chromatography, and the two compounds overlap although the band shape (oblong double spot) indicate the presence of two compounds of very similar Rf's, but not equal (thus with a Rf difference of 1% to 3%).

Table 2   Separated Products 1 and 2 run on chromatography (0.16mm thick)

Band	Rf Values	Average Rf
1	0.62 0.67	0.65
2	0.53 0.56	0.55
phenol	–	0.78

Numerical results

The calculated numerical results are shown in Table 1. As paper chromatography at molecular level is a statistical effect the average of all the energy results at each Cellulose II crystallite site for each of the isomers of dihydroxydiphenylmethane had to be taken into consideration. The number of sites combinations chosen must still be considered as a physical approximation necessitated by the number of permutations involved. If a considerably greater number of sites of the elementary Cellulose II crystallite for each isomer could have been considered, more accurate relative proportions of their average interaction energies could have been obtained. In theory only an infinite number of sites could have given a truly accurate result. The calculated percentage differences between the average minimum energies of the *para—para* and *ortho—para* isomers was of 18.45% and between the *ortho—para* and *ortho—ortho* isomers of 2.02%. Their correspondence with the experimental results appeared to be acceptable, indicating that the number of site permutations chosen appeared to be already enough to give at least an approximate result.

Of interest was that that, experimentally, variations of solvent mix polarity did not change the relative position of the two bands Rf's on the paper chromatograms. In the numerical calculations, in which the dielectric constant of water was used, a correspondence to this can be found in the lack of sensitivity of the electrostatic component in Tables 1,2 and 3 to a change of dielectric constant (for different solvents).

CONCLUSIONS

The separation by paper chromatography of the three isomers of dihydroxydiphenylmethane identified by C^{13} and H^1 NMR were grouped into two chromatographic bands, the first of which corresponded to the pure *para—para* isomers and the second of which to a 2/3:1/3 mixture of the *ortho—para* and *ortho—ortho* isomers respectively. Paper chromatography separation of the three compounds gave Rfs separation of 18.2% between the two chromatographic bands whatever the solvent mixture used. The second band was formed by two bands partially overlapping indicating a Rf difference of between 1% and 3%. The calculated minimum conformational energies of dihydroxydiphenylmethane isomers, when interpreted as average minimum total energy over the Cellulose II elementary crystallite, predicted a separation of 18.48% between the first two isomers and of 2.02% between the second and third isomer on paper chromatogram appear to indicate correspondence between calculated and experimental results.

REFERENCES

[1]　　E.Alvira, I.Vega and C.Girardet, *Chem. Phys.* 118, 223 (1987).
[2]　　E.Alvira, V.Delgado, J.Plata and C.Girardet, *Chem. Phys.* 143, 395 (1990).
[3]　　E.Alvira, J.Breton, J.Plata and C.Girardet, *Chem. Phys.*, in press (1991)

[4]   N.J.L.Megson, Phenolic Resin Chemistry (Butterworth, London, 1958) p 1—19.
[5]   A.J.Kinloch, Adhesion and Adhesives (Chapman and Hall, London, 1987) p.78—79.
[6]   E.Sjöstrom, Wood Chemistry — Fundamentals and Applications, (Academic Press, New York, 1981).
[7]   A.Pizzi and N.Eaton, *J. Macromol. Sci., Chem.* A21, 1443 (1984).
[8]   A.Pizzi and G.de Sousa, *Chemical Physics*, in press (1992).
[9]   N.L.Allinger and Y.H.Yuh, Operating instructions for MM2 and MMP2 programmes, 1977.
[10]   K.Gundertofte, J.Palm, I.Petterson, A.Stamvik, *J. Comp. Chem.*, 12, 200 (1991).
[11]   A.Pizzi, R.M.Horak, D.Ferreira and D.G.Roux, *Cellulose Chem. Technol.*, 13, 753 (1979).
[12]   R.Smit, A.Pizzi, C.J.H.Schutte and S.O.Paul, *J. Macromol. Sci., Chem.*, A(26), 825 (1989).
[13]   A.Pizzi and N.J.Eaton, *J. Macromol. Sci., Chem.*, A(24), 901 (1987).
[14]   F.J.Kolpak, M.Weih and J.Blackwell, *Polymer*, 19, 123 (1978).
[15]   J.T.Ham and D.G.Williams, *Acta Crystallogr.*, B26, 1373 (1970).
[16]   E.J.W.Whittaker, *Acta Crystallogr.*, 6, 714 (1953).

66

Strength properties of composites from biobased and synthetic fibers

R.A. Young, R.M. Rowell, A. Sanadi and C. Clemons - Department of Forestry, University of Wisconsin-Madison and USDA Forest Products Laboratory, Madison, Wisconsin, 53705, USA.

ABSTRACT

Biobased composites have been produced from esterified aspen fibers with either a thermosetting (phenol-formaldehyde, PF) or thermoplastic (polypropylene) matrix material. The polypropylene was incorporated in fibrous form. The level of esterification of the wood fibers had the greatest effect on the strength properties of PF bonded fiberboards, while the type of esterifying agent, acetic anhydride, maleic anhydride or succinic anhydride, had little or no effect. Acetylation of aspen fibers did not appear to increase the compatibility of the aspen fibers with polypropylene.

INTRODUCTION

Interest in fiber-reinforced and fiber-based composites has increased dramatically in recent years. Particularly notable is the increasing emphasis on biobased fibers such as wood, pulp, bagasse, jute, etc. as major components of these composites because of the low cost and high performance characteristics of these fibers. However, there are several important problems with the biobased fiber-plastic composites which have not yet been solved; notably the poor compatibility between biobased and synthetic materials, the poor dimensional stability of the natural fibers and the very limited plasticity. We are addressing all of these problems in a broad-based cooperative project on biobased composites between our institutions. In this paper we will give a brief report on our work related to strength properties of mixed fiber composites and describe a new technique we have developed to evaluate the interfacial shear strength of wood-thermoplastic systems.

EXPERIMENTAL

Acetylation was performed on attrition-milled aspen fiber with neat acetic anhydride (AA) with the technique developed by Rowell (1). Modification of the fibers with maleic anhydride (MA) and succinic anhydride (SA) was performed in hot xylene at reflux temperature. Excess anhydride was removed by Soxhlet extraction with xylene for 4 hours as described previously (2). The analysis of the modified products was carried out using the techniques of Matsuda (3) and Clemons et al. (2). The molar gains for aspen fiber esterified with acetic, maleic and succinic anhydrides are shown in Figure 1.

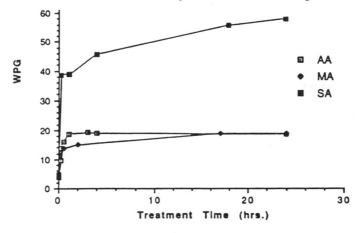

Figure 1. Weight Gains for Aspen Fiber Esterified with Acetic, Maleic, and Succinic Anhydrides.

Fiber mats were formed by sprinkling fiber onto a 15 x 15 cm screen. For treatment with a thermosetting resin, a liquid phenol-formaldehyde (PF) dry process hardboard resin (GP 2341, 50% aqueous solution) was sprayed on the fibers. Boards were made with a level of adhesive of 5, 8 or 12% based on the dry weight of the fiber. For incorporation of a thermoplastic material, a fibrillated form of polypropylene (PP) fiber from Hercules, Inc. (Pulpex, grade AD-H) was premixed with the wood fiber at levels of 0, 3, 10, 25 and 50% before sprinkling on the screen. The boards were bonded at elevated temperatures and pressures in a Carver press.

RESULTS AND DISCUSSION

Compression-shear tests were performed on fiberboards of esterified fiber bonded with phenol-formaldehyde resin. The compression shear test was used instead of the more traditional internal bond (IB) test because of the speed and simplicity of the procedure. The compression shear test has been shown to correlate well with the IB strength with similar or slightly lower variability (4).

As shown in Figure 2, the compression shear strength was found to increase when the extent of the esterification was increased, regardless of the type of esterifying reagent used for the modification. The type of fiber modification

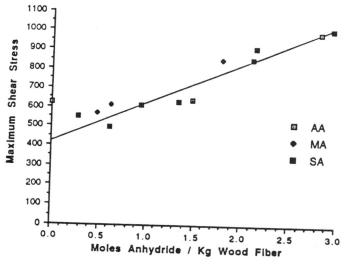

Figure 2. Compression-Shear Strength for Fiberboards of Acetic, Maleic, and Succinic Anhydride Modified Fiber and 5% Phenol Formaldehyde.

(AA, MA or SA) had no significant effect on the compression shear strength of the phenol-formaldehyde (5%) fiberboards. It was noted in our previous work (2) that at the higher levels of esterification the surfaces of the fibers became thermoplastic and exhibited flow at elevated temperatures and pressures. The increase in the maximum shear stress at higher levels of substitution could be due to the increased thermoplasticity of the esterified fiber since this would improve the fiber-to-fiber contact and subsequently the adhesion of the fibers in the boards.

An increase in the amount of PF resin deposited on the fibers was also found to increase the compressive shear stress of the fiberboards. As shown in Figures 3 and 4, when the PF resin content was raised from 5% to 12%, there was a corresponding increase in the strength. However the increase in board strength as a result of increased levels of PF resin (22-32%) were not nearly as great as the strength improvement realized from increased levels of esterification (135%). Thus, increased levels of esterification rather than further additions of PF resins are recommended for strength enhancement of these types of composites.

The equilibrium moisture contents (EMC) of the fiberboards made from aspen fibers modified with AA, MA and SA were reduced as compared to control boards; but similar to the strength properties of control boards, there was little effect of the type of esterifying agent on the final EMC. The same effect was noted in previous work for modified aspen fibers which had not been bonded into fiberboards (2).

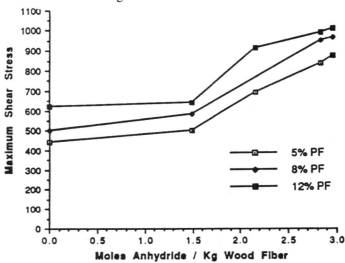

Figure 3. Compression-Shear Strength for Fiberboards of Acetylated Aspen Fiber and 5, 8, and 12% Phenol Formaldehyde.

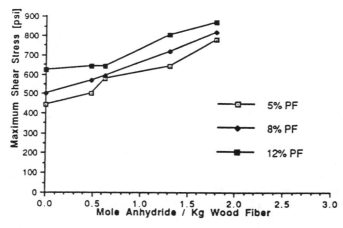

Figure 4. Compression-Shear Strength for Fiberboards of Maleic Anhydride Modified Aspen Fiber and 5, 8, and 12% Phenol Formaldehyde.

Composite boards from combinations of modified wood fibers and polypropylene were also evaluated. It was hoped that acetylation of the hydrophilic wood fibers would improve the compatibility with the polypropylene to give improved properties to the final composite board. Figures 5 and 6 show the density and internal bond strength of the composites from both unmodified and acetylated aspen fiber and polypropylene. Since the IB strength is very dependent on the density, it is not surprising that the two figures show curves of the same shape. The lower IB strength and density of the acetylated fiber composites at low levels of PP (0-10%) is probably due to both the poor compatibility adhesion of the modified wood fibers with PP and the increased stiffness of the acetylated fiber. Since water acts as a plasticizer for wood fibers and acetylated wood fiber has a lower EMC, the result is a stiffer AA modified fiber with reduced IB strengths.

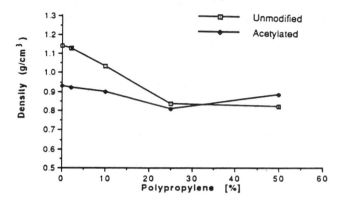

Figure 5.   Densities of Aspen Fiber or Acetylated Aspen Fiber and Polypropylene Composites.

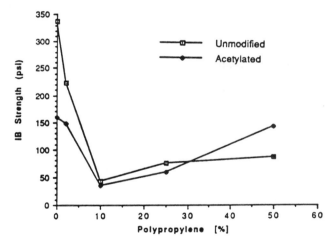

Figure 6.   IB Strengths of Aspen Fiber or Acetylated Aspen Fiber and Polypropylene Composites.

The acetylation treatment provided no improvement in the strength properties of the composites. The increase in density and strength of the composites at PP contents greater than 10% were attributed to the flow and film formation of the PP in the composite structure. As reported previously, both an increase in the PP content and acetylation of the wood fibers resulted in significant improvements in the EMC of the composites.

A new technique for determination of interfacial shear strength (IFSS) has also been developed to aid in the evaluation of strength improvement modifications for composites. The method involves measurement of the force required to pull a small wooden dowel from a thermoplastic matrix formed as a small plug around the dowel. The IFSS of unmodified and acetylated wood dowels in low molecular weight polyethylene was measured with the macroscopic pull-out test and the results showed no significant improvement in the pull-out force for the acetylated wooden dowel.

The effectiveness of the acetylation on the stability of the pull-out specimens was gauged by immersion of the specimens in distilled water for 48 hr. followed by a 2 hr. immersion in hot water at 80°C. The unacetylated dowels in polyethylene expanded and cracked the matrix such that an IFSS could not be obtained; however, with the acetylated dowels there was no change in the IFSS as compared to the dry condition test. A detailed description of this new macroscopic IFSS test will be given in a subsequent paper.

ACKNOWLEDGEMENTS

Financial support from DOE-SERI is gratefully acknowledged.

REFERENCES

1. Rowell, R. M.; Tillman, A.-M.; Simonson, R. *J. Wood Chem. Technol.,* **1986**, *6*, 427.

2. Clemons, C.; Young, R. A.; Rowell, R. M. *Wood Fiber,* **1991**, in press.

3. Matsuda, H. *Wood Sci. Technol.,* **1987**, *21*, 75.

4. Hall, H.; Haygreen, J. *Forest Prod. J.,* **1983**, *33*, 29; **1984**, *34*, 52.

67

New lightweight composites: synergistic interaction of balsa wood and ethyl α-(hydroxymethyl)acrylate

J. Randy Wright and Lon J. Mathias - Department of Polymer Science, University of Southern Mississippi, Hattiesburg, MS 39406-0076.

ABSTRACT

Wood-polymer composites (WPCs) from Southern pine and balsa wood have been prepared by impregnating whole wood samples with comonomer solutions of ethyl α-(hydroxymethyl)acrylate (EHMA) and styrene containing a radical initiator. The monomers were polymerized *in situ* using thermal activation or a combination of thermal and microwave cure. The WPCs were evaluated for mechanical properties compared to untreated wood. Good correlation between wood and WPC compression modulus with sample density was observed although increased data scatter developed with density increase. The latter suggests defect-limited physical property enhancement for all WPCs based on solid wood of density > 0.3-0.5 g/cm^3. Balsa WPCs were found to offer some of the best overall properties of any examined to date. Solid state ^{13}C NMR verified complete conversion of monomers with difference spectra showing strong interaction between the wood and polymer components. Scanning electron microscopy confirmed good polymer-cell wall interaction.

INTRODUCTION

Wood-polymer composites (WPCs) have been attracting renewed attention in the laboratory recently.[1],[2],[3]   A major family of WPCs is produced by impregnating solid wood with a vinyl monomer and polymerizing the monomer *in situ*. The most commonly used vinyl monomers are methyl methacrylate and styrene which are effective in filling the lumens of wood, but do not usually penetrate the cell walls to any extent.[4]   This results in a <u>mixture</u> of two components rather than a composite in which strong component interaction exists.   This can actually be detrimental since the interface between these components can provide a pathway for water penetration and fracture. A true wood-polymer composite should possess the ability to efficiently transfer stress from the polymer matrix to the rigid cellulose fibers while also increasing the stability and longevity of all components.

Ethyl α-(hydroxymethyl)acrylate (EHMA) is a multifunctional monomer containing an ester group, a primary hydroxyl group, and a polymerizable double bond. Due to the presence of the hydroxyl group, the polarity of EHMA is greater than methyl methacrylate. This increased polarity, plus the ability to hydrogen bond to wood component hydroxyl groups, enables EHMA to penetrate the cell walls and adhere strongly to all components of wood. It has been shown that EHMA alone,[5] or copolymers of EHMA and styrene or vinyl azlactone increase the strength and dimensional stability of whole-wood samples of Southern pine.[6],[7]

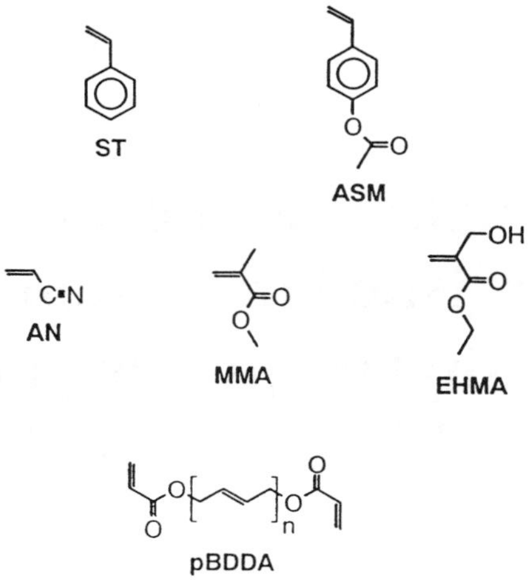

ST

ASM

AN

MMA

EHMA

pBDDA

Described here is an extension of this work to the impregnation of balsa wood with EHMA plus comonomers and a crosslinker, polybutadiene diacrylate (Figure 1). Balsa wood was chosen

ST =	Styrene
ASM =	para-Acetoxystyrene
AN =	Acrylonitrile
MMA =	Methyl methacrylate
EHMA =	Ethyl (α-hydroxymethyl)acrylate
pBDDA =	Polybutadiene diacrylate

Figure 1. Monomers and crosslinking agent.

because of its very low density $(0.10\text{-}0.30 \text{ g/cm}^3)$ which results in the production of lightweight composites with high specific strength. Balsa wood is used commercially as sound insulation, core material in boat hulls, and in aircraft sandwich panels for insulation and internal walls.[8]

EXPERIMENTAL

Materials

V-30 initiator was obtained as a donation from Wako Chemical Co. EHMA was synthesized by reaction of ethyl acrylate and paraformaldehyde in the presence of 1,4-diazabicyclo-[2,2,2]-octane (DABCO).[9] Styrene (ST) and methyl methacrylate (MMA) were purchased from Aldrich Chemical Co. Para-acetoxystyrene (ASM) was donated by Hoechst-Celanese. Polybutadiene diacrylate (pBDDA) was synthesized by reacting hydroxyl terminated polybutadiene with acryloyl chloride in the presence of triethyl amine and 4-dimethylaminopyridine. Balsa wood samples with the initial dimensions 1/16" x 1/2" x 5.0" were cut in half (perpendicular to the long axis). One half of each set was treated and the other used as a control for property measurements.

Impregnation and Curing

Wood samples were placed in a vacuum chamber (<3 mm Hg) for 20 minutes. While the samples were still under vacuum, the comonomer mixture containing crosslinker and initiator was introduced into the vacuum chamber until the samples were completely submerged. The vacuum was then slowly released and the samples were allowed to soak for 18 to 24 hours at 65 to 70 °C. After impregnation, samples were placed in separate test tubes which were subsequently fitted with septa. Each test tube was then purged with nitrogen and the samples were thermally cured for 24 to 36 hours or cured in a Cober LBM 1.2A microwave/convection oven. The cure conditions for the microwave/convection oven were as follows: 2 hours at 120 watts, then 2 to 4 additional hours at 240 watts. The thermal cure temperature was 55 °C for monomer containing AIBN initiator and 80 °C for monomer containing V-30 initiator.

Characterization

The compression modulus and toughness of the WPCs were measured by the buckled plate test[10] using a 1020C Instron. The compression modulus (E) was calculated from the following equation:

$$E = 12P_cl^2/\pi^2wh^3$$

where P_c = maximum force, w = width of wood sample, l = length of sample, and h = thickness of sample. The toughness was measured as

$$G = 0.82\ Eh^2(l-x)/l^2$$

where x = chord length of the buckled sample, E = compression modulus, l = length of sample, and h = thickness of sample.

Solid state ^{13}C CP/MAS NMR spectra were obtained on a Bruker MSL-200 spectrometer with a resonance frequency of 50.32 MHz. Spectra were acquired at 300K (27 °C) using a recycle delay of 3 sec, contact time of 5 msec, and a spinning rate of 4.8-5.0 KHz. An Electroscan environmental scanning electron microscope (ESEM) was used to evaluate the WPCs for polymer loading and polymer-cell wall interaction.

RESULTS AND DISCUSSION

The NMR spectra in Figure 2 affirm the presence of EHMA polymer within the balsa wood. While the spectrum of the wood-polymer composite clearly displays a combination of the peaks found in the individual spectra of poly-EHMA and untreated balsa wood, the spectral subtraction (WPC - wood) does not remove all the wood peaks which indicates strong interaction between the polymer and wood components causing chemical shift changes that do not cleanly subtract.

The dimensional stability of the WPCs is illustrated in Figure 3. This bar graph displays the amount of volume increase for various samples due to water uptake after soaking for 5 days. The sample containing pure EHMA polymer demonstrated the greatest dimensional stability. In general, as the concentration of EHMA in copolymers decreased, the dimensional stability of the WPCs decreased. This is probably due to a lower percentage of total polymer being incorporated into the cell walls, which increases the amount of water uptake possible. Styrene impregnated samples swelled more than twice as much as the EHMA impregnated samples while MMA impregnated samples swelled more than 3 times as much as the EHMA WPCs.

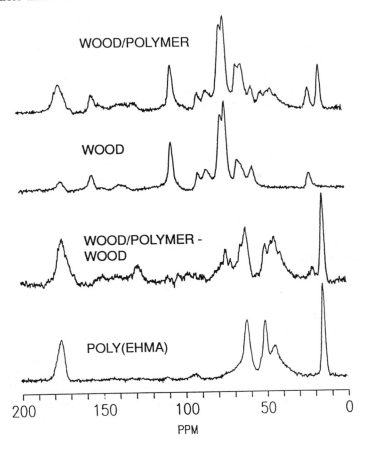

Figure 2. Solid state ^{13}C-NMR spectra.

Dynamic mechanical analysis was used to determine the improvement in mechanical properties of the WPCs over those of untreated wood. For this particular test, the same sample is analyzed before and after treatment. The DMA spectra in Figure 4 indicate that the storage modulus increased as the styrene content increased with the greatest improvement in modulus achieved using pure styrene. Based on these results it seemed reasonable to speculate that the most favorable properties would be attained by using EHMA as a comonomer with styrene or similar vinyl monomers.

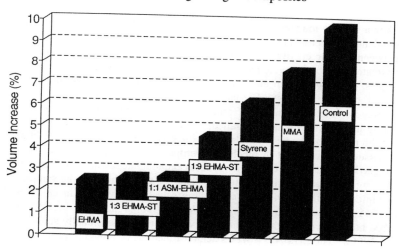

Figure 3. Dimensional stability of WPCs represented as degree of swelling due to water uptake.

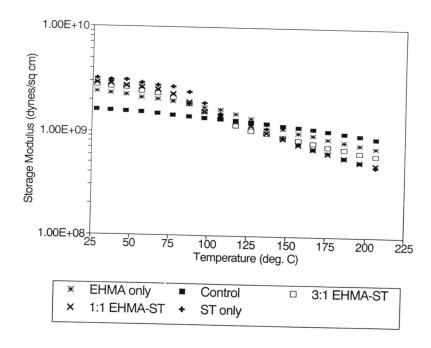

Figure 4. Dynamic mechanical analysis spectra of wood-polymer composites and untreated control sample.

It has been shown that the toughness of wood is directly related to its specific gravity and is independent of species.[11]   We have determined that there is also a strong correlation between specific gravity and modulus as shown in Figure 5. The moduli of untreated and treated balsa wood were plotted as a function of specific gravity. For the untreated samples, it is obvious that there is much better correlation at lower densities with more scatter at higher densities. For the treated samples the scatter became even greater. This may be due to variation in treatments for the samples, and suggests that physical property improvement is defect-limited; i.e., defects present before treatment. Table I shows the results of the buckled plate test for various sets of different WPCs. In general, a greater improvement in mechanical properties was attained with lower initial specific gravity. The improvements in modulus and toughness were much greater for samples having an initial specific gravity less than 0.2 g/cm^3 than for the samples having greater initial specific gravity. Overall, the greatest improvement in modulus and toughness was for the WPCs containing EHMA-ST copolymer and the WPCs containing ASM homopolymer, although the latter were not as dimensionally stable as the EHMA-ST WPCs. The EHMA-ST WPCs demonstrated the best overall properties.

In order for balsa WPCs to compete with other composites, the *specific* properties of the material should be improved by the treatment. The specific modulus (or specific toughness) is the modulus (or toughness) divided by the specific gravity, and gives a qualitative indication of component synergism. Significant improvements in *absolute* mechanical properties were achieved for all treatments, although the *specific* mechanical properties decreased with the exception of one type of treatment. Only for the EHMA-ST-pBDDA WPCs was there a substantial improvement in the specific modulus and specific toughness. To our knowledge, an improvement in specific properties of wood has never been reported previously.

In Table II these results are compared to specific property data reported in the literature. For the two sets of MMA impregnated wood samples there was a decrease in specific properties from 8.4% to 48.1%. These decreases are similar to those we obtained for the other treatments. The average improvement in specific modulus and specific toughness was 27.4% and 6.7%, respectively, for the three sets of samples impregnated with EHMA-ST-pBDDA. The improvement in specific modulus for the WPCs is graphically represented in Figure 6. The treated samples maintained a similar linear relationship to the untreated control samples, although the slope was somewhat greater. This indicates that an increase in density due to polymer is <u>more</u> effective in increasing the modulus than an increase in density from wood components. **This is a true synergistic effect; ie, the WPC properties are better than either component alone or combined in a linear manner**.

A major criterion for specific property improvement is control of the amount of weight gain due to impregnation. Most of the EHMA-ST-pBDDA WPCs had weight gains of 30-40%. Beyond a certain point, the increase in density will be greater than the increase in mechanical properties, thus causing a decrease in <u>specific</u> mechanical properties. In order to control the weight gain, the samples were weighed before and after impregnation. If the weight gain was acceptable, the samples were cured at atmospheric pressure. If the weight gain was too high, the samples were cured under high vacuum ($\cong$1 mmHg) to remove excess comonomer from the cell lumens. Polymer-filled lumens are <u>not</u> necessary for some property improvement since the major increase

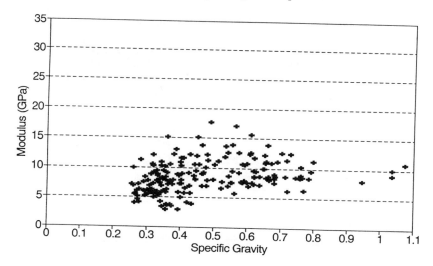

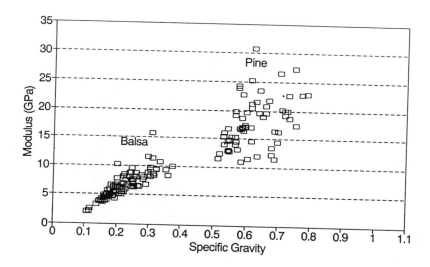

Figure 5. Sample modulus plotted as a function of specific gravity for untreated balsa and pine wood (lower plot) and balsa WPCs (upper plot).

Table I. Modulus and toughness data for wood-polymer composites.

Treatment	Number Tested	Average SG₁	Average SG Inc. (%)	Average Δ Modulus (%)	Average Δ Toughness (%)
EHMA only	12	0.268	68.9	+7.5	-6.0
MMA only	6	0.266	53.2	+9.7	+16.1
1:3 EHMA-MMA	12	0.214	61.2	+22.3	+2.5
EHMA-MMA-pBDDA	6	0.191	162.1	+40.0	+52.8
ASM only	6	0.184	410.8	+68.5	+75.4
1:1 ASM-AN	6	0.189	59.0	+26.3	+17.0
1:1 ASM-EHMA	6	0.192	105.0	+21.4	+12.9
1:1 ASM-ST	6	0.090	324.2	+180.5	+135.8
EHMA-ST-pBDDA	36	0.219	34.0	+32.6	+15.0
9:1 EHMA-ST	6	0.316	73.1	+28.2	+5.9
3:1 EHMA-ST	6	0.309	92.7	-1.3	+25.7
1:1 EHMA-ST	18	0.256	159.8	+43.4	+64.6
1:3 EHMA-ST	6	0.177	226.5	+85.2	+61.8
1:9 EHMA-ST	24	0.206	226.9	+34.8	+91.2
Styrene only	6	0.316	73.1	+28.2	+5.9

Table II. Comparison of absolute properties and *specific* properties.

Wood/Treatment	No. Samples	Average Toughness (kJ/m²)	Average ΔToughness (%)	Average Specific Toughness	Average ΔSpecific Toughness
Basswood/PMMA[1]	14	8.1	+50.0	8.1	-44.5
Basswood/control	14	5.4		14.6	
Sugar maple/PMMA[2]	9	30.9	+43.1	27.2	-8.4
Sugar maple/control	10	21.6		29.7	
Balsa/EHMA-ST-pBDDA[3]	18	8.9	+21.9	27.1	+6.7
Balsa/control	18	7.3		25.4	

Wood/Treatment	No. Samples	Average Modulus (GPa)	Average ΔModulus (%)	Average Specific Modulus	Average ΔSpecific Modulus
Basswood/PMMA[1]	14	11.6	+48.7	11.6	-45.0
Basswood/control	14	7.8		21.1	
Balsa/EHMA-ST-pBDDA[3]	18	10.7	+32.1	32.1	+27.4
Balsa/control	18	8.1		25.2	

[1]Modulus was determined from the static bending test using 2 x 2 x 30 cm samples. Toughness was determined using a toughness testing machine designed by the USDA Forest Products Laboratoy with 2 x 2 x 21 cm samples. (Langwig, J. E. *Forest Products Journal*, **1968**, *18*, 33.)
[2]Samples were 2 x 2 x 24 cm and testing was carried out on a toughness testing machine designed by the USDA Forest Products Laboratory. (Schneider, M. H. *Forest Products Laboratory*, **1989**, *39*, 11.)
[3]Samples were 1/16 x 1/2 x 2.5 inches. Toughness and modulus were determined by the buckled plate test.

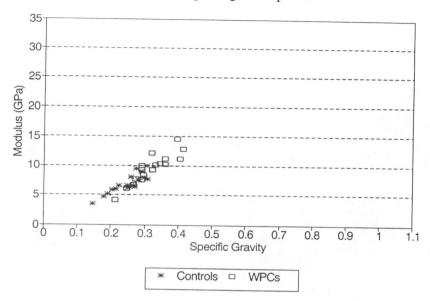

Figure 6. Modulus of balsa/EHMA-ST-pBDDA WPCs and control samples plotted as a function of specific gravity.

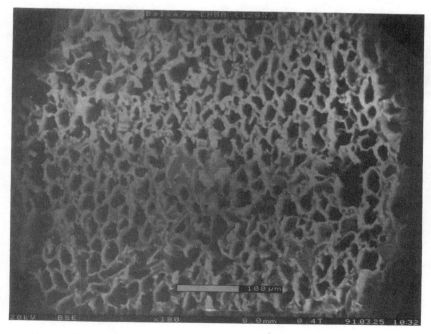

Figure 7. SEM micrograph of cross-section of balsa/EHMA WPC.

Figure 8. SEM micrograph of a split radial surface of balsa/EHMA-ST-pBDDA WPC.

in strength comes from cell wall reinforcement with polymer. However, if the lumens *are* filled with polymer, the cell walls must also contain that polymer for maximum property improvement and it should adhere to and penetrate the cell walls.

Scanning electron microscopy (SEM) was used to examine polymer distribution within the WPCs and to confirm polymer-cell wall interaction. Figure 7 gives a micrograph of a typical cross-section of a balsa/EHMA composite. This WPC is 55% polymer by weight with most of the cell lumens empty. This indicates that most of the polymer must be contained within the wood cell walls. This intimate relationship between poly(EHMA) and the cell wall is further demonstrated in Figure 8. Good adhesion of the polymer to the cell wall is indicated by the polymer that remained attached to the cell wall after fracture, indicating cohesive rather than adhesive (interfacial) failure.

CONCLUSION

It is clear that EHMA is an excellent monomer for increasing the dimensional stability of balsa wood. Copolymers containing EHMA, and especially EHMA-ST mixtures, are able to improve the dimensional stability *and* the mechanical properties of balsa wood. **Synergistically improved properties were achieved using EHMA-ST with pBDDA as crosslinker and toughening agent**. Scanning electron microscopy indicated that good interaction of monomers and copolymer with wood cell wall components is occurring, consistent with property improvements.

ACKNOWLEDGEMENTS

We are grateful to the Solar Energy Research Institute (Golden, CO) for partial support of this research through a contract from the Department of Energy-Energy Conversion and Utilization Technology/Biomass Materials Program; to Wako Chemicals and Hoechst Celanese for donation of chemicals; to Dr. W. L Jarrett and C. G. Johnson for obtaining the solid state ^{13}C-NMR spectra; and to the Office of Naval Research for a Department of Defense instrumentation grant to purchase our Bruker MSL spectrometer.

REFERENCES

1. Hoffmann, G.; Poliszko, S. *J. Appl. Polym. Sci.*, **1990**, *39*, 153.
2. Raj, R. G., Kokta, B. V., Daneault, C. *Intern. J. Polym. Mater.*, **1989**, *12*, 239.
3. Schneider, M. H., Phillips, J. G., Tingly, D. A., Brebner, K. I. *Forest Prod. J.*, **1990**, *40*, 37.
4. R. M. Rowell and P. Konkol, Gen. Tech. Rep. FPL-GTR-55, U. S. Dept. of Agriculture, Forest Service, Forest Products Laboratory, Madison, WI, 1987.
5. Wright, J. R.; Mathias, L. J. *Polym. Prepr.*, **1989**, *30*, 233.
6. Mathias, L. J.; Lee, S.; Wright, Warren, S. C. *Polym. Prepr.*, **1990**, *31*, 646.6
7. Mathias, L. J.; Lee, S.; Wright, J. R.; Warren, S. C. *J. Appl. Polym. Sci.*, **1990**, *42*, 55.
8. Kohn, J. *Forest Products Journal*, **1958**, *8*, 27-A.
9. Wright, J. R.; Mathias, L. J. *Polym. Prepr.*, **1989**, *30*, 233.
10. Brady, R. L.; Porter, R. S.; Donovan, J. A. *J. Mater. Sci.*, **1989**, *24*, 4138.
11. Wengert, E. M. *Wood Science*, **1979**, *11*, 233.

68

Studies on the influence of high frequency field on composites

S.O. Gladkov - Institute of Chemical Physics, USSR Academy of Sciences, 117977 Kosygin str. 4, Moscow, USSR.

ABSTRACT

Dielectric properties of a metal + glass compound placed in external alternating superhigh frequency (SHF) fields are considered. The structure of the composite is represented in the form of glass balls chaotically distributed in the volume of the metal with a certain function of the balls distribution by sizes.

It is shown that the imaginary part of the dielectric permeability $\varepsilon''_i(\omega)$ of such a compound differs strongly from the Debay dependence.

The composite ander considaration (metall+glass) is a simplified hypothetical model of such complex disordered substance as cellulose.We shall consider these composite below.

The investigation into the properties of substances placed in external high frequency fields are of great interest in view of discovering new unknown possibilities for these substances. For instance, as was shown in ref /1/, an external high frequency field alters strongly an exchange interaction of magnetic atoms and results in changing the temperature of the magnetic phase transition T_c . The value of T_c' starts to depend on the field frequency ω according to the law

$$\mathcal{J}_{ex}(\omega) = T_c = T_c(\omega) = \left(|t_{12}|^2 \Big/ 2\pi\hbar\omega\right)\sum_{n=-\infty}^{\infty}\mathcal{J}_n(A_d)\left[\ln\left(1+\frac{n\hbar\omega}{2\mathcal{E}_d}\right)\right]\Big/n$$

It is evident, therefore, that due to the SHF field one can control the temperature of the phase transition and bring about (at one's own discretion) either an early phase transition (at low temperatures) or a late one. Indeed, the practical importance of the possibility of phase transition temperature controlling is very great. As regards the physical aspect of this problem, it is perfectly clear. In fact, since the exchange interaction is accounted for the electronic mixing up of d -electrons of the magnetic atoms outer shells, then, on condition that the interaction of the SHF field with electrons proceeds according to the scheme

$$\hat{V}_{SHF} = -\frac{e}{\rho mc}\left(\vec{P}\vec{A} + \vec{A}\vec{P}\right)$$

where $\vec{P}$ is the electron impulse operator, $\vec{A}$ is the external vector potential, with $\vec{A}(t) = \frac{ic}{\omega}\vec{E}(t)$

where m is the mass of an electron, hence, thanks to $\hat{V}_{SHF}$, the energy of electrons becomes dependent on time and it finally results, by the invariant theory of disturbances, in the dependence $\mathcal{J}_{ex} = \mathcal{J}_{ex}(\omega)$ (note that $T_c \sim \mathcal{J}_{ex}$).

As concerns the composites (e.g. metal+glass compound), it is very important to introduce a correct definition of the effective dielectric permeability of such a complex compound. Take the following example. Let have a melted metal with liquid glass added. The glass concentration x_g is optional. We are now going to answer the question how the properties of the metal can change depending on x_g . Assume, first, that $x_g \ll 1$. In this case the molecular compound SiO_2 in the process of crystallization will somewhat chaotically be distributed over the whole volume of the metal. It is clear that physical properties of the composite will change depending on the amount of SiO_2 . Suppose, for instance, that electric current is passed through such a compound. Then, on condition $x_g \ll 1$, the current depends on x_g as $j(x_g) = j_{el} - j_o x_g$. With the concentration increasing, when x_g tends to a certain critical value x_{cr} , the current $j(x_g) = j(x_{cr} - x_g)^d$ where j is a constant. At $x_g \geqslant x_{cr}$ there is no current. It is a natural result because at the glass concentration of x_{cr} the composite becomes a non-conductor. The value x_{cr} , as is easily understood, defines the percolation threshold. As regards an index of a power d , it can be calculated by applying the scale theory (or, in other words, renormgroup).

We have been distracted a little from the main goal of our investigation, i.e. dielectric permeability of the composite. So, just when we are going to focus on the dielectric permeability of the metal+glass compound let us confine ourselves, in the meantime, to the very simplest case when x_g is small ($x_g \ll 1$) and an additional phase of SiO_2 is derived from spherical inclusions with a certain function of distribution $f(r)$ by sizes of these spheres. In this case, according to /2/ (p.69), we have for ε_{ef}:

$$\varepsilon_{ef} = \varepsilon_M + 3x_g \varepsilon_M (\varepsilon_g - \varepsilon_M)/\varepsilon_g + 2\varepsilon_M \quad (1)$$

where ε_M is the dielectric permeability of the metal, ε_g is the dielectric permeability of a glass ball.

It should be noted that expr.(1) is valid only for small x_g . The value x_g may mean, for instance, a voluminous share of glass inclusions:

$$x_g = V_g / V_M + V_g$$

V_M is the volume of metal, V_g is the volume of glass, or a massive one: $x_g = m_g / m_M + m_g$.

With the increase of the concentration x_g the ratio (1) becomes invalid and such a simple dependence does not work in any way to describe the dielectric permeability.

We represent the frequency dependence of the dielectric permeability in accordance with the Debay formula: $\varepsilon(\omega) = \varepsilon'(\omega) + i\,\varepsilon''(\omega)$
where

$$\varepsilon'(\omega) = \varepsilon_0/(1+\omega^2\tau^2) \;,\; \varepsilon''(\omega) = \varepsilon_0\,\omega\tau/(1+\omega^2\tau^2)$$

where τ is the relaxation time . The composite absorption of the SHF field, described by an imaginary part of dielectric permeability, in accordance with expr.(1) is

$$\varepsilon''_{ef}(\omega) = \varepsilon''_M + 3x_g\{[\varepsilon''_M (\varepsilon''_g - \varepsilon''_M) - \varepsilon'_M (\varepsilon'_g - \varepsilon'_M)](\varepsilon'_g + 2\varepsilon''_M)$$
$$+ [(\varepsilon'_g - \varepsilon''_M)\varepsilon'_M + \varepsilon''_M(\varepsilon'_g - \varepsilon'_M)](\varepsilon'_g + 2\varepsilon'_M)\}/|\varepsilon_g + 2\varepsilon_M|^2 \quad (2)$$

where

$$\varepsilon'_M = \varepsilon_{0M}/(1+\omega^2\tau_M^2), \; \varepsilon''_M = \varepsilon_{0M}\,\omega\tau_M/(1+\omega^2\tau_M^2),$$

$$\varepsilon'_g = \varepsilon_{0g}/(1+\omega^2\tau_g^2), \; \varepsilon''_g = \varepsilon_{0g}\,\omega\tau_g/(1+\omega^2\tau_g^2),$$

where τ_M and τ_g are relaxation times in the metal and glass respectively.

Now we shall analyze the frequency dependence $\varepsilon''_{ef}(\omega)$. For this purpose we examine some ultimate cases:

1. $\omega \ll 1/\tau_M$, $1/\tau_g$
2. $1/\tau_M \ll \omega \ll 1/\tau_g$
3. $1/\tau_g \ll \omega \ll 1/\tau_M$
4. $\omega \gg 1/\tau_g$, $1/\tau_M$.

In the first case the dependence $\varepsilon''_{ef}(\omega)$ is given by expr.

$$\varepsilon''_{ef}(\omega) = \varepsilon_{oM}\,\omega\,\tau_M + 3x_g\,\varepsilon_{oM}\,\omega\,\tilde{\tau} \tag{3}$$

where

$$\tilde{\tau} = \left\{(\varepsilon_{oM}-\varepsilon_{og})(\varepsilon_{og}\,\tau_g + 2\varepsilon_{oM}\,\tau_M) + [\varepsilon_{og}(\tau_g+\tau_M) - 2\varepsilon_{oM}\,\tau_M]\cdot\right.$$
$$\left. \cdot (\varepsilon_{og} + 2\varepsilon_{oM})\right\} \Big/ (\varepsilon_{og} + 2\varepsilon_{oM})^2 \tag{4}$$

In the second frequency range

$$\varepsilon''_{ef}(\omega) = \left(\varepsilon_{oM}/\omega\tau_M\right) + 3x_g\,\varepsilon_{oM}\left(\omega\tilde{\tau}_g^2/\tau_M + 1/\tau_M\omega\right) \tag{5}$$

In the third

$$\varepsilon''_{ef}(\omega) = \varepsilon_{oM}\,\omega\,\tau_M + 3x_g\left\{(\varepsilon_{oM}+\varepsilon_{og}\tilde{\tau}_M/\tau_g)(2\varepsilon_{oM}\omega\tau_M + \right.$$
$$\left. + \varepsilon_{og}/\omega\tau_g) + 2\varepsilon_{oM}([\varepsilon_{og}/\omega\tau_g]-2\varepsilon_{oM}\omega\tilde{\tau}_M)\right\}/4\varepsilon_{oM} \tag{6}$$

And, finally, in the fourth

$$\varepsilon''_{ef}(\omega) = \left(\varepsilon_{oM}/\omega\tau_M\right) + 3x_g\varepsilon_{oM}/\omega\tau^* \tag{7}$$

where

$$1/\tau^* = \frac{1}{\tau_M}\left(\varepsilon_{og}-\varepsilon_{oM}\tau_g/\tau_M\right)\left(\varepsilon_{og}+2\varepsilon_{cM}\tau_g/\tau_M\right)\Big/\left(\varepsilon_{og}+2\varepsilon_{oM}\tilde{\tau}_g/\tau_M\right)^2 \tag{8}$$

    The graph of changing $\varepsilon''_{ef}(\omega)$ is shown in Fig.1. As is seen from the figure, the difference of ab-sorbing capacity of a composite from that of a pure metal is very great. In our opinion, the most interesting is the possibility of practical use of a composite which absorbs a minimal part of the energy in a certain frequency range. This frequen-cy range is defined by the vicinity of frequencies $\omega_1^* \simeq (1+3x_g)^{1/2}/x_g^{1/2}\tilde{\tau}_g$ (at $\tau_M \gg \tau_g$) or $\omega_2^* \simeq (x_g/\tau_M\tau_g)^{1/2}$. (at $\tau_M \ll \tau_g$). By varying the concentration of glass we can vary the absorption range as well. It is a very interesting peculiarity.

    It is important to note that the temperature dependence involves the relaxation times τ_M and τ_g. The time τ_M is defined, as a rule, by an electro-nic mechanism of interaction. In fact, there are three main mechanisms of relaxation: electron-electron, electron-impurity and electron-phonon. Each of these mechanisms is characterized by its individual function of temperature, depending on a temperature range (see Review /3/), which is of practical use for us. As concerns the time τ_g, it characterizes, as was mentioned above, the time of glass relaxation. It is to be stressed that in the region of low temperatures τ_g is defined by the

mechanism of phonons scattering on a two-well potential, but in case of a high temperature range τ_g
is defined by the four- or three-phonon mechanism
of scattering.

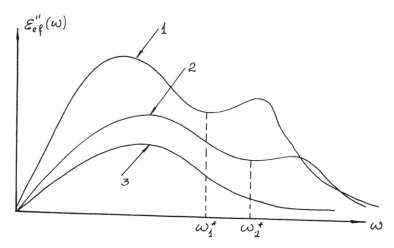

Fig.1

Curve 1 shows the dependence of dielectric
permeability of a composite for the frequency range
in the region 2. Curve 2 shows the frequency range
3 and Curve 3 relates to pure metal.

Within the scope of our present work it is of
interest to find connection between the sound velocity in a composite and the concentration of
glass balls. For that purpose we shall get the following way to define $c_s(x_g)$. As the energy of a
sound wave is

$$E = \frac{1}{2} \int \rho |\dot{u}|^2 dv = \frac{1}{2} \int_{V-V_2} \rho \omega^2 u_o^2 dv + \frac{1}{2} \int_{V_2} \rho \omega^2 u_o^2 dv$$

where $\omega = c_s \cdot K$.   (9)

Since the sound vector $\vec{K}$ remains after wave
hitting the media boundary($K_1 = K_2$) eq(9)
can be rewritten differently:

$$E = \frac{K^2}{2} \left\{ \rho_1 c_{1s}^2 u_{o1}^2 (V-V_2) + \rho_2 c_{2s}^2 u_{o2}^2 V_2 \right\} =$$

$$= \frac{K^2 V}{2} \left\{ \rho_1 c_{1s}^2 (1-x_g) u_{o1}^2 + \rho_2 c_{2s}^2 u_{o2}^2 x_g \right\} .$$

(10)

On the other hand, since the effective sound velo-
city can be determined by the ratio $E = \rho K^2 c_s^2 V u_{o1}^2 / 2$
or otherwise $c_s^2 = 2(\partial \varepsilon / \partial v) / \rho u_{o1}^2 K^2$

hence, according to (10) we have

$$c_s^2 = \left\{ \rho_1 c_{1s}^2 (1-x_g) |u_{o1}|^2 + \rho_2 c_{2s}^2 |u_{o2}|^2 x_g \right\} / \rho |u_{o1}|^2$$

On the assumption that
$$1/\rho = (x_g/\rho_2) + (1-x_g)/\rho_1 \quad \text{and} \quad |u_{02}|^2/|u_{01}|^2 = (1-\sqrt{R})^2 c_{2s}^2/c_{1s}^2$$
where the coefficient of reflexion in normal incidence

$$R = \left[(\rho_2 c_{2s} - \rho_1 c_{1s})/(\rho_2 c_{2s} + \rho_1 c_{1s}) \right]^2$$

we obtain
$$c_s^2 = (1 - x_g + x_g \rho_1/\rho_2)\left[c_{1s}^2(1-x_g) + 4 x_g \rho_1 c_{2s}^4/\rho_2(c_{2s} + c_{1s}\rho_1/\rho_2)^2 \right] \quad (11)$$
If we introduce a variable $y = \rho_1/\rho_2$, the behaviour $c_s(y)$ can be illustrated in Fig.2.

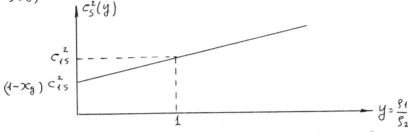

The expr.(11) is valid only for the values of concentration $x = x_{cr}$. At $x > x_{cr}$ the given ratio is invalid and an analogous formula should be used with index 1 changed for 2, that is at
$$c_s^2 = (1 - x_M + x_M \rho_2/\rho_1)\left[c_{2s}^2(1-x_M) + 4 x_M \rho_2 c_{1s}^4/\rho_1(c_{2s} + c_{1s}\rho_1/\rho_2)^2 \right] \quad (12)$$
So, the examinations of physical properties of a composite (metal+glass) are of great practical significance, because they permit to use in practice the predicted results. Indeed, for measuring losses of EM energy in such systems one can check minima in the function $\varepsilon''_{e\rho}(\omega)$. For measuring the sound velocity the functional behaviour specifity of $c_s(x, y)$ is of interest.

REFERENCES

1. S.O.Gladkov, Phys.Lett., 1992,v.163,p.460.
2. L.D.Landau, E.M.Lifshits, Electrodynamics of solid media, v.8, M.Nauka,1981 (in Russian).
3. S.O.Gladkov, Phys.Reports, 1989, v.182,p.211.

69

Cellulosic materials as a humidity sensor

Kenichiro Arai - Faculty of Engineering, Gunma University, Kiryu, Gunma 376, Japan.

INTRODUCTION

Recently, many new types of humidity sensor have been developed to measure the humidity of atmosphere more quickly and accurately. The information obtained by the sensor is converted to an electric signal to be utilized for regulation of the humidity in industrial and agricultural applications. They are based on the change in the characteristics, such as electric resistance, of a certain material with the changing humidity, as shown in Table 1. The used materials may be classified into inorganic and organic ones. As the former, for example, porous sintered blocks of alumina or silica, and as the latter typically polyelectrolytes such as polystyrenesulfonates have been used. However, each type of the sensor materials has inherent disadvantages.

We have investigated the salts of cellulose monosulfate(CS) as the humidity sensor materials for the system measuring the change in electric resistance with the changing humidity. In the preliminary experiment using polyelectrolytes, including various salts of CS and poly(styrenesulfonic acid)(PSS), in film form, isopropylamine salt of CS was found to be best in respect to the sensitivity to the humidity, among the examined polyelectrolyte salts. However, it needed a longer time to obtain an equilibrium value of the electric resistance at a relative humidity(RH), and showed a lower reproducibility of the values at the higher RH's.

In the present experiment, in order to overcome these problems, and to develop a simple and expedient humidity sensor,

Table 1. Classification of modern humidity sensors

Characteristics	Materials
Electric resistance or electrostatic capacitance	Inorganics; ceramics based on alumina or silica Organics; polyelectrolytes such as polystyrenesulfonates
Electromagnetic wave absorption	Organic polymers, with infrared ray or microwaves
Thermal conductivity	A couple of a dry and a wet bulbs, with a thermometer
Color	Dyes
Length	Hair, Polymer films

the polyelectrolyte salts are impregnated to the cellulose paper, and the performances of the paper as the humidity sensor are investigated.

EXPERIMENTAL

CS was prepared from dissolving pulp according to the method reported by Schweiger[1) as follows;

$$Cell-OH \xrightarrow{\quad N_2O_4/Dimethylformamide \quad} Cell-O-N=O$$

$$Cellulose \xrightarrow{\quad SO_3 \quad} Cell-O-SO_3H \; (DS \doteqdot 1.0)$$

The CS and commercially available PSS were neutralized with sodium hydroxide, ethylamine, or isopropylamine aqueous solutions. The salts obtained are abbreviated as CS-Na, CS-EA, and CS-IPA for CS salts, and PSS-Na, PSS-EA, and PSS-IPA for PSS salts, respectively.

A thin($14g/m^2$) and porous rayon paper was immersed in a given concentration of the polyelectrolyte salt aqueous solution. After taken out, it was squeezed lightly and then air-dried to obtain the polyelectrolyte salt- impregnated paper.

A 10mm square piece of the paper was held between two electrodes in a hygrothermostat at 25°C as shown in Fig.1. The electric resistance of the paper was measured by using a high resistance meter(Hewlett·Packard 4329).

The constant RH was obtained by using inorganic salt-saturated aqueous solution; $CaCl_2$ for RH 31%, $Mg(NO_3)_2$ for RH 52%, a mixture of NH_4Cl and KNO_3 (1:1 in mole) for RH 71%, and $BaCl_2$ for RH 88%.

RESULTS AND DISCUSSION

Fig.2 shows the electric resistance of the polyelectrolyte

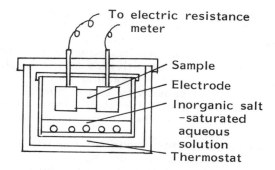

Fig.1. Apparatus for measurement of electric resistance of the sample at a constant RH

salt-impregnated papers in equilibrium at different RH's.   Upper plots are for PSS salts and lower ones are for CS salts. The ordinate is in logarithm of electric resistance value.   The equilibrium values were obtained after the sample had stood at the RH for 24h.   In the figure, even the untreated original paper itself is found to show a change in the electric resistance with the changing humidity.   However, the sensitivity to the humidity, which is expressed by the difference between

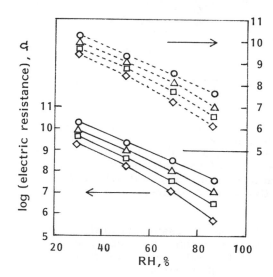

Fig.2. Electric resistance of polyelectrolyte salt-impregnated papers in equilibrium at different RH's. Polyelectrolyte: -----,PSS; ———,CS.   Salt:△,Na; □,EA;◇,IPA.   ○,untreated original paper

the electric resistance values at RH of, for example, 31 and
88%, is increased by the impregnation with the polyelectrolyte
salts in order of Na < EA < IPA.   On comparison of the lower
plots with the upper ones, the papers impregnated with the
CS salt are found to show higher sensitivities to the humidity
than those impregnated with the corresponding PSS salt.
Thus, on the viewpoint of the sensitivity to the humidity, CS
-IPA is considered to be most preferable as the impregnated
polyelectrolyte salt.   However, further higher amine salts of
CS did not give better results.

Fig.3 shows the influence of the amount of the IPA salts of
CS and PSS impregnated to the paper, on the electric resistance at different RH's.   The sensitivities to the humidity are

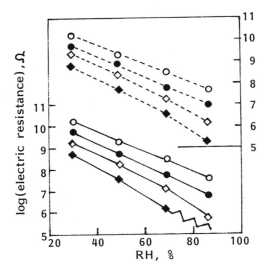

Fig.3. Influence of amount of IPA salts impregnated to
the paper on electric resistance of the paper in equilibrium at different RH's.   Polyelectrolyte: - - - - ,PSS;
———— ,CS.   Amount of the salt in wt%: ●,◇ , and ◆
are 0.9, 3.0, and 9.9 for PSS, and 0.8, 3.0, and 9.6
for CS, respectively.   ○,Untreated original paper

found to increase with the increasing amount of the salts for
both polyelectrolytes.   However, the papers impregnated with
too large an amount of IPA salt show a tendency to give lower
reproducibilities at the higher RH's.

The IPA salts impregnated papers in equilibrium at RH 31
and 88% were transferred to RH 88 and 31%, respectively, and
the changes in the electric resistance values were measured
with time.   The results are shown in Fig.4.   Upper curves

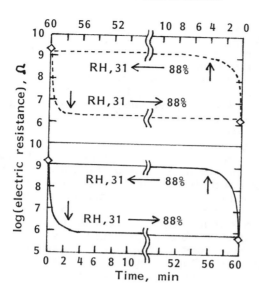

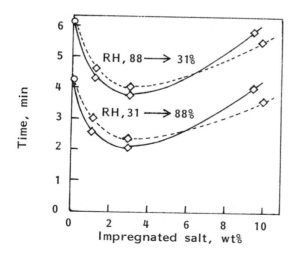

Fig.4. Changes in electric resistance of CS-IPA and PSS-IPA impregnated papers with time, after they have transferred from RH 31 to 88% and from 88 to 31%. Polyelectrolyte: – – – –, PSS; ————, CS.   ◇, electric resistance values in equilibrium at RH 31 and 88%.   Amount of impregnated IPA salt: 3.0wt%.

Fig.5. Plots of responsive time to obtain the 90% change in log(electric resistance) values to the corresponding equilibrium ones against amount of impregnated IPA salt to the paper.   Polyelectrolyte: – – – –, PSS; ————, CS

are for PSS-IPA and the lower ones are for CS-IPA.   In the humidification, shown by an arrow as RH 31→88%, both the salt-impregnated papers are found to need about 2.5 minutes to obtain the 90% change in logarithm of electric resistance value to the corresponding equilibrium one, and in the desiccation process, shown as from RH 88 to 31%, they need about 4 minutes.   These responsive times are considered to be allowable for some practical applications.   But, even after 60 minutes, the electric resistance values do not perfectly reach the equilibrium ones; hysteresis still remains.

Fig.5 shows plots of the responsive time to obtain the 90% change against the amount of impregnated IPA salts to the paper.   In comparison with the original paper, the responsive times are found to be improved considerably by the impregnation of the salt, but they show an optimum value at a certain amount of each salt impregnated; too large an amount of the salt impregnated leads to a longer responsive time.

Table 2 shows the hysteresis in the electric resistance value remaining 60 minutes after changing RH from 88 to 31% and from 31 to 88%, respectively.   The hystereses are found

Table 2. Hysteresis in electric resistance after 60 minutes.   Impregnated IPA: 3.0wt%

Paper	Transfer	Difference in log-(electric resistance)
PSS-IPA	RH 88 →31%	0.14
	RH 31 →88%	0.12
CS-IPA	RH 88 →31%	0.15
	RH 31 →88%	0.13
Original paper	RH 88 →31%	0.12
	RH 31 →88%	0.10

to be in the range of 0.10 to 0.15, which values are less than 4% of the total change.   Again they are larger for desiccation than humidification.

CONCLUSION

From the above results, CS-IPA may be concluded to be equally or more useful as an organic humidity sensor material than PSS salts which have been commercially used.   And CS-IPA impregnated paper has a possibility to be utilized as a simple and expedient humidity sensor, although some factors, such as durability, remain for further investigation.

REFERENCE

1. R.G.Schweiger, Tappi, 57, 86 (1974)

70

Langmuir-Blodgett films from cellulose ester derivatives

Y. Tsujii, T. Itoh, S. Ito*, and T. Miyamoto - Institute for Chemical Research, Kyoto University, Uji, Kyoto 611, Japan. *Department of Polymer Chemistry, Faculty of Engineering, Kyoto University, Sakyo-ku, Kyoto 606, Japan.

ABSTRACT

Mono– and multi–layer films prepared from cellulose octadecanoate derivatives were studied by electron microscopy and fluorescence spectroscopy. By electron microscopy it was found that a cellulose tri(octadecanoate) forms a perfectly homogeneous monolayer at π = 10 mN/m. This homogeneous monolayer on the water surface was repeatedly transferred onto a substrate by a horizontal lifting method to form a Y–type multilayer film. The surface structure was also observed to be homogeneous by a dark–field technique, and the film thickness was found to be proportional to the number of layers. This suggests a good regularity of the film structure. The in situ fluorescent data directly observed for pyrene–labeled cellulose octadecanoate monolayer at the air/water interface were consistent with the film structures. Analysis of the fluorescent quenching efficiency in built–up films revealed that the pyrenyl groups are two–dimensionally randomly distributed within each layer. Furthermore, in the multilayer film with high pyrenyl concentration the formation of two types of excimers was demonstrated by time–resolved fluorescence spectroscopy.

INTRODUCTION

Langmuir–Blodgett (LB) technique is very useful to architect high–performance multilayer films.[1,2] The fine structure of a surface film is a critical factor determining the quality of the relevant LB films.[3] Transmission electron microscopy (TEM) gives direct information on the fine structure of LB films. Recently, polymer LB films have attracted much attention for their many potential applicabilities as candidates with thermal and mechanical stability. By introducing a hydrophobic side chain, cellulose can be converted into an amphiphilic polymer that may possibly form an LB film. In this report, we prepared mono– and multi–layer films prepared from cellulose tri(octadecanoate) which would be expected to be thermally stable owing to the semiflexibility of the cellulosic main chain and studied the fine structure of the LB films by TEM.

For preparing a highly functional LB film, a functional group must be introduced without disturbing the monolayer characteristics. It is also very important to control the distribution of the introduced functional group. One method to introduce a functional group into an LB film is to mix amphiphilic chromophores with amphiphiles such as fatty acids to form a surface monolayer. However, the mixed LB film of fatty acid often causes aggregation (phase separation) and have a disadvantage of thermal and mechanical instability.[4-6] We have prepared cellulose octadecanoates partially labelled with a fluorescent probe, a (1–pyrenyl)butyroyl group, and studied by fluorescence spectroscopy the influence of the chromophore on the monolayer characteristics and the distribution of the chromophore within the LB film.

$$R = CO(CH_2)_{16}CH_3$$

$$CO(CH_2)_3$$

EXPERIMENTAL

Cellulose tri(octadecanoate) (P0) was prepared according to the acid chloride – pyridine procedure.[7] For fluorescence measurements, four samples (P1 – P4) of pyrene–labeled derivatives were prepared by reacting partially substituted cellulose octadecanoates with 1–pyrenebutyric acid.[8,9] The characteristics of the polymers prepared are listed in Table I.

Monolayer spreading on the surface of pure water was carried out from a 0.01 wt% benzene solution. The temperature of the subphase was kept at 293 K. The monolayer was compressed with a Teflon bar at a rate of 5–10 mm min^{-1}. The surface monolayer was transferred onto a substrate by the horizontal lifting method without a frame.[10] During the deposition, the surface pressure was kept constant.

The surface structure of mono– and multi–layer films transferred on the carbon film was observed on a JEOL JEM 200–CX or a Hitachi H–300 by applying a dark–field imaging mode of TEM. The film thickness was measured by a folding method; the

Table I Overall Degree of Substitution (DS), Degree of Substitution of Pyrene Units (DS$_{Py}$), and Limiting Areas per Glucose Unit (A$_0$)

Sample	DS	DS$_{Py}$	A$_0$ / nm^2
P0	3.0	0	0.71
P1	3.0	0.007	0.70
P2	3.0	0.038	0.70
P3	3.0	0.21	0.70
P4	3.0	1.1	0.84

mono– and multi–layer films transferred onto the platinum film were folded into two with the lining inside, and the fold edge was observed in the bright–field imaging mode of TEM.

Fluorescence spectra were recorded on a Hitachi 850 spectrophotofluorometer. A single photon counting method was applied for the measurements of time–resolved fluorescence spectra. In situ fluorescence measurements at the air/water interface were performed through two quartz optical fibers set on the middle of the trough.[9a] The fluorescence measurements were made for the chromophoric multilayers which were sandwiched between the two bilayers of P0 to suppress possible effect of the substrate and air interface. The deposition was made at the surface pressure of 10 mN/m.

π – A ISOTHERMS

Figure 1 shows surface pressure (π) – area (A) isotherms of the cellulose derivatives. P0 gives a typical condensed–type/isotherm, and the limiting area per glucose unit estimated by extrapolating the steepest tangent of the isotherm to zero surface pressure is 0.71 nm^2. Pyrene–labeled derivatives, P1 – P3, give almost the same π – A isotherms as unlabeled P0, and the limiting area, 0.70 nm^2, is also similar to that of P0. This suggests that the introduction of pyrenyl group as much as 7 % (by side–chain mole base) does not essentially alter the monolayer characteristics, giving a condensed monolayer similar to that of P0. On the other hand, P4 (DS$_{py}$ = 1.1) having a pyrenyl content of 38 % (by side–chain mole base) shows a larger limiting area (0.84 nm^2) and a small pressure of collapse (ca. 20 mN/m) because of too many bulky pyrenyl groups.

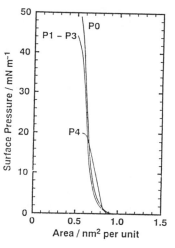

Fig. 1. Surface pressure (π) – area (A) isotherms measured at 293 K.

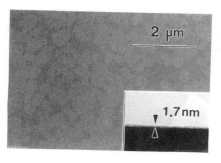

Fig. 2. Dark- and bright-field images of the P0 monolayer deposited at $\pi = 0$ mN m^{-1}.

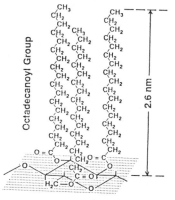

Fig. 3. Monolayer structure of cellulose tri(octadecanoate).

ELECTRON MICROSCOPY

Fine Structure of Monolayer Film Observed by TEM

Electron microscopy provides direct information on the morphology of surface films. Figure 2 shows a surface structure (dark field image) and film thickness (bright field image) of the P0 monolayer transferred at 0 mN/m. The surface film takes the so-called "island structure"; a heterogeneous film consisting of aggregates of non uniform size can be seen. With increasing compression, the film structure becomes more and more uniform and a perfectly homogeneous monolayer is obtained at 10 mN/m. The homogeneous monolayer has a thickness of 2.4 nm, which approximately suggests a molecular model illustrated in Figure 3: this model assumes that the glucopyranose rings lie flat in the layer boundary, and the alkyl side chains, fully extended, stand normal to the rings, hence to the film surface. Surprisingly, the film thickness at 0 mN/m was measured to be 1.7nm, not so different from that of the homogeneous monolayer. This presumably shows that, at 0 mN/m, the alkyl chains in the "island" already take a similar conformation to that in the monolayer.

Preparation and Structure of Multilayer Film

By horizontal lifting method homogeneous monolayers on the water surface at $\pi = 10$mN/m were repeatedly transferred onto the hydrophobic substrate with a constant transfer ratio of 2.0. When the plate is horizontally brought down to just touch the surface, the first layer will be deposited on it with the hydrophobic alkyl groups attached to the plate surface and the hydrophilic glucopyranosic groups appearing on the film surface. As the plate is brought up, the surrounding monolayer will be drawn up, folded, and deposited on top of the first layer with the hydrophilic groups inside and the hydrophobic groups outside. Namely, a bilayer (Y–type) film is formed through one operation.

In Figure 4, the thickness L of the multilayer films measured with a bright-field imaging mode of TEM is given as a function of the number of layers. The figure shows that L is proportional to the number of layers, indicating that each layer is built up regularly. From the slope of the line, the thickness per layer is calculated to be 2.3 nm, in good agreement with the thickness of the monolayer, 2.4 nm. The electron microscopic observation of the P0 10-layer film gave no characteristic texture, indicating a completely homogeneous surface structure of the built–up film.

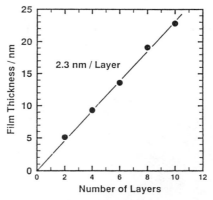

Fig.4 Thickness of multilayer films with differing number of layers.

The good transferability of the surface monolayer onto the substrate was also checked by measuring the fluorescence intensity of P1 and P2 as a function of the number of layers; the monomer fluorescence intensity (at 377 nm) increases proportionally to the number of layers.[9b)]

FLUORESCENCE SPECTROSCOPY

Fluorescence Properties at the Air/Water Interface

Figure 5 shows the fluorescence spectra of the P3 film as a function of the surface pressures, in which the intensity is normalized at the maximum peak height. The spectra are almost independent of surface pressures ranging from 0 to 10 mN/m. This indicates that the microscopic environment around the pyrene probe remains unchanged throughout the compression process between 0 and 10 mN/m, which is consistent with the fine structure of the surface monolayer observed by electron microscopy. On the other hand, when the surface monolayer collapses, the excimer emission intensity increases. This is

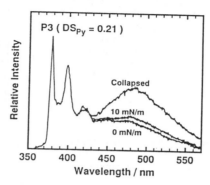

Fig.5 Fluorescence spectra of the P3 monolayer on the water surface.

because the collapse causes an increase in excimer site density. This observation suggests that the excimer emission can be a useful measure of monolayer structural orders.

The fluorescence intensity at 377 nm (monomer emission) was measured under nitrogen atmosphere as a function of surface concentration of sample P2.[9a] Below the surface concentration at which the monolayer begins to collapse, the intensity is proportional to the surface concentration. With a further compression, the intensity deviates from the proportional line. This is presumably because the collapse proceeds from weak positions in the film.

Excimer Formation and Chromophore Distribution in Multilayer Film

Figure 6 shows the fluorescence spectra of 10–layer films of P1 to P4. Samples of low pyrenyl content, P1 and P2, give virtually the monomer fluorescence only. With increasing pyrenyl content, the emission intensity from an excimer increases relatively to the monomer fluorescence. The excimer formation in the multilayer films of these polymers is highly suppressed compared with that in an LB film of mixed fatty acids having the same surface concentration of pyrenyl groups. This is clearly because the pyrenyl group is distributed homogeneously in the polymer multilayer film, whereas it easily aggregates in an LB film of mixed fatty acids.

The randomness of the spatial distribution of pyrenyl groups was confirmed by the following simulation. In Figure 7, the experimental data

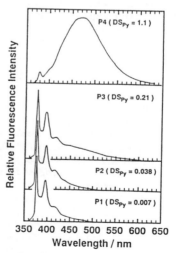

Fig.6 Fluorescence spectra of 10–layer films.

(closed circles) of a total quenching efficiency are simulated. The simulation line is calculated assuming a statically random distribution of pyrenyl group within a layer. The good fit of the experimental data to the simulation line suggests that the pyrenyl groups are two–dimensionally randomly distributed within each layer, directly reflecting the chemical randomness of the chromophoric side–chain distribution along the cellulosic main chain. This homogeneity is characteristic of a polymer LB film.

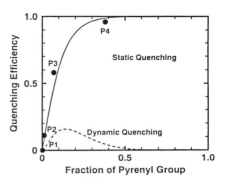

Fig.7 Quenching efficiency as a function of pyrenyl fraction.

Time–Resolved Fluorescence Spectroscopy

Figure 8 shows the time–resolved fluorescence spectra of 2–layer film of P4. A broad emission band observed in a long wavelength region exhibits a red–shift with time. This broad band is ascribed to two types of excimers; one is a normal excimer with a sandwich arrangement observed at around 480 nm, and the other is a partially overlapped excimer observed at around 430 nm. E1 is less stable and has a shorter lifetime than E2. The existence of the partially overlapped excimer indicates that the motion of pyrenyl groups in a condensed monolayer should be highly restricted and there are two types of excimer–forming sites in a layer.

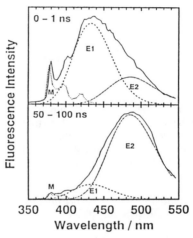

Fig.8 Time–resolved fluorescence spectra of 2–layer film of P4.
The symbols M, E1, and E2 indicate the emitting components due to the monomer excited state, partially overlapped excimer, and sandwich excimer, respectively.

REFERENCES

1) Roberts, G. G. *Adv. Phys.* **1985**, *34*, 475.
2) Kuhn, H.; Möbius, D.; Bücher, H. In *Physical Methods of Chemistry*; Weissberger A.; Rossiter, B. W., Eds.; Wiley; New York, **1972**; Vol. 1, Part 3B, p. 577.
3) Ries, H. E.; Walker, D. C. *J. Colloid Sci.* **1961**, *16*, 361.
4) Yamazaki, I.; Tamai, N.; Yamazaki, T. *J. Phys. Chem.* **1987**, *91*, 3572.
5) Blodgett, K. B. *Phys. Rev.* **1939**, *55*, 391.
6) Fukui, T.; Sugi, M.; Iizima, S. *Phys. Rev. B* **1980**, *22*, 4898.
7) Malm, C. J.; Mench, J. W.; Kendall, D. L.; Hiatt, G. D. *Ind. Eng. Chem.* **1951**, *43*, 684.
8) Shimizu, Y.; Hayashi, J. *Cellulose Chem. Tech.* **1989**, *23*, 661.
9) (a) Itoh, T.; Tsujii, Y.; Fukuda, T.; Miyamoto, T.; Ito, S.; Asada, T.; Yamamoto, M. *Langmuir*, in press. (b) Tsujii, Y.; Itoh, T.; Fukuda, T.; Miyamoto, T.; Ito, S.; Yamamoto, M. *Langmuir*, in press.
10)(a) Kamata, T.; Umemura, J.; Takenaka, T. *Chem. Lett.* **1988**, 1231. (b) Umemura, J.; Hishino, Y.; Kawai, T.; Takenaka, T.; Gotoh, Y.; Fujihira, M. *Thin Solid Films* **1989**, *178*, 281.

71

Processing and characterization of polyamide 6,6 – cellulose blends

M. Garcia Ramirez, J.Y. Cavaillé, J. Desbrières, D. Dupeyre and A. Péguy -
Centre de Recherches sur les Macromolécules Végétales, CNRS, BP 53X 38041
Grenoble Cedex, France.

ABSTRACT

Polyamide 6,6/cellulose blends were obtained using N-methyl morpholine N-oxide and phenol as solvents, and methanol as coagulant. The process used results mainly in a non miscible system, where PA6,6 spherulites are embedded in a cellulose matrix. Achieved blends were characterized by differential scanning calorimetry, dynamic mechanical measurements, scanning electron microscopy and wide angle X-ray scattering. From the comparison of mechanical data and theoretical predictions used for composite systems, it appears that, at the interface between PA6,6 and cellulose domains, a certain miscibility exists.

INTRODUCTION

For several years, the interest in polymer blends has been increasing, but in the particular case of cellulose as parent polymer, only a few studies have been published due to difficulties in processing this polymer. The two principal processes, are (i) chemical modification of cellulose allowing films or fibers production, followed by a regeneration step, or (ii) dissolution in an appropriate solvent, followed by a precipitation (or coagulation) step. We have chosen the second method, which necessitates finding a solvent common to each parent polymer of the expected blend. On the other hand, the choice of a partner for cellulose is made on the basis of the best interactions which can be expected between both polymers. For this reason, polymers with a high capability for hydrogen bonding are generally considered in such studies [1-5]. For our work, we have chosen polyamide 6,6 (PA66), which is known for its good mechanical properties. We have used N-methyl morpholine N-oxide (NMMO) and phenol as co-solvent of cellulose and PA66.

EXPERIMENTAL SECTION

Materials.

The cellulose used was supplied by Buckeye (England) with a degree of polymerization (DP) of 600. Polyamide 6,6 (PA66) was provided by Rhône-Poulenc (France) with a (DP) of 180. It was dried for 15 hours at 100°C. Solvents employed were N-methyl morpholine N-oxide (NMMO) and phenol from TEXACO (USA) and Prolabo (France), respectively.

Preparation of samples.

In order to dissolve PA66, phenol was used as co-solvent with NMMO. The water/cellulose ratio has been kept constant and the total amount of polymer was 5 wt% for all the blends. From the dissolution diagrams (fig.1) between NMMO and phenol in the presence of cellulose and in the presence of PA66, it was found that the optimal ratio between these solvents is 80/20 (NMMO/phenol) (w/w) at the final solution composition. Solutions were separately prepared and mixed in a Mini Max Molder (CS-183 MMX, from Custom Scientific Instruments, Inc) in appropriate ratios to produce blends with compositions ranging from 50/50 to 90/10 (cellulose/PA66) (w/w), see table I. The different mixtures were poured into an injection spinning device to obtain monofibers. Then the materials were precipitated using methanol as coagulant. Samples were finally dried at room temperature under vacuum and stored in a desiccator until used. Mass spectroscopy was used to confirm the absence of solvents.

TABLE I : Composition of solutions

Ref.	Blends	NMMO wt %	Phenol wt %	Cel. wt %	PA66 wt %	Water wt %
A	cellulose	62.4	18	10	0	9.6
B	90/10	66	19	4.5	0.5	10
C	80/20	66	19	4	1	10
D	65/35	68.7	19	3.25	1.75	7.3
E	50/50	72	16.7	2.5	2.5	6.3

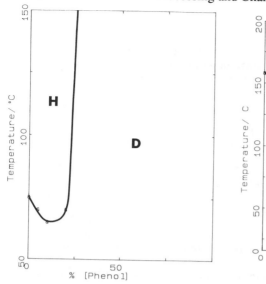

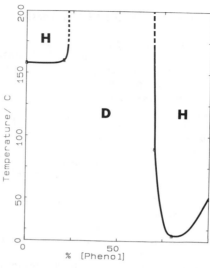

Figure 1. Dissolution diagrams of NMMO/Phenol ; on the left, in the presence of 5 wt % of cellulose, and on the right, in the presence of 5 wt % of PA66. H : homogeneous ; D : heterogeneous

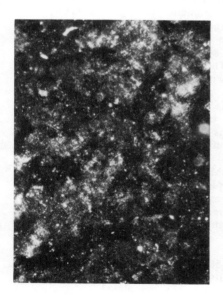
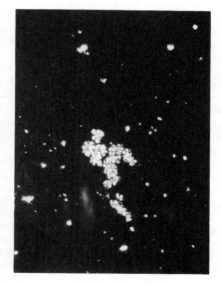

Figure 2. Optical micrographs of 50/50 cellulose/PA66 in a NMMO/ phenol solution; on the left, at 6.3 wt % of water and on the right, at 10.6 wt % of water.

EXPERIMENTAL RESULTS AND DISCUSSION

An optical microscope fitted with a METTLER heater was used to build the dissolution diagrams of NMMO/phenol in the presence of cellulose, and in the presence of PA66 . When PA66 and cellulose solutions are mixed, PA66 appears to precipitate, leading to its crystallization as spherulites with an average diameter of 1 to 5 µm, (fig.2a).

For comparison, another 50/50 blend was prepared with a total amount of water of 10.6 wt%. It results in PA66 domains of about 50 µm (fig.2b).

Differential Scanning Calorimetry (DSC) tests were performed with a Perkin Elmer DSC7. The scans were run at a heating rate of 20°C/min from -60 up to 110°C in the presence of nitrogen. In order to avoid the effect of the remaining water, samples were kept for 10 minutes at this temperature. Then the second scan was run from —60 to 300°C. The melting endotherm of PA66 can be observed even for the 90/10 blend, and the temperature of the peaks remains about constant at increasing PA66 content.

Dynamic mechanical measurements were carried out in a Mecanalyser (Metravib Instruments) at three frequencies, namely 0.01, 0.1 and 1 Hz in the temperature range -173 to 177 °C (100 K to 450 K), temperature being raised at 0.25 °C/min. All these tests were performed in the presence of dried nitrogen. Samples were previously heated at 140°C for one hour. From the experimental data, activation energy could be determined for each relaxation process, allowing detection of the presence of PA66 domains through its main relaxation (or α) process, which is characterized by a very high apparent activation energy (>300kJ/mol). On the other hand, we have used two models developed for composite systems in order to predict the mechanical behavior of our blends. The first one consists of a mean field approximation and is known as the Kerner-Dickie approach [6,7]. The second one is based on a percolation approach [8]. Both of these models lead to the same data. The main point is that the half-width of the a relaxation of PA66 in each blend is much larger for experimental data than for predictions (fig. 3). This probably results from the existence of a composition gradient at the interface between PA66 and cellulose domains.

A 6100 Scanning Electron Microscope (SEM) from Jeol was used to observe fractured surfaces. Two sets of samples were fractured at 77 K (liquid nitrogen), one for direct observations, the other one after an etching treatment by phenol, in order to remove PA66 domains from the blends. These observations (i) confirm that PA66 and cellulose are mainly inmiscible, and (ii) clearly show that PA66 forms inclusions in a continuous cellulose matrix (fig. 4). In addition, we have determined the loss of weight of 50/50 samples during their immersion in pure phenol at 60°C. Results show that most of the PA66 is dissolved, indicating that PA66 domains are percolated.

Wide angle X-Ray scattering (WAXS) patterns were obtained on a Philipps diffractometer. X-Ray diffractograms exhibit two main peaks at 20° and 24° which correspond to those of pure PA66 after crystallization from a solution of PA66 in phenol. In the case of the cellulose 90/10 wt%, the peak at 24° appears as a shoulder on the

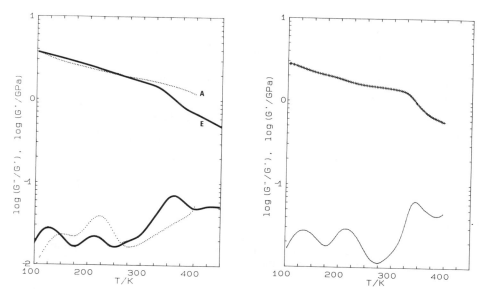

Figure 3. Dynamic mechanical behaviour at 0.1 Hz ; log(G'/GPa) and log(G"/G'); on the left, experimental data for cellulose (A) and 50/50 blend (E) ; on the right, theoretical.

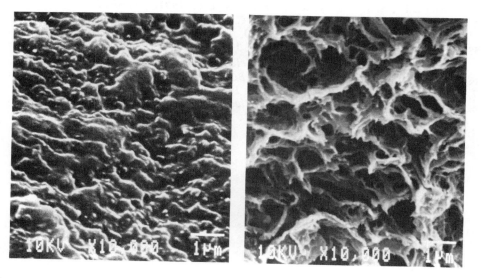

Figure 4. SEM micrographs obtained after an etching treatment; on the left, pure cellulose ; on the right, 50/50 blend.

amorphous bump of cellulose.

CONCLUSION

Following our conditions, crystallization of PA66 occurs during the mixing of solutions of cellulose and PA66. This phase separation is detected by different techniques. It results in a rather thin homogeneous dispersion, of PA domains in a matrix of cellulose (1 μm for 50/50 blend, 0.1μm for 80/20). However from the comparison between theoretical predictions and experimental data, it seems that a small part of PA66 is homogeneously blended with cellulose. This can occur at the interface between cellulose and PA66 and results in a composition gradient. The average glass transition temperature Tg seems to be increased and the glass transition range enlarged.
Finally our processing conditions allows us to drive the phase separation of PA66 and cellulose during the mixing step.

ACKNOWLEDGEMENT

We greatly acknowledge CONACYT (Mexico) for the financial support of one of us, M. Garcia Ramirez. This work is a part of her PhD Thesis.

REFERENCES

1. Nishio, Y ; St J. Manley, R. *Macromolecules* **1988**, *21*, 1270.
2. Nishio, Y ; St J. Manley R. *Polym. Eng. Sci.* **1990**, *30*:, 71.
3. Masson, J.F ; St J. Manley, R. *Macromolecules* **1991**, *24*, 6670.
4. Shibayama, M ; Yamamoto, T ; Xiao, C.F ; Sakurai, S ; Hayami, A. ; Nomura, S ; *Polymer* **1991**, *32*:, 1010.
5. Paillet, M ; Cavaillé, J.Y ; Desbrières, J ; Dupeyre, D ; Péguy, A ; submitted to *Colloid and Polymer Science.*
6. Kerner, E.H. *Proc. Phys. Soc.* **1956**, *B69*, 808.
7. Dickie, R.A. *J . Appl. Polym. Sc.* **1973**, *17*, 45.
8. Ouali, N ; Cavaillé, J.Y ; Pérez, J. *Plastics Rubber Composites Processing and Applications* **1991**, *16*, 55.

72

Morphologies in cellulose based blends

W. Berger, B. Morgenstern and H.W. Kammer - Dresden University of Technology, Institute of Macromoleculare Chemistry and Textile Chemistry, Mommsenstr. 13, O-8027 Dresden, Germany.

ABSTRACT

Phase morphologies arising from blend solutions which contain besides cellulose, aramide or poly(acrylonitrile) depend on the course of phase separation and are also effected by the initial total polymer concentration and the blend ratio of the polymers. Regular bicontinuous phase structures are favoured only when their pinning-down time is shorter than the "life-time" of the regular structures and when the growth rate of domain size is sufficiently small.

INTRODUCTION

Cellulose based polymer blends can be prepared via solution state only. The phase behavior of mixtures of cellulose with synthetic polymers was studied to some extent [1-4]. Generally, processing of a solution of incompatible polymers such as cellulose and synthetic polymers results in two-phase morphologies. However, this fact conveys the possibility to manipulate and to control the phase structure in polymer blends. From a practical point of view, a largely open field is offered for studies of morphology and structure-property relationships. It involves, from an academic view point, the fundamental problem of morphology control at different spatial levels under non-equilibrium conditions. Basic knowledge is necessary to understand and to master structure evolution in blend solutions by solvent evaporation, coagulation or thermal agitation.

This paper reports on evolution of phase morphologies in solutions and blends of cellulose and flexible coil or rigid-rod molecules subjected to different phase separation courses. Systems under discussion are blend solutions which contain besides cellulose, aramide or poly(acrylonitrile). The results are qualitatively interpreted in terms of spinodal decomposition in the early and late stage.

EXPERIMENTAL

The polymers used in our studies are commercial products. The cellulose samples (Heweten-201® and Avicel PH-101®) originate from degraded cotton linters having a cuoxam-DP of 230 and 170, respectively. Poly(acrylonitrile) (PAN) having a molecular weight of 100,000 g.mol^{-1} was supplied by Märkische Faser AG Premnitz. Aramide is the fiber HM-50 provided by Teijin Co.

The preparation of blend solutions using lithium chloride/dimethylacetamide (LiCl/DMAc) as solvent is described elsewhere [4,5]. The LiCl content of the blend solutions was kept constant and amounted to 4 wt%. The solutions contained the dissimiliar polymers in different ratios at 2 to 4 wt% of total polymer.

The solutions were poured onto glass slides and spread out to films of about 0.1 mm thickness. To study the evolution of phase morphologies the following course was chosen: After keeping the samples for different periods of time ranging from 10 s to 15 min at room temperature they were heated rapidly and annealed at constant temperatures ranging from 60 to 130°C until no further change in the resultant morphology could be detected. Isothermal phase structure formation was analyzed by light microscopy.

RESULTS AND DISCUSSION

Phase morphologies as they evolved from 50/50 cellulose/PAN solutions 50/50 cellulose/HM50 solutions, respectively, of initially 3 wt% of total polymers using different courses of solvent evaporation are presented in Figures 1 and 2. As can be seen, quite different phase morphologies emanate, e.g., in Figure 1a an irregular morphology of droplet-matrix type occurs while in Figure 1b a regular, cocontinuous morphology can be seen. This behavior can be discussed in terms of spinodal decomposition and subsequent coarsening processes.

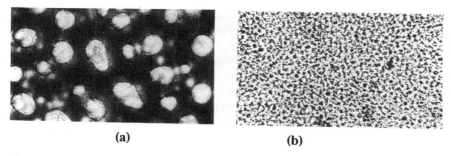

(a) **(b)**

Figure 1. Optical micrographs of decomposition structures evolved
from cellulose/PAN/DMAc solutions by keeping the cast films for
10 s (a) and 15 min (b) at room temperature followed by annealing
at 60°C. (⊢————⊣ 25 μm).

As the experimental results suggest a cocontinuous, network-like pattern of the
growing morphologies exists only over a certain period of time. Afterwards,
the regular, cocontinuous structures decay into droplet-matrix morphologies.
This is equivalent to the existence of a critical domain size d_c. One may say,
the phase-separated structure grows self-similarly over a range of length scales
bounded by an upper cut-off length which is determined by inner parameters
of the system. Above d_c it breaks up into an irregularly shaped morphology.

If one assumes that domain growth proceeds below d_c in conformity with the
Lifschitz-Slyozov process [6], one can estimate the period of time t_c in which
the cocontinuous morphology exists after initiation of the phase separation
process

$$t_c \sim \frac{\eta}{kT} d_c^3 \tag{1}$$

where η denotes the viscosity and kT the thermal energy. The critical length
scale d_c can be expressed by terms of the total polymer concentration ϕ_p [4].
The quantity d_c has been calculated in Ref. [4]. Therefrom, one gets

$$t_c \sim \frac{\eta}{kT} (\Phi_p / \Phi_p^c)^{-3/4} (\Phi_p - \Phi_p^c)^{-3/4} \tag{2}$$

The results presented in more detail in the following can be discussed at least
qualitatively in terms of Equation (2) when one compares t_c and the pinning-
down time t_p which is governed by the rate of solvent evaporation. The lower
the rate of solvent evaporation the lower the polymer concentration ϕ_p where
phase separation begins. On the other hand, the pinning-down time after which
phase separation ceases due to enhanced viscosity is inversely proportional to
the evaporation rate. The thermodynamic driving force for phase separation is
related to the difference $(\phi_p - \phi_p^c)$. Consequently, a low evaporation rate is

accompanied by a low thermodynamic driving force. When the thermodynamic driving force is low, regular phase patterns may grow self-similarly until they are pinned-down. In the opposite case, at high evaporation rates, the driving force is high, but, the pinning-down time is shortened to such an extent that again regular phase morphologies are preserved. Thus, there exists a delicate interplay of two characteristic periods of time - the life-time t_c of the regular morphology and the externally imposed pinning-down time t_p. Only when (t_c-t_p) is positive regular morphologies can be seen. This behavior is documented in Figure 2 showing examples of morphology evolution in 50/50 blends of cellulose and aramide.

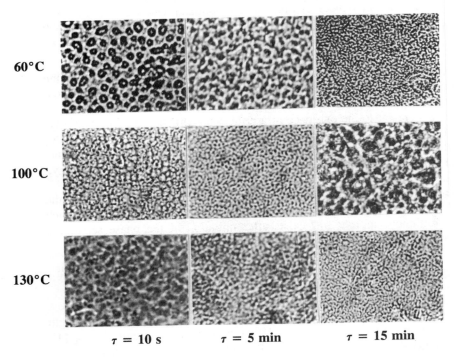

Figure 2. Optical micrographs of phase-separated morphologies obtained from 50/50 cellulose/HM50 solutions containing initially 3 wt% of total polymer. Storage periods τ at room temperature and annealing temperatures as indicated. (⊢――――⊣ 20 μm).

Looking for example at solutions kept for 5 min at room temperature and annealed afterwards at 60 and 130°C, respectively, phase separation starts in the former case at lower polymer concentration than in the latter. However, owing to the shortened pinning-down time at 130°C a regular morphology can be detected at higher temperature but not at the lower. Moreover, inspection of the blends annealed at 60 and 130°C, respectively, reveals that with increasing storage period τ at room temperature the phase structure becomes increasingly finer.

The effect of the storage period τ turns out to be different for different annealing temperatures. At low annealing temperatures the evaporation rate is sufficiently low. Therefore, the system can follow the course of evaporation over an extended period of time which means that phase separation starts at lower polymer concentrations with increasing τ. As a result, the phase structures become coarser more and more with decreasing τ. At high annealing temperatures the polymer concentration where phase separation starts is approximately independent of storage period τ at least for $\tau \geq 5$ min. But, the period of time for phase unmixing is significantly reduced resulting in finer phase structures as in the former case.

It is interesting to note that a divergent behavior is found for blends annealed at 100°C. Owing to the higher rate of evaporation at an annealing temperature of 100°C in comparison to lower temperatures the initial polymer concentration of phase separation is enhanced with increasing storage period at least again for $\tau \geq 5$ min. This results in finer structures with decreasing storage periods.

One should stress the point that the initially more dilute solution ends up in an irregular phase structure while more regular structures emanate from the concentrated solution. This can clearly be recognized in Figure 4. The initially reduced mobility of the chain molecules in more concentrated solutions means that phase separation is terminated earlier than in the case of diluted solutions and in almost all cases bicontinuous morphologies are preserved. The differences in the morphologies resulting from the various courses are less pronounced for solutions with initial polymer concentrations in the vicinity of the critical concentration.

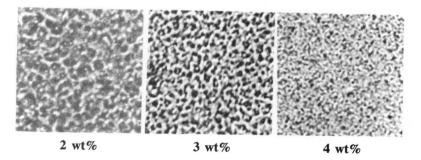

2 wt%   3 wt%   4 wt%

Figure 3. Optical micrographs of phase-separated morphologies obtained from 50/50 cellulose/HM50 solutions of given initial polymer concentrations. Storage period at room temperature, 5 min; annealing temperature, 60°C. (⊢————⊣ 20 μm).

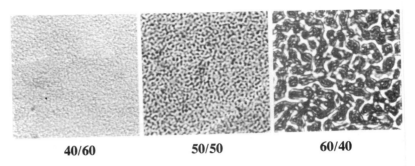

40/60   **50/50**   **60/40**

Figure 4. Optical micrographs of phase-separated morphologies obtained from cellulose/HM50 solutions of the indicated polymer ratios containing initially 3 wt% of total polymer. Storage period at room temperature, 5 min; annealing temperarture, 100°C. (⊢————⊣ 20 μm).

More diluted solutions tend to more coarse-grained structures than the concentrated solutions. The same effect can be seen when one varies the blending ratio as Figure 4 shows for cellulose/aramide solutions containing initially 3 wt% of total polymer and annealed at 100°C. This behavior is related to the asymmetry of the phase diagram [4]. The cellulose/aramide 60/40 blend solution is equivalent to initially more dilute solutions than the 40/60 blend solution. Consequently, the latter display finer structures than the former.

In conclusion, we studied phase morphologies arising from solution cast films of cellulose-containing polymer blends in the late stage of demixing. These morphologies can be characterized as being either regular bicontinuous or irregular. The conditions for the existence of regular phase structures in the late stage of phase decomposition can be specified as follows: Periodic structures can be detected only when the pinning-down time t_p is shorter than the "life-time" t_c of the regular structures: t_c-t_p>0. The growth rate of domain size which is proportional to kT/η must be sufficiently small.

REFERENCES

1. Savard, S.; Levesque, D.; Prud'homme, R.E. *J. Appl. Polym. Sci.* **1979**, *23*, 1943.
2. Tager, A.A.; Adamova, L.V.; Kryakunov, A.A.; Grinsphan, D.D.; Savitskaya, T.A.; Kaputskii, F.N. *Vysokomol. Soed.* **1985,** *B27*, 1943.
3. Morgenstern, B.; Kammer, H.W. *Polym. Bull.* **1989**, *22*, 265.
4. Schöne, A.; Morgenstern, B.; Berger, W.; Kammer, H.W. *Polym. Networks Blends* **1991**, *1*, 109.
5. Morgenstern, B.; Keck, M.; Berger, W. *Acta Polym.* **1990**, *41*, 86.
6. Lifschitz, I.M.; Slyozov, V.V. *J. Phys. Chem. Solids* **1961**, *19*, 35

73

Reinforcing wood particleboards by glass or flax fibres

F. Tröger and G. Wegener - Institute for Wood Research, University of Munich, D-8000 München 40, Winzerestrasse 45, Germany.

ABSTRACT

The aim of the present investigation was to effect an improvement of conventional phenol resin particleboards by reinforcing the surface layers. The surface layer materials consist of very fine as well as slender particles. Phenol resin impregnated glass fibre mats and laboratory isocyanate glued flax straw mats were used for reinforcing. An increase in bending strength of 40 % of the non-reinforced control was achieved with slender as opposed to fine surface layer chips. A comparison with the control showed a strength increase due to glass fibre reinforcement of 30 % and 20 % for fine and slender surface layer chips respectively. The E-modulus in bending of the non-reinforced control varied for slender and fine surface layer chips and was 25 % higher for slender chips. Glass fibre reinforcement improved E-modulus in bending by 19 % and 16 % for fine and slender surface layer chips respectively. Reinforcing particleboards with glued flax straw mats led to a marked improvement in bending properties. Compared with the control, the reinforcing resulted in an increase in bending strength and in E-modulus in bending of 60 % and 40 % respectively.

INTRODUCTION

The future of particleboard application will depend increasingly on the development of speciality type particleboards. Existing particleboards for special applications such as OSB, waferboard, veneer panels, molded parts from reconstituted wood and boards for structural purposes provide ample proof for the fact that the "breeding" of wood based materials with special properties is opening new fields of application.

The present investigation deals with the improvement of three-ply particleboards by reinforcing the surface layers. This also led to a definite increase in bending strength. The reinforcing materials used were different glass fibre mats and flax straw mats.

Platform floors, a field of application for special particleboards

Platform floors consist of adjustable metal pedestals (steel, aluminium) glued or fastened onto the subfloor by means of screws. Larger horizontal shear loads are counteracted by specific pedestal reinforcements. Standard distance between pedestals is 600 mm, i.e. they are arranged in a pattern of 600 x 600 mm. This substructure serves as support for the finished platform floor. The space inbetween can be used for installations (electricity, water, gas, air vacuum cleaning devices) and is easily accessible anywhere by removal of individual platform elements.

The following materials are used for modern platform floors:
- wood based panels (mainly high density particleboards)
- mineral materials (gypsum bonded fibreboards, cement bonded particleboards)
- metals (steel, aluminium)
- combinations of the above

MATERIAL AND METHODS

Glass fibre mats

Glass fibres, in particular the E-glass types, are known for their high tensile and low stretch properties and for their low density. They are therefore often used as reinforcing material for different composites. In this investigation three different types of glass fibre mats, A, B and C, were tested (Table 1). These mats had been pre-condensed with phenolic resin.

Table 1: Characteristic values of pre-impregnated glass fibre mats

glass fibre mats type	A	B	C
mesh width [mm]	0.45	0.6	0.75
thickness [mm]	0.63	0.92	1.80
filament thickness [μm]	5	8	13
area weight [g/m^2]	300	477	865
resin impregnation [g/m^2]	102	157	260
tensile strength [N/mm^2]	350	720	1400

Flax straw

A high cellulose content and very long fibres account for the much greater tensile strength of flax as compared with e.g. wood (Table 2). Flax suggested itself, therefore, as suitable material for the reinforcement of particleboards. Flax is generally used in the form of retted flax or flax straw.

Table 2: Comparison of characteristic values of flax and spruce

	flax	spruce
cellulose content [%]	65	42
polyoses content [%]	16	24
lignin content [%]	2.5	28
fibre length [mm]	20 - 40	3.4
fibre thickness [μm]	26	31
breaking length [km]	73[1)	19.5

[1) Breaking length of unretted flax; that of retted flax is 2 to 3 times as high

Particles for surface layer and core

Surface layer
- very fine particles from beech wood
- very slender, sieved softwood particles

Core
- typical core material from beechwood

Laboratory particleboards and testing
Production and method of reinforcement

- surface layer to core ratio   35 : 65
- resin type   phenol-formaldehyde
- resin content: - surface layer   12
                          - core   10.5 %
- catalyst   0.7 %
- wax   0.3 %
- hot pressing: - time   17s/mm
                          - temperature   180°C
- thickness of boards: - with glass fibre reinforcement   38 mm
                          - with flax straw reinforcement   20 mm
- density of boards: - with glass fibre reinforcement   820 kg/m³
                          - with flax straw reinforcement   710 kg/m³

Board testing according to German standards:

density   DIN 52 361
surface soundness   DIN 52 366
internal bond   DIN 53 365
bending strength   DIN 52 361
E-modulus in bending   DIN 52 361

Reinforcement with glass fibre mats

For the investigation with glass fibre mats as reinforcing material two different surface layer particle types (fine particles from beech wood and slender softwood particles) were used in combinations with mats A, B and C (Table 1).
For comparison, one surface layer particle type each was used to produce control specimens without glass fibre mat reinforcement. For maximum improvement of bending properties the glass fibre mats were placed within the uppermost part of the surface layer.

However, to prevent damage to the glass fibre mats during sanding of the finished product, mats A and B had to be covered on both sides by a 25 % portion by weight of the surface layers.

For mat C, the thickest one, which was used in combination with slender softwood surface layer particles, a higher covering layer was required, effected by placing a 40 % portion by weight of the surface layer on top of the glass fibre mat.

Reinforcement with flax straw mats

Flax stems were placed together to form mats (Fig. 1) to which isocyanate was applied on both sides prior to use in the particleboard. Table 3 provides a survey of the different flax straw reinforced particleboards. The flax straw mats were incorporated in the particleboard mat and were covered on both sides by 20 % portions by weight of the surface layers.

Table 3: Different types of particleboard reinforcement with flax straw mats

	surface layer particle type	flax portion related to dry weight particle-board [%]	orientation of flax stems in relation to particleboard axis		position of mat within particleboard cross-section	
0 Fl	softwood slender	non-reinforced control				
1 Fl	softwood slender	3	parallel		center of surface layer	
2 Fl	beech wood fine	3	parallel		center of surface layer	
3 Fl	softwood slender	6	mat 1: parallel	mat 2: across	mat 1: center of surface layer	mat 2: between surface layer and core

Figure 1:
Laboratory flax
straw mat before
gluing

RESULTS

Improving bending properties of particleboards by reinforcing with glass fibre mats

The high density particleboards show extremely high values as regards internal bond and surface soundness (Table 4) [1, 2].
However, smooth rupture areas observed after the surface soundness tests indicate that the bonding of resinated glass fibre mats with wood particles can still be improved.
Differences between the two controls (0) produced with the two types of surface layer particles confirm the positive effect, often described in literature, of slender surface layer particles on bending strength as well as on E-modulus in bending. Glass fibre reinforcement with mat A of the surface layer led to an improvement by 20 % for the slender particles as compared with the control and by 30 % for the beechwood surface layer particles. The stronger glass fibre mats B and C produced no significantly greater improvement in bending strength, as shown in Table 4.

Table 4: Properties of 38 mm laboratory particleboards reinforced with glass fibre mats (A, B, C) and non-reinforced controls (0)

surface layer particle type	density [kg/m^3]	internal bond [N/mm^2] V 20	surface soundness [N/mm^2]	bending strength [N/mm^2]	E-modulus in bending [N/mm^2]
slender softwood part.					
0	820	1.12	2.54	33.8	4200
A	835	1.18	2.47	40.4	4500
B	835	1.12	2.35	42.5	4700
C	810	1)	1)	42.4	4900
fine beech wood part.					
0	815	1.04	1.96	24.1	3300
A	835	1.07	1.93	31.6	3700
B	830	1.04	1.82	34.2	3900
C	790	1)	1)	34.7	4000

1) not tested

Improving bending properties of particleboards by reinforcing with flax straw material

The incorporation of flax straw mats into the surface layers had no adverse effects on internal bond and on surface soundness (Table 5) [3].
Similar to glass fibre reinforcement, surface soundness samples of particleboards reinforced with flax straw show smooth glueline ruptures and only few material ruptures. This implies that the bonding of flax straw and wood particles needs improving.
Reinforcing particleboard with isocyanate impregnated flax straw mats led to a marked improvement in bending properties. This goes in particular for specimen Fl 1 which, compared with the control Fl 0, produced an increase in bending strength and in E-modulus in bending of 60 % and 40 % respectively.

Table 5: Properties of 20 mm laboratory particleboards reinforced with flax straw mats in surface layers (Fl 1 to Fl 3) and of non-reinforced controls (Fl 0)

surface layer material	density [kg/m^3]	internal bond [N/mm^2]	surface soundness [N/mm^2]	bending strength [N/mm^2]	E-modulus in bending [N/mm^2]
Fl 0	710	0.85	1.52	24.8	3200
Fl 1	715	0.67	1.31	39.6	4500
Fl 2	710	0.69	1.32	36.8	3900
Fl 3	725	0.70	1.42	40.5	4700

CONCLUSION

1. All reinforced materials applied in this investigation contribute effectively towards a considerable improvement in bending strength and E-modulus.
2. Surface soundness tests, in particular, revealed that the bonding of wood particles with the reinforcing materials (glass fibre mats and flax straw mats) needs improving.
3. The bending strength and elasticity levels achieved to date with glass fibre and flax straw reinforced particleboards for special applications are clearly superior to the bending properties of waferboard while they do not, as yet, equal those of OSB.

REFERENCES

1. Tröger, F.; Wörner, H., *Holz Roh-Werkstoff* **1991**, 49 (10) 405-409.

2. Wörner, H. *Diplomarbeit 1990, Forstwissenschaftl. Fakultät der Universität München.*

3. Ullrich, M. *Diplomarbeit 1991, Fachhochschule Rosenheim.*

74

The effect of compatibilizers on interfacial bonding in lignocellulosic fiber/polyethylene composites

Wen-Chung Tai*, Stephen L. Quarles* and Timothy G. Rials** - *University of California, Forest Products Laboratory, 1301 S. 46th Street, Richmond, California 94804. **US Forest Service, Southern Forest Experiment Station, 2500 Shreveport Highway, Pineville, Louisiana 71360 USA.

ABSTRACT

Douglas-fir (*Pseudotsuga menziesii*) bark fibers were treated with a maleic-anhydride-modified polypropylene (MAPP) compatibilizing agent to final loading levels of 0%, 4%, and 20% by weight. In order to evaluate the ability of the bark fiber treatments to enhance interfacial properties of the fiber/polyethylene composites, single- and multiple-fiber specimens were isolated from compression molded samples. Results from the photoelastic observations from fracture tests using transmitted polarized light and fracture surface evaluation using scanning electron microscopy indicated that adhesion between the matrix and fiber was better at a 4% MAPP loading level compared to either 0% or 20%. Differences in acoustic emission generated during fracture in specimens with bark fibers having different MAPP loading levels were not evident.

INTRODUCTION

One of the dominant factors that influences performance properties of multi-phase polymer systems is the strength and quality of the interfacial region [1]. This is particularly true for fiber reinforced plastics (FRP's) in that the level of adhesion between the two dissimilar phases determines the stress transfer efficiency from the polymer matrix to the load-bearing fiber. The use of lignocellulosic fibers for reinforcement has met with only limited success due largely to deficiencies in interfacial quality, particularly with polyolefin matrix materials such as polyethylene and polypropylene. As a result, research in this area has focused on developing potential coupling or compatibilizing agents, designed to interact with both fiber and matrix, to improve phase compatibility and composite properties.

Recent reports in the literature [2-5] have identified maleic anhydride-modified polypropylene (MAPP) as one of the more effective compatibilizing agents for wood fiber/polyolefin composites, resulting in the improved ultimate mechanical properties. While mechanical testing has indicated a true reinforcing effect for the polyethylene and polypropylene, the exact origin of this improvement remains somewhat ambiguous. Felix and Gatenholm [6] recently confirmed the presence of covalent bonds between the MAPP and wood fiber; however, the nature of interaction with the polymer matrix is unclear. In an effort to further define the interfacial characteristics of this composite system, Douglas-fir (*Pseudotsuga menziesii*) bark fiber was modified with a MAPP, and single-fiber specimens prepared with a low-density polyethylene matrix. This report presents preliminary results on the use of stress optical and acoustic emission analysis to evaluate interfacial quality.

EXPERIMENTAL

Douglas-fir bark fiber was obtained from Weyerhaeuser Company, Federal Way, Washington and was used as received. Typical fiber dimensions were 2.0 mm in length and 0.12 mm in width ($l/d = 16.7$). A commercially available MAPP, Epolene-43 (Eastman Chemical Products, Inc.), was used to modify the fibers; this material had an average molecular weight of 4,500 and an acid number of 47. The matrix polymer was a commercially available low-density polyethylene obtained from Scientific Polymer Products, Inc., and had a number average molecular weight of approximately 50,000 g/mole, and a density of 0.92.

Douglas-fir fibers were immersed in a dilute solution of MAPP in hot (65 C) cyclohexane and treated to selected final loadings. Fibers incorporated in the 0% MAPP specimens were still immersed in hot cyclohexane. After drying, the fibers were placed between two sheets of polyethylene and compression molded at a press temperature of 150 C. The pressure was stepped up from 0.7 MPa to 1.8 MPa to 2.9 MPa in 30 s intervals (90 s press cycle). The final thickness of the composite was approximately 0.3 mm. Single-fiber specimens were then isolated from the molded plaque by punching dog-bone (for fracture testing) or rectangular (for AE testing) specimens.

Stress optical micrographs were taken at different strain levels using a micro-tensile stage mounted on an Olympus BH-2 microscope with a polarizing attachment. Fracture surface analysis was accomplished using a Hitachi Model S-2300 scanning electron microscope (SEM). Specimens used in the SEM analysis were obtained by making a razor cut perpendicular to the fiber at mid-length in a dog-bone specimen after tension testing.

AE was monitored during tension testing. The rectangular specimen used for AE testing was approximately 50 mm in length by 10 mm in width, and was prepared as a double-edge-notched fracture specimen, each notch being 1 mm long, and located at mid-fiber-length at either edge of the specimen. AE time-domain data was collected using an AET 5500 general purpose monitoring system (Babcock and Wilcox, Inc.), and data to be used for frequency-domain analysis was acquired using a Sonotek 825 transient recorder (Sonix, Inc.). A PICO transducer (Physical Acoustics Corp.), resonant at 500 kHz, was used for all AE tests. A hot melt adhesive was used to couple the transducer to the composite.

RESULTS AND DISCUSSION

Photoelastic results for 0% and 4% MAPP loading fiber composites are given in Figures 1 and 2. These figures show the fiber tip in the polyethylene matrix at a 3 mm elongation during tension testing. As is evident in Figure 1 (0% MAPP), a significant photostress pattern is not observable at this elongation, and there is a noticeable cavity beyond the fiber tip. This observation was consistent with an earlier report [7]. Figure 2 (4% MAPP) shows a much more clearly defined photostress pattern and much less of a cavity at the fiber tip. Micrographs of the 20% MAPP composite at the 3 mm elongation were very similar to the 0% MAPP composite. These results indicate that at the 4% MAPP loading level, the adhesion between the fiber and matrix material was more effective, and consequently so was the stress transfer to the fiber.

Scanning electron microscopy results were similar to those obtained in the photoelastic analysis. The micrograph of the fiber/matrix interface for the 0% and 4% MAPP loading are given in Figures 3 and 4. Lack of adhesion between the fiber and matrix was evident at all loading levels, but comparing Figures 3 and 4, it was clear that the quality of the interface containing the 4% MAPP fiber was better than that with the 0% MAPP fiber. The interface of the composite containing the 20% MAPP fiber (not shown here) was worse than that containing the 0% MAPP fiber, possibly due to a weak boundary layer created by the high volume of the low molecular weight MAPP material.

The results of the AE analysis were not as conclusive as those resulting from the photoelastic and SEM analyses. Time- and frequence-domain features of the AE data generated during tension testing were very similar regardless of

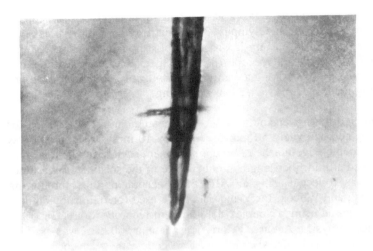

Figure 1. Tip of bark fiber with 0% MAPP. The micrograph was taken at 3 mm elongation and 10x magnification.

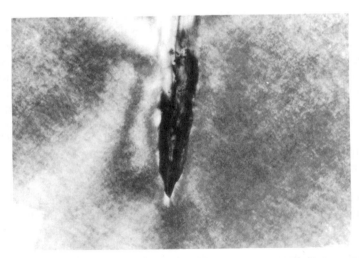

Figure 2. Tip of bark fiber with 4% MAPP. The micrograph was taken at 3 mm elongation and 10x magnification.

MAPP loading. For example, the peak amplitude of approximately 95% of all AE events for all specimens were less than 59 dB (considering total system gain, this level can be considered a low peak amplitude). The low peak amplitude events were most likely associated with bark fiber/matrix frictional forces occurring during tension testing. Frequency domain analysis consistently showed two dominant peaks. The first was approximately at 150 kHz, and the second was at 500 kHz. The 500 kHz peak could be expected because it was the same as the resonant frequency of the transducer. The lack of distinction in AE between specimens with different MAPP loadings indicted that the failure mechanism for all specimens was the same.

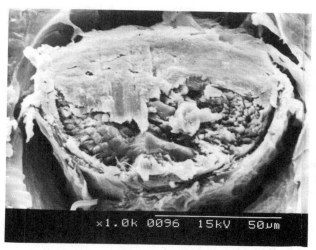

Figure 3. Scanning electron micrograph of bark fiber composite with 0% MAPP. Micrograph taken after tension testing.

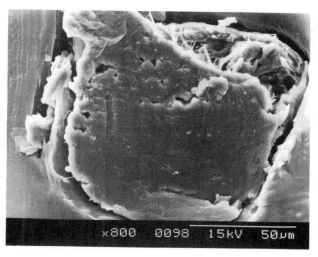

Figure 4. Scanning electron micrograph of bark fiber composite with 4% MAPP. Micrograph taken after tension testing.

CONCLUSIONS

Results from the photoelastic tests and scanning electron microscopy indicated that a 4% loading level was superior to either the 0% or 20% MAPP loading levels. Results from the AE tests were inconclusive. All specimens, regardless of MAPP loading level, generated predominately low peak amplitude events. Spectral analysis consistently showed two peaks, one at approximately 150 kHz, and the other at 500 kHz. The AE results indicated that failure mechanisms were similar regardless of the MAPP loading level. It is clear that with the current experimental conditions, insufficient stress transfer existed between the polyethylene and MAPP treated bark fibers to cause fiber failure.

ACKNOWLEDGEMENTS

The authors would like to thank Weyerhaeuser Company for providing the bark fiber used in this study. The use of trade names does not constitute endorsement by the USDA-Forest Service.

REFERENCES

1.   Theocaris, P.S. *Colloid & Polymer Sci.*, 265:461-480 (1987).

2.   Raj, R.G., B.V. Kokta, and C. Daneault. *Makromol. Chem., Macromol. Symp.*, 28:187-202 (1989).

3.   Takase, S. and N. Shiraishi. *J. Appl. Polym. Sci.* 37:645-659 (1989).

4.   Dalvag, H., C. Klason, and H.-E. Stromvall. *Intern. J. Polymeric Mater.*, 11:9-38 (1985).

5.   Woodhams, R.T., G.T. Thomas, and D.K. Rodgers. *Polym. Eng. Sci.*, 24:1166-1171 (1984).

6.   Felix, J. and P. Gatenholm. *J. Appl. Polym. Sci.*, 42:609-620 (1991).

7.   Johnson, J.A. and W.T. Nearn. in Jayne, B.A. (Ed.) *Theory and Design of Wood and Fiber Composite Materials*, pp. 371-400, Syracuse University Press, Syracuse, NY, (1972).

75

Changes in wood polymers during the pressing of wood-composites

D.J. Gardner*, D.w. Gunnells*, M.P. Wolcott* and L. Amos** - *West Virginia University, Division of Forestry, Morgantown, WV 26506-6215 USA. **Weyerhaeuser Technology Center, 32901 Weyerhaeuser Way S., Federal Way, WA 98003 USA.

ABSTRACT

This study describes changes in polymer structure of wood flakes when heated under conditions relevant to the hot-pressing of wood composites. The techniques of Dynamic Contact Angle (DCA) analysis, Carbon 13 Nuclear Magnetic Resonance Spectrometry with cross polarization and magic angle spinning (C-13 NMR CP/MAS), X-ray Photoelectron Spectroscopy (XPS), and X-ray Diffraction were used to describe the chemical nature of the wood polymers in situ. Resulting data indicate that the cellulose crystallinity in the wood increases slightly in response to heat treatment while aryl-alkyl linkages are cleaved in the lignin polymer. The surface elemental composition of the heated wood does not appear to be altered to any great extent. Surface free energy as measured by DCA exhibits a step decrease above the glass transition temperature of lignin.

INTRODUCTION

During the hot-pressing of wood composites, the wood component is exposed to rapidly changing temperature and humidity conditions. In addition, the wood consolidates and densifies under the extreme mechanical forces applied during pressing. Previous studies have shown that the physical properties of the wood component change during the pressing cycle [1-3]. In general, low pressing temperatures result in a decrease in flake strength and stiffness, whereas, high pressing temperatures result in an increase in flake strength and stiffness.

Dynamic mechanical analysis (DMA) was used to non-destructively determine stiffness changes in wood flakes after hot-pressing [3]. Flakes pressed at different moisture and temperature conditions showed varying changes in specific modulus (ratio of modulus to density). Depending on the temperature and moisture conditions encountered during hot-pressing, different degrees of cell wall damage will occur [4]. This damage primarily results when lignin is in the glassy region.

Moisture has been shown to lower the glass transition temperature (Tg) of the wood polymers [5]. After hot-pressing flakes at zero percent moisture content and 220°C, substantial increases in specific modulus occur [3]. Recovery of the wood cellular structure decreases with increased pressing temperature unless moisture is present [4]. These findings are corroborated by the decreases in thickness swell that result when hot-pressed panels are treated under pressure for extended periods of time [6]. These results suggest that changes in wood polymer structure or morphology are occurring with heat treatments. Furthermore, these changes are likely to be influenced by the presence of moisture.

EXPERIMENTAL

Yellow-poplar (*Liriodendron tulipifera*) sapwood flakes were used for the testing. Wood chemical compositions were obtained using standard methodologies. Flakes were pressed at a stress of 22 kN for 12 minutes at three environmental conditions: 1. 23°C, 2. 220°C dry, and 3. 120°C steam. Dynamic contact angle (DCA) analysis was used to measure wood surface free energy changes, C-13 NMR CP/MAS spectroscopy was used to measure bulk wood polymer chemical functionality changes, X-ray photoelectron spectroscopy (XPS) was used to measure surface elemental composition changes and X-ray diffraction was used to measure cellulose crystallinity changes resulting from hot-pressing. Further information on the experimental procedures used in this study can be found in the literature [7].

RESULTS AND DISCUSSION

POLYMER STRUCTURE

The chemical composition of the yellow-poplar sapwood was 18.5% lignin, 47.6% alpha cellulose, 34.1% hemicellulose, 2.4% extractives, and 0.12% ash.

The crystallinity index of the pressed yellow-poplar sapwood was 63.4% at 23°C, 65.9% at 220°C, and 68.7% at 120°C in steam. Increases in cellulose crystallinity for both heat-treated and steam-treated flakes correspond to increases in elastic modulus described in the literature [4]. Crystallinity increases are likely to occur when amorphous polymers exceed the glass transition temperature. In the rubbery phase, the increased mobility of the matrix polymers, i.e. lignin and hemicellulose, can permit reorientation and crystallization of the cellulose microfibrils. This effect has been noted in

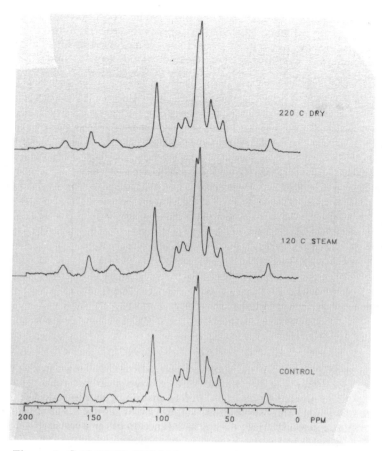

Figure 1 C-13 NMR CP/MAS spectra of yellow-poplar sapwood.

cellulose derivative and lignin blends [8]. However, crystallinity increases alone can not account for increased mechanical properties of hot-pressed flakes because the largest stiffness increases occur at 220°C and dry conditions, whereas the largest increases in crystallinity index occur in 120°C/steam pressed flakes.

The C-13 NMR CP/MAS spectra for the pressed yellow-poplar samples are shown in Figure 1. When the wood was heated without water present, cleavage of the alkyl-aryl ether bonds in lignin were noted by the shift in the C-13 NMR CP/MAS spectra from the 158-151 ppm to the 150-145 ppm frequency range indicating formation of free phenolic groups in the lignin polymer structure. These effects were not evident in the steam heated wood flakes. It is theorized that under the high temperature pressing conditions, some of the hemicelluloses can decompose to acetic acid and furfural compounds which can react with the free phenolic lignin moieties to auto-crosslink the amorphous wood polymer components. Although more evidence is needed to support this theory, auto-crosslinking can explain the low recovery of wood flakes compressed at high temperatures, the increased stability of heat-treated wood panels, and disproportionate increases in specific modulus from dry heat-treatment.

Table 1 Surface elemental composition as determined by XPS.

Sample	% Atom Composition		
	Carbon (C)	Oxygen (O)	O/C Ratio
Untreated	80.0	20.0	0.25
Pressed @ 25°C	85.3	14.7	0.17
Pressed @ 220°C	85.2	14.8	0.17
Pressed @ 120°C/Steam	86.7	13.3	0.15

SURFACE CHEMISTRY

The surface elemental composition of the yellow-poplar sapwood is shown in Table 1. Although a difference exists in the surface oxygen/carbon ratio between the untreated and pressed flakes, the different pressing conditions do not alter the surface elemental composition of the yellow-poplar sapwood. The changes in yellow-poplar sapwood surface energy as a function of temperature are shown in Figure 2. A step decrease in surface free energy was noted at approximately 65°C. This temperature was confirmed as the glass transition of lignin using dynamic scanning calorimetry [9]. Minimization of surface free energy above Tg might occur either by reorientation of hydrophobic functionalities on the amorphous polymer components from increased molecular mobility or by a migration of extractive molecules to the surface from the increase in diffusion coefficient at Tg. Because the XPS samples were cooled slowly after pressing, there may have been sufficient time for hydrophobic groups to reorient back to original conditions. Additional research is required to discern the exact mechanism controlling surface energy changes. Surface free energy and chemical functionality are important factors governing interactions between wood and synthetic polymer adhesives.

CONCLUSIONS

Changes in the mechanical and physical properties of wood flakes have been shown to occur with hot pressing [1-3]. The increased cellulose crystallinity and theorized lignin crosslinking will produce physical property changes consistent

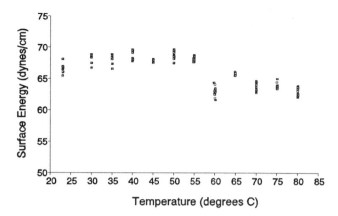

Figure 2 Changes in wood surface energy with temperature for yellow-poplar sapwood.

with those observed in previous studies. The importance of the glass transition temperature to wood polymer structure has been shown through the observed changes in cellulose crystallinity and surface energy. The relation of the glass transition temperature to densification has previously been shown [10]. The results presented here reinforce the importance of wood-polymer phase changes to composite manufacture and behavior.

ACKNOWLEDGEMENTS

Funding for this research was provide by Michigan State University/ USDA:CSRS Eastern Hardwood Special Research Grant Programs Nos. 90-34158-4989 and 90-34158-4990 and West Virginia Agriculture and Forestry Experiment Station, McIntire Stennis Grant 955. The authors thank Gene Shutler for providing mechanical testing information and Steve Kelley for providing cellulose crystallinity data.

REFERENCES

1.   Geimer, R. L.; Mahoney, R. J.; Loehnertz, S. P.; Meyer, R. W. *Forest Products Lab. Res. Paper #463*, **1985**.
2.   Price, E. W. *Forest Products J.* **1976**, 26(10), 50.
3.   Kamke, F.A.; Wolcott, M. P.; Casey, L. J.; Brady, D. E. *Final Report for USDA Competitive Grants Program for Forest and Rangeland Renewable Resources, Project No. FP-86-0813*, **1988**.
4.   Shutler, E. Master's Thesis, *West Virginia University, Division of Forestry, Morgantown, WV, 1992*.
5.   Kelley, S. S.; Rials, T. G.; Glasser, W. *J. Materials Sci.* **1987**, 22, 617.

6.   Hsu, W. E. *Proceedings from the Washington State University Particleboard Symposium* **1987**.

7.   Gardner, D. J. *Proceedings of the 6th International Symposium on Wood and Pulping Chemistry, Vol. I* **1991**, 345.

8.   Rials, T. G.; Glasser, W. G. *Wood and Fiber Sci.* **1989**, 21, 80.

9.   Gunnells, D. W. Master's Thesis, *West Virginia University, Division of Forestry, Morgantown, WV, 1992*.

10.   Wolcott, M. P.; Kamke, F. A.; Dillard, D. A. *Wood and Fiber Sci.* **1990**, 22(4), 345.

76

Comparison of the performance of various cellulosic reinforcements in thermoplastic composites

D. Maldas and B.V. Kokta - Centre de Recherche en Pâtes et Papiers, Université du Québec, C.P. 500, Trois-Rivières, Québec G9A 5H7, Canada.

ABSTRACT

In the current study, the performance of MA-modified cellulosic materials (e.g. chemithermomechanical pulp (CTMP) of hardwood aspen, wood flour of hardwood aspen and nutshell flour) in polyethylene and polystyrene composites has been compared by evaluating the mechanical properties of the composite materials. Experimental results indicate that the performance of three different cellulosic materials are not the same, and it also varies with concentration of fillers and nature of thermoplastics.

INTRODUCTION

In the recent years, the demand for reinforcing, low-cost and renewable resource fillers in thermoplastic composites has increased tremendously. In this respect, natural fillers, particularly cellulosics, offer a number of potential advantages, e.g., low-cost, renewable source, flexibility during processing, light weight, no health hazard, etc. On the contrary, poor wettability of the polar cellulose surface by the non-polar polymer surface produces poor dispersion of the cellulose fiber and low interface adhesion between polymer and filler. This often results in poor physical and mechanical properties in the end-product. However, modifying the filler's surface by selecting suitable coupling agents in order to improve the compatibility level, enhances filler use (1-4). Moreover, a higher filler concentration can also be used with improved physical and mechanical properties.

A survey of the literature, as well as our previous reports, reveal that maleic anhydride offers greater adhesion between cellulose fibers and thermoplastics in the composites (4-9). In the present study, the mechanical properties of MA-modified nutshell and wood fiber-filled polyethylene and polystyrene composites have been evaluated.

MATERIALS

Two types of thermoplastics, e.g. high-impact polystyrene PS 525 [HIPS] (supplied by Polysar Limited, Canada) and high density polyethylene (HRSN 8907) [HDPE] (supplied by Novacor Co., Canada) were used in this study. Hardwood aspen was used in the form of wood flour (sawdust) and chemithermomechanical pulp (CTMP). CTMP and chips used to make sawdust were dried and then ground to a mesh size 60 mixture. Blends of pecan shell and peanut hull flour of mesh size 200 was supplied by Southeastern Reduction Co., Valdosta, Georgia, U.S.A. The fillers were oven-dried by circulating air at 55 °C for 5-7 days. Both maleic anhydride (MA) and dicumyl peroxide (DCP) were supplied by Anachemia, Montréal, Canada.

EXPERIMENTAL

Fillers were surface modified with HDPE/HIPS which is the same as the one used in composites (5% by weight of filler), maleic anhydride (1% by weight of filler) with the help of a Laboratory Roll Mill. The surface modified fillers were passed through a screen of mesh size 20 in a grinder. Mixture of polymer and surface modified filler (10-30% by weight of composites) were mixed in the roll mill and ground once more to mesh size 20. The mixtures were then molded into shoulder-shaped test specimens (ASTM D-638, Type V).

The mechanical properties (e.g. modulus at 0.1% strain, tensile strength at yield point and corresponding elongation and energy) of all the samples were measured with an Instron Tester (Model 4201) following ASTM D-638. The mechanical properties were automatically calculated with a HP-86B computer. Un-notched Izod impact strength was tested following ASTM D-256 with an Impact Tester (Model TMI, No. 43-01). The statistical average of the measurements of at least 5 specimens was taken to obtain a reliable average and standard deviation.

RESULTS AND DISCUSSION

Figures 1-5 compare the mechanical properties of both HDPE and HIPS composites filled with non-treated (NT) and 3% MA-coated (T) cellulose fibers (e.g. nutshell, CTMP aspen and sawdust aspen). It is obvious from these figures that properties improve due to the addition of MA and they change according to filler content. Optimum fiber levels where maximum improvement in properties takes place change with the nature of polymer and fiber, and overall with the quality of the mechanical properties. For example, in comparison to virgin polymers, tensile strength for MA-coated fiber-filled composites improve in most cases up to 20 wt. % of fiber and rarely up to 30 wt. %. Modulus always increases linearly following the

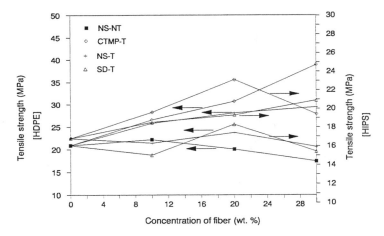

Fig. 1. Variation of tensile strength with the concentration and nature of cellulose fiber for polyethylene (HDPE) and polystyrene (HIPS) composites.

concentration of fiber. The impact strength of the composites diminishes in comparison with that of virgin polymers. Generally, impact strength decreases at the initial level with the rise in fiber content, and then either level off or increase with the further increase of fiber content. Elongation of PE-based composites is always less than that of virgin polymer. Energy for the same composites follow more or less similar trends, except for CTMP-filled composites where the opposite is observed, i.e. it

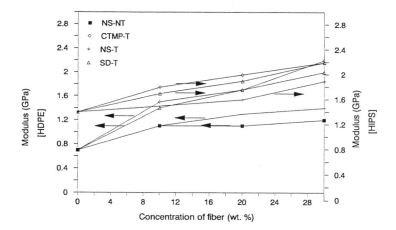

Fig. 2. Variation of modulus with the concentration and nature of cellulose fiber for polyethylene (HDPE) and polystyrene (HIPS) composites.

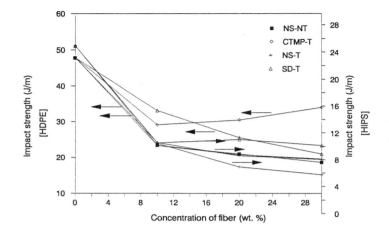

Fig. 3. Variation of Izod impact strength with the concentration and nature of cellulose fiber for polyethylene (HDPE) and polystyrene (HIPS) composites.

drives upwards with the upgrading in fiber content. The energy value exceeds that of unfilled polymer even at 20 wt. % of fiber level. Both elongation and energy of nutshell-filled PS composites remains more or less unchanged compared to those of virgin polystyrene, while for CTMP and sawdust-filled PS composites, the same properties improve with respect to unfilled polymer, up to 20 wt. % of fiber level.

Concerning the tensile strength of PE composites, when one judges the effectiveness of all three fibers according to the high-

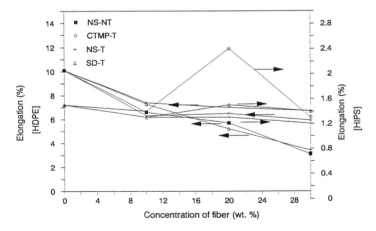

Fig. 4. Variation of elongation with the concentration and nature of cellulose fiber for polyethylene (HDPE) and polystyrene (HIPS) composites.

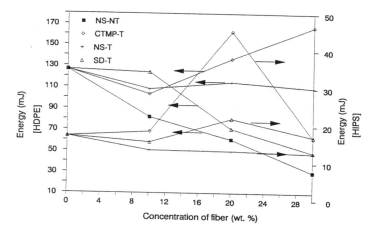

Fig. 5. Variation of energy with the concentration and nature of cellulose fiber for polyethylene (HDPE) and polystyrene (HIPS) composites.

est property performance, it can be noticed that CTMP aspen > nutshell > sawdust aspen. CTMP aspen ≈ sawdust aspen > nutshell for modulus of PS composites, while sawdust aspen > nutshell > CTMP aspen for elongation of PE composites. On the other hand, the order of performance for all other remaining cases are CTMP > sawdust aspen > nutshell.

The mechanical properties improved due to the addition of MA to the composites because the thermoplastic, cellulose and MA in the presence of an initiator form a block copolymer containing a succinic half ester bridge between cellulose and thermoplastic, e.g. polyethylene or polystyrene (4,6,7). Moreover, the -OH group of cellulose also has the ability of forming hydrogen bonds with the -COOH group of the MA segment. In this way, MA develops an overlapping interface area between fiber and polymer matrices. Prior grafting of the fiber with polymer and MA produces a soft film of hydrophobic material on the surface of the hydrophilic fiber. As a result, the phase separation between the two different matrices may reduce. In addition, strong fiber-fiber interaction due to intermolecular hydrogen bonding has also been diluted, which leads to better dispersion of the fibers.

It is also obvious that the performance of three different fibers is not the same. In fact, the performance of cellulosics as the reinforcing fiber for plastics depends on their quality, e.g. fiber length, fiber-making technique, morphology, inherent physical and mechanical properties, origin, and lignin content, etc. (10-12). Properties are also affected by the fiber-making process, even within the same species. Different pulping techniques offer various ways of separating fibers from the chips or raw materials, e.g. sawdust and nutshell particles were prepared by mechanical means, while both chemical and mechanical methods were used to prepare CTMP. As a result, fibers of CTMP are more separated and they are more flexible than the other two cellulose fibers. However, our experimental results are consistent with the fact that CTMP is without doubt the best pulp as far as mechanical properties of composites are concerned.

It is a well-established fact that fiber length is a critical parameter in the evaluation of the composites' properties (12). In the present study, both CTMP and sawdust were used with mesh size 60, while nutshell required mesh size 200. Therefore, it is also very difficult to compare the performance of cellulose fibers with different mesh sizes.

CONCLUSION

Coating treatments show some positive influence on the mechanical properties of the composites over that of uncoated fiber-filled composites and of virgin polymers. Among the three different fibers, CTMP seemed best with regard to mechanical properties. As far as cost is concerned, sawdust and nutshell were the most logical selections as fillers for thermoplastic composites after being submitted to surface modifications.

ACKNOWLEDGEMENT

The authors wish to thank the NSERC of Canada, and the FCAR and the CQVB of Québec for their financial support.

REFERENCES

1.   A. Y. Coran and R. Patel, U.S. Pat., 4 323 625 (April 6, 1982).
2.   P. Zadorecki and A. J. Michel, Polym. Composites, 10(2), 69 (1989).
3.   D. Maldas and B. V. Kokta, J. Appl. Polym. Sci., 41, 185 (1990).
4.   N. G. Gaylord, U.S. Pat., 3 645 939 (Feb. 29, 1972).
5.   H. Kishi, M. Yoshioka, A. Yamanoi and N. Shiraishi, J. Japan Wood Res. Soc., 34(2), 133 (1988).
6.   G. S. Han, H. Ichinose, S. Takase and N. Shiraishi, J. Japan Wood Res. Soc., 35(12), 1100 (1989).
7.   D. Maldas and B. V. Kokta, Intern. J. Polymeric Mater., 14(3), 165 (1990).
8.   S. Takase and N. Shiraishi, J. Appl. Polym. Sci., 37, 645 (1989).
9.   A. A. Pohlman, M. S. Dissertation, Univ. of Calif., Berkeley, 1974.
10. J. A. Clark, "Pulp Technology and Treatment for Paper", 2nd Ed., Ch. 9, Miller Freeman Publications, Inc., San Francisco, 184 (1985).
11. B. V. Kokta, J. L. Valade and C. Daneault, Pulp and Paper Canada, Transactions, TR 59 (Sept., 1979).
12. D. Maldas, B. V. Kokta and C. Daneault, J. Appl. Polym. Sci., 38, 413 (1989).

77

New types of polyurethanes derived from lignocellulose and saccharides

H. Hatakeyama, S. Hirose, K. Nakamura* and T. Hatakeyama** - Industrial
Products Research Institute, 1-1-4 Higashi, Tsukuba, Ibaraki 305. *Otsuma Women's
University, 12 Sanbancho, Chiyoda-ku, Tokyo 102. **Research Institute for
Polymers and Textiles, 1-1-4 Higashi, Tsukuba, Ibaraki 305, Japan.

ABSTRACT

New types of polyurethanes, such as films, foams and composites,
were prepared from kraft lignin, solvolysis lignin, wood tar
residue, woodmeal and molasses by polymerization with
polyethylene glycol, polypropylene glycol and diphenylmethane
diisocyanate. It was found that the glass transition temperature,
tensile and compression strength, and also Young's modulus of the
polyurethanes increased with increasing content of the above
plant materials. However, the degradation temperature of the
polyurethanes decreased with increasing content of the lignins
and the wood tar residue. The above fact suggest that lignin and
saccharide residues act as the hard segments in the
polyurethanes, and also that the phenolic groups in the lignin
residue have the effect of decreasing the degradation
temperature.

1. Introduction

It is generally recognized that polyurethane (PU) is one of the
most useful three-dimensional polymers, since PU has unique
features: for example, various forms of materials such as sheets,
foams, adhesives and paints can be obtained from PU, and their
properties can easily be controlled. Accordingly, many attempts
to use lignocelluloses as raw materials for PU synthesis have
been made, since natural polymers having more than two hydroxyl
groups per molecule can be used as polyols for polyurethane
preparation if the polyols from natural polymers can be reacted

efficiently with isocyanates.

Recently, Glasser et al. prepared PU's from hydroxyalkylated lignins and studied the relationship between their structures and properties [1,2]. They also studied adhesive properties of PU's derived from lignins [3].

Yoshida et al. [4,5] and also Reimann et al. [6] prepared polyurethane from non-modified kraft lignin. Hirose et al. [7] prepared heat-resistant polyurethanes from solvolysis lignin. Yano et al. [8] prepared new types of polyurethanes from wood meal and coffee bean parchment and studied their mechanical properties. Nakamura et al. [9] prepared polyurethanes from softwood kraft lignin and also from hardwood solvolysis lignin, studying the mechanical properties of the obtained polyurethanes.

The present paper reports our recent studies on the thermal and mechanical properties of polyurethanes prepared from various plant resources such as kraft lignin, solvolysis lignin, wood tar residue, woodmeal and molasses by polymerization with polyethylene glycol (PEG), polypropylene glycol (PPG) and diphenylmethane diisocyanate (MDI). Preparation and mechanical properties of molecular composites consisting of polyurethane and cellulose will also be reported.

2. Materials

Industrial kraft lignin separated from black liquor was used as kraft lignin (KL) sample. Hardwood solvolysis lignin (SL) used in this study was obtained as a by-product in organosolve-pulping of Japanese beech (Fagus crenata). The SL was supplied by the Japan Pulp and Paper Research Institute Co. The Eucalyptus kraft lignin (EL) was supplied by Litchem Limited, U. K. The wood meal (WM) from pine was supplied by Miki Sangyo Co. Ltd. The particle size was 60-90 um. Wood tar residue (WTR), which was obtained from Brazilian eucalyptus, was supplied by Nissho Iwai Corporation. Molasses (ML) were supplied by Tropical Technology Center Ltd., Okinawa. PEG, PPG and MDI were commercially obtained.

3. Preparation of Polyurethane

Prior to obtaining polyurethane, it was necessary to dissolve lignocelluloses (KL, SL, EL, WM and WTR), and saccharides (ML) in polyols such as PEG and PPG in order to prepare polyol solutions. The obtained polyol solutions were mixed with MDI and plasticizer (PEG or PPG) at room temperature, and precured polyurethanes were prepared. Each of the precured polyurethanes was heat-pressed and a PU sheet was prepared. In order to prepare PU foam, first the above polyol solution was mixed with plasticizer, surfactant (silicone oil), and catalyst (di-n-butyltin dilaurate), and then MDI was added. This mixture was vigorously stirred with a droplet of water which was added as a foaming agent. Scheme 1 shows the above preparation processes.

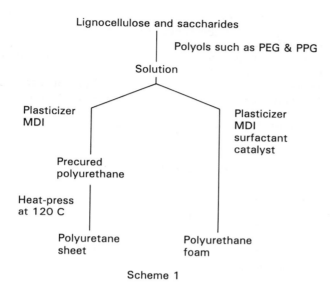

Lignocellulose and saccharides

Polyols such as PEG & PPG

Solution

Plasticizer
MDI

Plasticizer
MDI
surfactant
catalyst

Precured
polyurethane

Heat-press
at 120 C

Polyuretane
sheet

Polyurethane
foam

Scheme 1

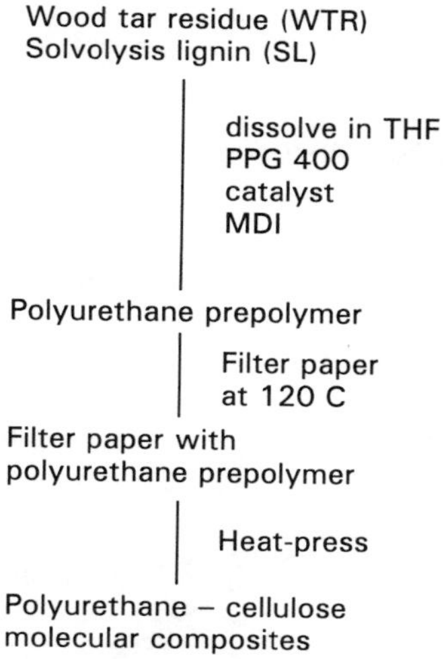

Wood tar residue (WTR)
Solvolysis lignin (SL)

dissolve in THF
PPG 400
catalyst
MDI

Polyurethane prepolymer

Filter paper
at 120 C

Filter paper with
polyurethane prepolymer

Heat-press

Polyurethane − cellulose
molecular composites

Scheme 2

Polyurethane composites were prepared by dipping filter paper (Toyo filter paper No.2) into a tetrahydrofuran solution of precured polyurethane, and then by drying and heat-pressing the filter paper with precured polyurethane. In order to increase the amount of precured polyurethane reacted with filter paper, the above process was repeated. Scheme 2 shows the preparation process for polyurethane composites.

In the above processes, the NCO/OH ratio, the weight of starting materials and the contents of lignocellulose and saccharides (shown as LC in following equations), polyols such as PEG and PPG (shown as PEG in the following equation), and MDI were calculated as follows:

$$NCO/OH = M_{MDI} \times W_{MDI} / (M_{LC} \times W_{SL} + M_{PEG} \times W_{PEG})$$

$$W_t \ (g) = W_{LC} + W_{PEG} + W_{MDI}$$

$$LC \ content \ (\%) = (W_{LC} / W_t) \times 100$$

$$PEG \ content \ (\%) = (W_{PEG} / W_t) \times 100$$

$$MDI \ content \ (\%) = (W_{MDI} / W_t) \times 100$$

where NCO/OH is the molar ratio of isocyanate and hydroxyl groups, M_{MDI} the number of moles of isocyanate groups per gram of MDI (8.0 mmol/g), W_{MDI} the weight of MDI, M_{LC} the number of moles of hydroxyl groups per gram of LC, W_{LC} the weight of LC, M_{PEG} the number of moles of hydroxyl groups per gram of PEG, W_{PEG} the weight of PEG, and where W_t is the total weight of the components in the PU system.

Table 1 shows the relationship between the above plant resources and the prepared polyurethanes.

Table 1. Polyurethanes from Various Lignocelluloses and
           Saccharides

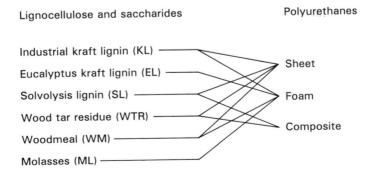

4. Macromolecular Structure of Polyurethanes

The chemical structure of prepared polyurethanes are dependent on
the plant raw materials. The polyurethanes consist of core
structures of lignin and saccharides linked by urethane bonding.
Schemes 3 and 4 show structures of saccharide and lignin with
urethane bonding. Accordingly, we may assume that the obtained
polyurethanes are essentially block copolymers having three
dimensional networks of urethane bonding which combine saccharide
and lignin components. Scheme 5 shows a schematic macromolecular

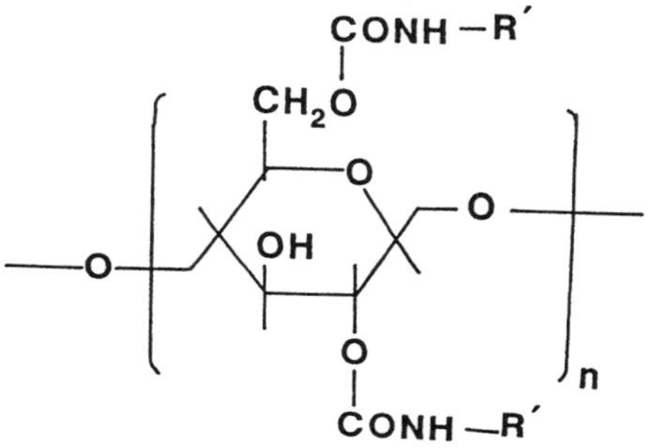

Scheme 3

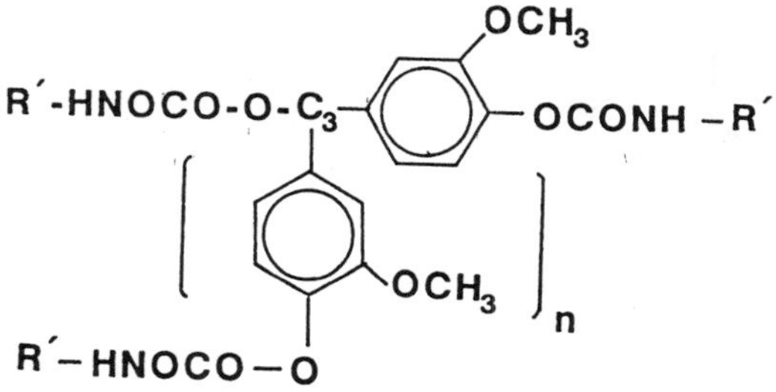

Scheme 4

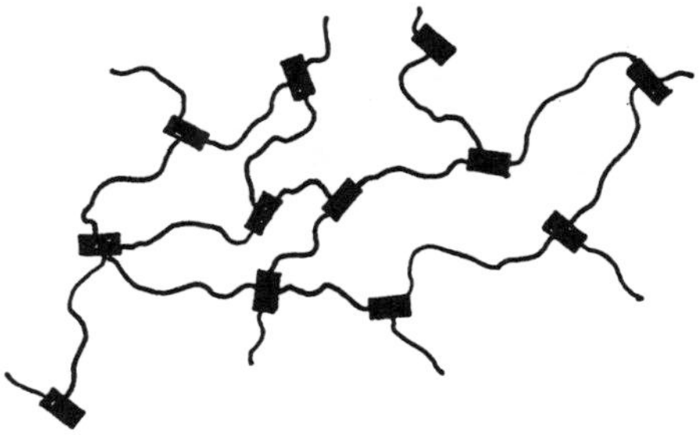

Scheme 5

network structure of the obtained polyurethanes. Saccharide and lignin components, which are shown as rectangular shapes, are connected with either polyethylene chains (from PEG) or polypropylene (from PPG) chains by urethane bonding, which are shown as real lines.

5. Thermal Properties of polyurethanes

It was found that glass transition temperature (Tg) and degradation temperature (Td) of prepared PU's were mostly dependent on the content of KL, SL, EL, WM, WTR and ML. Tg increased with increasing content of phenyl groups from lignins and WTR, since phenyl groups work as hard segments in PU network. However, Td decreased with increasing content of lignin and WTR, since the dissociation of urethane bonds between isocyanate groups and phenolic hydroxyl groups occurs at the temperature region lower than the dissociation of urethane bonds between isocyanate groups and alcoholic hydroxyl groups [10-13].

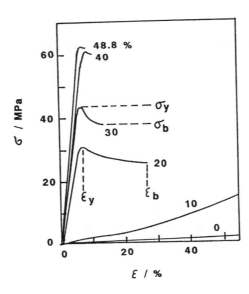

Figure 1. Stress (σ) – strain (ε) curves of polyurethane sheets from solvolysis lignin (SL).

6. Mechanical Properties of Polyurethane Sheets

Figure 1 shows the stress-strain curves of PU's with various SL contents ranging from ca. 10 to 50%. In the figure, ε (%) is elongation; e_y, elongation (%) at the yielding point; ε_b, elongation (%) at the breaking point; σ_y, stress (MPa) at the yielding point and σ_b, stress (MPa) at the breaking point. The σ values of PU without SL are small when the ε value is large. The elongation decreases with the increase in SL content. Yielding was observed in the samples with the SL content higher than 20 %. The breaking strength becomes larger with increasing SL content within the amount used in this experiment.

Figure 2 shows the relationship between tensile properties and SL content. As can be seen from the figure, σ_b of the PU increases with increasing SL content. The effect of SL content on σ_b is apparent in the SL content lower than 40 %, although it is necessary to consider that the PU with SL content lower than 20 % is in the rubbery state at room temperature. Young's moduli (E's) for the PU's also increases with increasing SL content, although the E value of PU with SL content lower than 20% is quite small since the PU is in the rubbery state. The above results suggest that the tensile properties of the PU are markedly improved with the addition of an appropriate amount of SL.

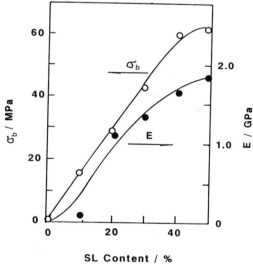

Figure 2. Change of breaking strength (σ_b) and Young's modulus (E) of polyurethane sheets with SL contents.

Figure 3 shows changes of breaking stress (σ_b), strain (ε) and Young's modulus (E) of polyurethanes with the WTR/polyol (PPG) ratio. The above polyurethanes were prepared by reacting the mixture of PPG and WTR with MDI. As shown in Figure 3, σ_b and E values increase with increasing WTR/polyol ratio. This suggests that the lignin degradation components having aromatic rings in WTR act as hard segments in the polyurethanes, since most parts of WTR components are lignin degradation products having the aromatic ring structure [11].

7. Mechanical Properties of Polyurethane Foams

Figure 4 shows the change of compression strength (σ) and Young's modulus (E) in PU foams which were prepared from EL, PEG and MDI. In this relationship, the MDI/polyol ratio is maintained as 1.2. The values of both σ and E increase with increasing amount of EL. This fact suggests that EL acts as the hard segments in the PU foam.

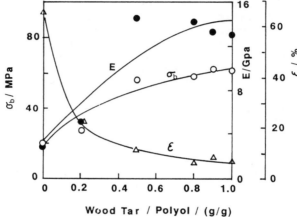

Figure 3.   Change of σ_b , ε and E of polyurethane sheets with wood tar residue/polyol (WTR/PPG) ratio.

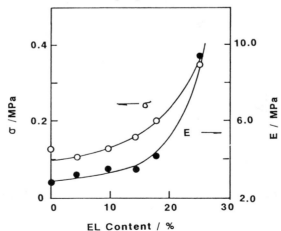

Figure 4.   Change of compression strength (σ) and compression modulus (E) of polyurethane foams with Eucalyptus lignin (EL).

Figure 5 shows the change of s and E with the density (ρ) of
the PU foam having the EL content of 30 % and the MDI/polyol
ratio 1.2. As clearly seen from the figure, the   σ   and E of
the PU foam increase with increasing ρ . This suggests that the
value is important in estimating the mechanical strength of the
PU foam, since the thickness of each cell wall of the PU foam is
dependent on the   ρ   value, when the pore size of each cell is
assumed to be almost the same.

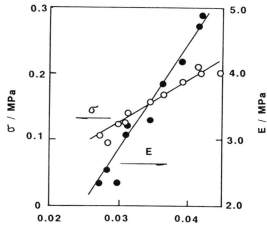

Figure 5.   Change of (σ) and modulus (E) of polyurethane foams
with apparent density (ρ).

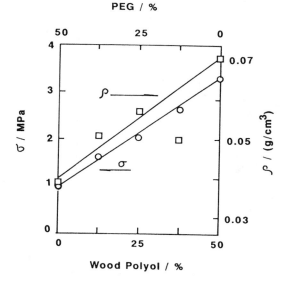

Figure 6. Change of σ   and   ρ   of polyurethane foams with
woodmeal (WM) contents.

Figure 6 shows the relationship between σ and WM content in polyurethane foams, and also the relationship between ρ and WM content. It is clear that both σ and ρ of polyurethanes increase with increasing WM content. The above results suggest that WM components in polyurethane contribute to the increase of compression strength of polyurethane foams, because of furanose, pyranose and aromatic rings of cellulose, hemicellulose and lignin in WM.

8. Mechanical Properties of Polyurethane Molecular Composites

Polyurethane molecular composites were prepared according to Scheme 2. In the molecular composites, cellulosic components of filter paper, as well as SL and WTR components, are connected with each other through polyethylene or polypropylene networks by urethane bonding.

Figure 7 shows breaking strength (σ_b) of the composites with polyurethane (PU)/filter paper (FP) ratio. The σ_b values of composites increased with increasing PU/FP ratio. The σ_b values of FP-WTR-MDI composites (curve A) are higher than those of the FP-WTR-PPG-MDI composites (curve C). However, the σ_b values of FP-SL-PPG-MDI composites (curve B) are higher than those of the FP-SL-MDI composites. The differences between the FP-WTR-PPG-MDI (and FP-WTR-MDI) composites and the FP-SL-PPG-MDI (and FP-SL-MDI) composites are probably dependent on the compatibility of WTR and SL with FP, although more detailed studies are needed to give a clear explanation concerning the relationship between mechanical properties and compatibility of the above composites.

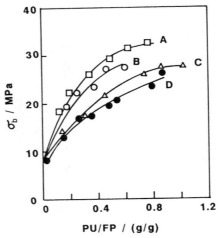

Figure 7.   Change of breaking strength (σ_b) of polyurethane molecular composites with polyurethane/filter paper (PU/FP) ratio.

9. Conclusion

Table 2 shows the influence of KP, EL, SL, WTR, WM and ML on thermal and mechanical properties, such as T_g, T_d, σ, ε and E', on polyurethane sheets, foams and composites. The T_g, σ and E' of polyurethanes increase with increasing lignin and saccharide components, while T_d decreases with increasing amount of lignin components. The decrease of T_d of polyurethanes with lignin components is probably caused by the dissociation of urethane bonding between the isocyanate group and the phenolic hydroxyl group in lignin components. This dissociation occurs at a temperature lower than that of the urethane bonding between the isocyanate group and the alcoholic hydroxyl group of saccharide components.

Table 2. Influence of Industrial Kraft Lignin (KL), Eucalyptus Kraft Lignin (EL), Solvolysis Lignin (SL), Wood Tar Residue (WTR), Woodmeal (WM) and Molasses (ML) on Thermal and Mechanical Properties of Polyurethane Sheets, Foams and Composites.

Polyurethane Components	Sheet					Foam					Composite					References
	T_g	T_d	σ_b	ε	E	T_g	T_d	σ	ε	E	T_g	T_d	σ_b	ε	E	
KL	+	-	+	-	+	+	-	+	-	+						4, 5, 6
EL						+	-	+	-	+						13, *1
SL	+	-	+	-	+						+	=	+			7, 9, *1
WTR	+	-	+	-	+						+	-	+			11, *1
WM	+	=	+	-	+	+	=	+	-	+						8
ML						+	-	+								*2

T_g: glass transtion temperature, T_d: thermal degradation temperature
σ_b: breaking strength, ε : elongation, E: Young's modulus or compression modulus
σ : compression strength
+ : increase, -: decrease, =: almost same
*1: this report, *2: our experimental data

The following conclusions may be drawn from the above results.

(1) Plant components such as lignin and saccharides act as stiff segments in polyurethanes.
(2) The influence of plant components on the molecular motion of polyurethanes is greater than that of other factors such as NCO/OH ratio and molecular weight of polyols.
(3) The introduction of polyols such as PEG and PPG into polyurethanes increases the flexibility of polyurethanes.

(4) Formation of cellulose-polyurethane composites improves the mechanical properties of polyurethane.
(5) The thermal and mechanical properties of polyurethane can be controlled by changing the contents of plant components and polyols.

Acknowledgment

The authors are grateful to Japan Pulp and Paper Research Institute for providing the solvolysis lignin samples, to Litchem Ltd. for the Eucalyptus kraft lignin samples, to Miki Sangyo Co. for wood meal, to Nissho Iwai Trading Industries for wood tar residue and to the Tropical Technology Center for molasses.

REFERENCES

[1] Saraf, V.P.; Glasser, W.G. J. Appl. Polym. Sci., **1984**, 29,1831.
[2] Saraf, V.P.; Glasser, W.G; Wilkes, G.L.; McGrath, J.E. J. Appl. Polym. Sci., **1985**, 30, 2207.
[3] Newman, W.H.; Glasser, W.G. Holzforschung, **1985**, 39, 345.
[4] Yoshida, H.; Morck; R., Kringstad, K. P.; Hatakeyama, H. J. Appl. Polym. Sci.**1987**, 34, 1198.
[5] Yoshida, H.; Morck, R.; Kringstad, K. P.; Hatakeyama, H. J. Appl. Polym. Sci.**1990**, 40, 1819.
[6] Reimann, A.; Morck, R.; Yoshida, H.; Hatakeyama, H.; Kringstad, K. P. J. Appl. Polym. Sci.**1990**, 41, 39.
[7] Hirose, S.; Yano, S.; Hatakeyama, T.; Hatakeyama, H. in Glasser, W. G.; Sarkanen, S.(Eds.) Lignin: Properties and Materials, ACS Symposium Series 397, American Chemical Society, Washington DC., **1989**, 382.
[8] Yano, S., Hirose, S. ; Hatakeyama, H. in Kennedy, J. F.; Phillips, G. O.; Williams, P. A. (Eds.) Wood Processing and Utilization, Chichester, Ellis-Horwood, **1989**, 263.
[9] Nakamura, K.; Morck, R.; Kringstad, K. P.; Hatakeyama, H. in Kennedy, J.F.; Phillips, G. O.; Williams, P. A. (Eds.) Wood Processing and Utilization, Chichester, Ellis-Horwood, **1989**, 175.
[10] Saunders, J.H.; Frisch, K.C. Polyurethanes, Chemistry and Technology. in High Polymers, Vol. XVI, Part I, New York, Interscience Publishers, **1962**, 103.
[11] Nakamura, K.; Hatakeyama, T.; Hirose, S.; Hatakeyama, H. Sen-i Gakkai Reprints 1991, S-107, Tokyo.
[12] Hatakeyama, H.; Hirose, S.; Nakamura, K. Synthesis and Thermal Properties of Polyurethanes from Lignocellulose, 201st ACS National Meeting, Atlanta, Georgia, April, 1991.
[13] Hatakeyama, H.; Hirose, S; Nakamura, K; Yoshida, H. Proceedings, 6th International Symposium on Wood and Pulping Chemistry, **1991**, 1, 517, Melbourne.

78

Lignin biopolymers in the composite polypropylene films

B. Kosikova, V. Demianova* and M. Kacurakova - Slovak Academy of Sciences.
*Slovak Technical University, Bratislava, Czecho-Slovakia.

ABSTRACT

The lignin bipolymers, obtained as co-products of spruce wood organosolv pulping and/or of beech wood pre-hydrolysis, were tested as components of polypropylene films without any modification. A sufficient thermostability and tensile streghts of the polymers with lignin weight fractions ranging from 2 to 10% were observed. The structural changes of both lignin preparations in the process of film production were examined by difference IR spectroscopy. In addition, the rate of photo-oxidation of the films prepared in the presence or absence of stabilizers from the polypropylene blends containing lignin preparations was examined.

INTRODUCTION

The characteristics of kraft lignin as a polymer are utilized in the preparation of three-dimensional polymers, such as polyurethanes or phenol resins (1,2). Kraft lignin addition to rubber leads to an increase in hardness and to improvements in abrasion (3).

As to some drawbacks of kraft lignin, co-products of organosolv pulping and/or steam treatment of lignicellulosics have attracted great attention for

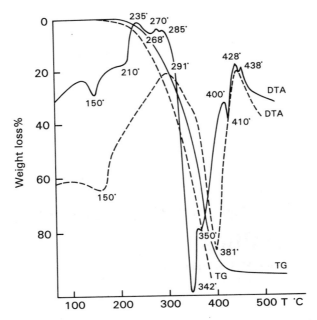

Fig. 1 DTA and TG curves of lignin-free polypropylene
blend (-----) and polypropylene blend containing 2%
wt of lignin (WL) (————)

Composition	Tensile strength δr /MPa/
Stabilized PP	70.9
PP + 2% wt WL	76.5
PP + 2% wt OL	76.0
PP + 4% wt OL	73.2
PP + 6% wt OL	72.7

Table 1 Tensile strengths of polypropylene films

The structural changes of lignin during film
processing were investigated by IR spectroscopy. The
quantitative changes in the intensity of the O-H band
at 3385 cm^{-1} in the spectra of lignin component in
polypropylene films investigated are shown in Fig.
2. The absorbance of this band related to 1510 cm^{-1}

their utilization in engineering plastics. The objective of the present paper was to examine the effect of both lignin preparations on physicochemical properties of polypropylene films including their stability towards photo-oxidation.

EXPERIMENTAL

The polymer blends were obtained by mixing polypropylene powder (TATREN,HPE,Slovnaft Bratislava) with organocell lignin (Gm 6H corporation, München), and/or water soluble pre-hydrolysis lignin (Bukóza,Vranov) in the presence or absence of 0.15% wt of 2,6-di-tert-buthyl-4-methylphenol (AO4K) and 0.15% wt calcium stearate. The content of lignin was 0,2,4,6,8 and 10% wt. Then the blends were homogenized in a twin Screw extruder at 200°C.

Photooxidation was carried out in equipment described elsewhere (4) using 125W mercury arc RVC (Tesla,Holešovice) at 60±5°C. The course of photo-oxidation was followed by monitoring the carbonyl products IR-spectroscopically in the region between 1700 and 1740 cm^{-1}. IR spectra were measured with a Perkin-Elmer 457 spectrometer.

RESULTS AND DISCUSSION

Both pre-hydrolysis lignin (WL) and organocell lignin (OL) seem to be attractive for incorporation into polymeric materials with regard to the average molecular weight (Mw = 2000 and 3300, respectively) and polydispersity (D = 1.2 and 3, respectively). The lignin containing polypropylene blends, possesing a good compatibility show sufficient thermal stability at 200°C (the temperature used for proceeding of PP films Fig. 1). The thermal effects which accompanied the interaction between polypropylene and lignin, are evident from the comparison of DTA curves of lignin-free polymer and that of containing 2% wt of lignin. The latter contains an exothermic peak starting at 200°C indicating chemical reaction between polypropylene and lignin during thermal processing of the composite blends.

To elucidate the effect of lignin addition, the films prepared from the blends containing lignin were compared with lignin-free polymer with the respect to mechanical properties. From table 1 it is seen that tensile strengths were not significantly influenced by addition of lignin.

is lower in all composite samples than that in the
original lignin sample. The decrease is most
pronounced in the films containing OL and/or WL at
2-3% wt which is probably the optimal regarding
lignin function as a stabilizer.

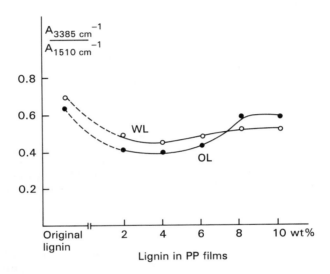

Fig. 2 Relationship between the OH stretching
vibrations and the weight % of lignin in
polypropylene films

The results of photo-oxidation of lignin
modified polypropylene films supported the above
described suggestion. Kinetic curves for the increase
in carbonyl absorption during the course of arc
irradiation of polypropylene films, monitoring by IR
spectroscopy, demonstrated photoiradiation protection
of polypropylene in composities under investigation
if WL is present at 2% wt (Fig. 3).

This revealed photostability of polypropylene
film containing 2% wt.WL is comparable with that of
lignin free polypropylene stabilized by commercial
stabilizer. The obtained results confirm the
suggestion (5) that phenolic structures in lignin
represent potential radical scavengers by terminating

chain reaction of polymers induced by oxygen, whereby
phenoxy radicals are formed. Higher lignin content
probably initiates competitive radical reactions

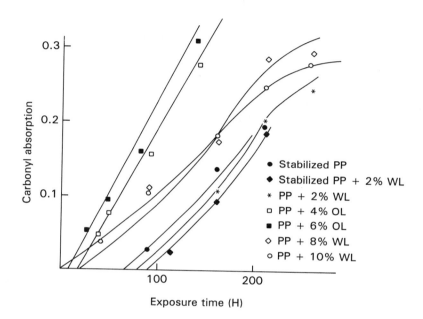

Fig. 3 Rates of photo-oxidation of polypropylene
films modified with lignin (OL, WL)

leading to oxidative degradation of polypropylene by
a hydroperoxide radical mechanism. Fig. 3 illustrates
that OL is more effective in this direction.

CONCLUSIONS

1. The optically transparent and homogeneous PP films
containing 2-10% wt WL or OL with sufficient
tensile strengths could be obtained in the absence of
commercial stabilizers.

2. The results of photo-oxidation of lignin
containing polypropylene films as well as the
structural changes of lignin component during
processing indicate that the effect of lignin
addition on the physico-chemical properties of films
vary with lignin content in the composite blends.

REFERENCES

1. Saraf,V.P., Glasser,W.G., Wilkes,G.L.,McGrath, J.E.: J.Appl.Polymer Sci. 30, 2207 (1985).

2. Yoshida,H., Mörck,R., Kringstadt,K.P., Hatakeyama, H.: J.Appl.Polymer Sci. 34, 1187 (1987).

3. Lubeskina, E.G.: USp. Khim. 52, 1116 (1983).

4. Chmela,Š., Hrdlovič,P.: Polymer Degradation and stability 27, 159-167 (1990).

5. Kratzl,K., Claus,P., Lonsky,W., Gratzl,J.S.: Wood Sci.Technol. 8, 35 (1974).

79

Reactive extrusion of polypropylene-lignocellulosic blend materials

Ramani Narayan *,**, **Chittoor K. Mohanakrishnan**** and **John D. Nizio***** -
*Michigan Biotechnology Institute, 3900 Collins Road, Lansing, MI 48910, USA.
**Department of Chemical Engineering, Michigan State University, East Lansing, MI 48824, USA. ***Southeastern Reduction Company, 309 S. Lee Street, Valdosta, GA 31601, USA.

ABSTRACT

Reactive extrusion processing was employed to fabricate blend matrices of a lignocellulosic material and modified polypropylene. A polypropylene-ethylene copolymer (PP) was functionalized with maleic anhydride (MA) in a twin screw extruder using dicumyl peroxide as an initiator. The modified polymer (MAPP) was alloyed with a lignocellulosic material (LDHW) in the extruder to form a blend material with good mechanical properties. Three types of lignocellulosics (LDHW 1, 2, and 3) with different average particle aspect ratios and two lignocellulosics contents (20 and 30% by weight) were evaluated as parameters that influenced the mechanical properties of the blend. The MAPP-LDHW blend system exhibited improved tensile properties as compared to both the virgin polymer and the maleated polymer, in contrast to a PP-LDHW blend system which suffered a reduction in tensile strength and elongation. This demonstrated the influence of adhesion between the blend components on the mechanical properties of the blend system.

INTRODUCTION

The use of fillers and reinforcements in thermoplastic compositions is a well-known practice in the polymer industry. The motivation behind the use of organic materials as fillers or reinforcements should not be just cost-driven considerations. In this work, the emphasis was on thermoplastic resin compositions with good performance properties to be used at the low end of the structural composites market such as the industrial and automotive markets (high volume, low cost). Different blend matrices consisting of recycled/reclaimed polymers (PET, PE, PP) or natural polymers (amylose, lignocellulosics) are currently being investigated as potential materials for composite fabrication. The use of recycled/reclaimed polymers not only lowers the cost of the composite but also opens up high value markets for the use of recycled resins, a national priority. Natural polymers like lignocellulosics reduce product cost, can be expected to improve creep resistance, and at the same time can aid in easy recyclability of the composite materials. Suitable chemical modifications to either the polymer matrix or the organic material or both can result in compositions with mechanical properties superior to or equivalent to other mineral and inorganic material-filled composites.

Two important factors in the processing of such compositions is the dispersion of the organic filler/fiber and the interfacial adhesion between the organic material and the matrix. Fiber agglomeration during processing results in composites with poor mechanical properties, as does the lack of interfacial adhesion.

There are three ways of improving the fiber-matrix adhesion. These are as follows-
  (i)   use of interfacial agents or coupling agents or additives
  (ii) fiber modification
  (iii) matrix modification

In this work, matrix modification using maleic anhydride was employed to alter the interfacial adhesion characteristics of the system. It involves reaction of the hydroxyl group on the cellulosic fiber/filler to a suitable functional group on the polymer chain, thereby generating interfacial adhesion. Chemical functionalities, specifically the anhydrides have been used as adhesion promoters in polymer systems[1-3]. In this work, an effort has been made to react the anhydride functionality with the components of the blend system, instead of simply mixing it in as an additive. Polymer modification using reactive extrusion produced maleic anhydride functionalities in a polypropylene-ethylene copolymer which were subsequently employed in the generation of in-situ grafts between the copolymer and the lignocellulosic component. The anhydride group on the synthetic polymer was reacted with the hydroxyl group of the lignocellulosic component by use of a suitable catalyst.

Also, the dispersion of the lignocellulosic fibers in the matrix was improved by the incorporation of ground pecan shells in the lignocellulosics blend material.

EXPERIMENTAL METHODS

Materials

PRO-FAX 8301 with a melt index of 11 (PP), supplied by Himont Advanced Materials, East Lansing, was used as the synthetic polymer component of the blend system. Three different types of low-density hardwood pulp blended with ground pecan shells (LDHW 1, 2, and 3), supplied by Southeastern Reduction Company, Valdosta, Georgia, were used as the natural polymer blend component. Ground pecan shells were incorporated with the wood pulp specifically to prevent fiber agglomeration, to improve dispersion of the particles during processing, and for surface smoothness of the extrudate.The average particle aspect ratio for each of the three lignocellulosic materials was determined using a video display microscope under a 100X magnification.

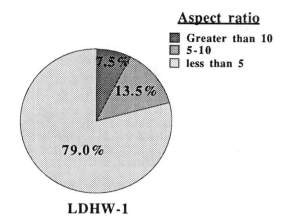

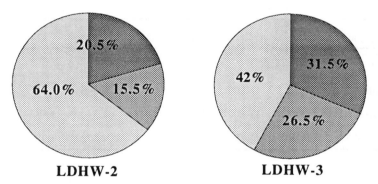

Figure 1.   Average particle aspect ratio distribution for the lignocellulosic
              materials

Slides were prepared using glycerine as a mounting agent and the size
measurements were made on 200 individual particles at random. The
distribution of aspect ratios for each of the lignocellulosic material types is
shown in Figure 1. Maleic anhydride (MA) in the form of briquettes, and the
initiator for the maleation reaction, dicumyl peroxide (DCP), were obtained
from Aldrich Chemical Company.

Biobased Materials Processing

Prior to compounding, the lignocellulosics were dried in a vacuum oven at
70°C for 24 hours. The biobased materials were fabricated using 20 and 30%
by weight of each of the three types of lignocellulosics. The formulations for
the blend matrices were first dry-mixed in the appropriate ratios and were
subsequently compounded in the extruder under a vacuum. The extrudate was
air-dried and pelletized. The unmodified formulations are simple blends of PP
and the desired amount of lignocellulosics (LDHW). The modified formulations

are blends of the maleated polymer (MAPP) and the desired amount of lignocellulosics. All the pelletized formulations were injection molded as ASTM Type I (D-638) specimens at a nozzle temperature of 170°C.

RESULTS AND DISCUSSION

The mechanical properties of the unmodified PP-LDHW blends are shown in Table 1. There was a significant decrease in tensile strength of the blend upon addition of the lignocellulosics. This was unlike the MAPP-LDHW blend matrix in which the addition of lignocellulosics resulted in an increase in the tensile strength and tensile modulus. Figures 2, 3, and 4 represent the tensile properties of the MAPP-LDHW blend matrices.

Table 1. Tensile properties of PP-LDHW blend materials.

Material	Tensile strength at yield (MPa)		% Elongation at yield		Initial tensile modulus (GPa)	
Fiber (wt%)	20	30	20	30	20	30
PP	18.48		13.34		0.913	
PP-LDHW 1	15.93	14.86	7.59	5.74	1.346	1.557
PP-LDHW 2	16.62	15.80	8.24	5.92	1.320	1.587
PP-LDHW 3	17.30	17.12	7.53	6.18	1.472	1.902

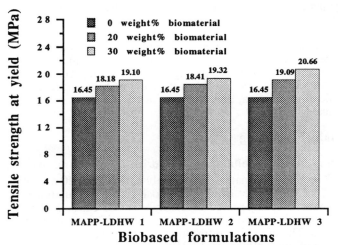

Figure 2. Tensile strength at yield data of MAPP-LDHW blends

The maleated polymer, MAPP suffered a 11% decrease in tensile strength, a 33% decrease in tensile modulus and a 56% increase in elongation at yield as compared to the starting polymer, PP. The maleation of the polymer in the melt state is accompanied by chain scission and a reduction in molecular weight, as discussed by Gaylord [4]. On account of the decreased molecular weight, the maleated polymer showed a reduction in tensile strength as compared to the virgin polymer.

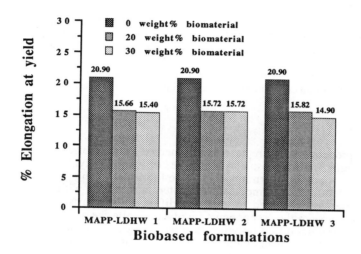

Figure 3. Elongation at yield data of MAPP-LDHW blend materials

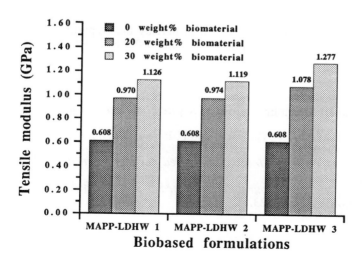

Figure 4. Tensile modulus data of MAPP-LDHW blend materials

The MAPP-LDHW blend matrices showed a maximum increase in tensile strength of 12% over the virgin polymer, PP, and a maximum increase in elongation of about 19%. The tensile modulus increased by a maximum of 40% as compared to the polymer, PP.

Compositions containing LDHW 2 and 3 exhibited the greatest improvement in properties. This was due to the higher aspect ratio of the particles of these lignocellulosic compositions. However, formulations containing LDHW 1 had a smooth exterior surface, while compositions with LDHW 2 and 3 had visible fiber specks seen under the surface. The MAPP-30% LDHW 3 blend material had a tensile strength (at yield) of 20.66 MPa as compared to 17.12 MPa for the PP-30% LDHW 3 material. The elongation (at yield) values were 6.18% for the PP-30% LDHW 3 blend and 14.90% for the MAPP-30% LDHW 3 blend material. In comparision, the PP-30% glass fiber material had a tensile strength at yield of 19.30 MPa, an elongation at yield of 6.53%, and an initial flexural modulus of 2.341 GPa.

CONCLUSIONS

Reactive extrusion processing is a simple, efficient method for the fabrication of polymer blend systems and composites. Composites of natural polymers like lignocellulosics with modified thermoplastics were fabricated using reactive extrusion. Modification of the thermoplastics resulted in materials with good mechanical properties as compared to the unmodified counterparts due to improved adhesion between the components involved. While the addition of lignocellulosics to the unmodified polymer resulted in a decrease in tensile strength, there was an improvement in tensile strength and elongation values in the case of polymers functionalized with maleic anhydride. The tensile strength at yield for the A-modified MAPP-30% LDHW 3 was higher than that of PP reinforced with 30% by weight glass fibers. The microstructure and morphology of these biobased materials are currently being investigated.

ACKNOWLEDGEMENTS

Financial support for this work was provided in part by the Department of Energy through SERI CEW-D071. Helpful discussions with Ms. Helena Chum of SERI is gratefully acknowledged.

REFERENCES

1. J.M. Felix and P.Gatenholm, *Journal of Applied Polymer Science*, **1991**, 42, 609.
2. B.V. Kokta, D. Maldas, C. Daneault, and P. Beland, *Polymer Composites*, **1990**, 11, 84.
3. S. Takase and N. Shiraishi, *Journal of Applied Polymer Science*, **1989**, 37, 645.
4. N.G. Gaylord, M. Mehta, *Journal of Polymer Science, Polymer Letters Edition*, **1983**, 21, 23.

80

Ligols – novel hydroxyl-containing multipurpose materials

M. Gromova*, A. Arshanitca, G. Telysheva and A. Sevastianova - Institute of Wood Chemistry, Latvian Academy of Sciences, Riga, Latvia.
*Present address: Institute of Wood Chemistry, 27 Akademijas st., 226006 Riga, Latvia.

ABSTRACT

Ligols - oligoethers based on technical lignosulphonates - were obtained by a hydroxypropylation method. High reaction ability of ligols in the reaction with isocyanates enable production of polyurethane materials with a wide application range, including rigid polyurethane foams and polyurethane binders. Ligols were successfully tested as additives to drilling fluids, surfactants and flotation reagents.

INTRODUCTION

Hydroxypropylation of lignin-like model compounds, acid hydrolysis lignin and lignosulphonates (LS) under alkaline conditions in inert and hydroxyl containing solvents at different temperatures was studied in previous investigations, using propylene oxide [1,2].

It was observed that reaction mechanism involved first order kinetics with regard to propylene oxide (PO) concentrations and that the propoxylation rate constant, K, greatly depends on acidity of hydroxyl containing compound increasing with the growth of acidity. It was shown that synthesys parameters, such as reaction medium, temperature and pressure, KOH concentration and propylene oxide to lignin hydroxyl groups ratio had influence on the chemical structure of hydroxypropyl lignin derivatives. These results were in accord with the findings of Glasser et al. [3].

EXPERIMENTAL

Commercial lignosulphonates were used throughout. LS had a methoxyl content 9.4...10.0%, a total OH content 9.6...10.3%, and elemental analysis values of 43.7...45.2% C, 4.7...5.6% H, 6.0...6.4% S, 4,6...5.3% Na, 0.4...0.5% Ca, 0.2 K% and 0.9...1.0% N. Ash and carbohydrate contents were 13.8...16.5% and 11.5...13.9%, respectively.

The reaction of PO and LS under alkaline conditions (KOH) in a medium of glycerol or propylene glycol was carried out in an industrial reactor with mechanical stirring at 110...150 C. After the reaction was completed, the reactor was cooled and KOH neutralized with acetic acid.

Depending on the properties of initial LS, solvent and temperature, the properties of ligols were as follows: hydroxyl number 300...660 mg KOH/g; density at 20 C 1.08...1.14 g/cub.cm; viscosity at 25 C 500...12000 MPa∗s; solubility in water 100%.

APPLICATION

High reaction ability of ligols in reaction with isocyanates was established (Figure 1). This property of ligols allows us to produce polyurethane materials, including rigid foams (Trade mark "Ligopor").

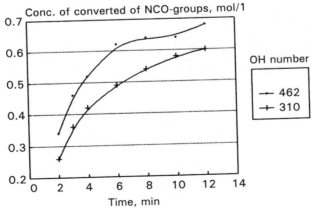

Figure 1. The reaction ability of ligols in the reaction with phenylisocyanate at 20 C.

Ligopors on the basis of ligols with definite characteristics (hydroxyl number 430...500 mg KOH/g and viscosity at 25 C 800...4100 mPa∗s) were utilized in the production of furniture frames.

The characteristics of Ligopors:

- density, kg/cub.m   46...50
- compression strength, MPa   0.30...0.35
- compression modulus, MPa   5.0...6.0
- tensile strength, MPa   0.41...0.51
- tensile modulus, MPa   14.0...15.3
- percentage elongation, %   3.9...5.3
- heat resistance, C   148...152
- closed cells, %   90
- water absorbtion per day,
  kg/sq.m   0.2
- natural color   light brown

The replacement of phenol-formaldehyde resins in the binders for foundry moulds and cores by ligol-based binders increases the mechanical properties of the articles (Figure 2) and improves labour conditions due to elimination of evolving toxic monomers.

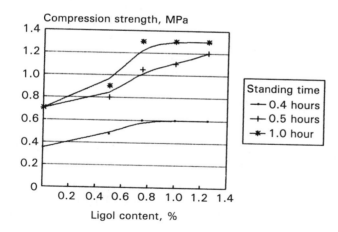

Figure 2. Compression strength of foundry moulds and cores depending on ligol content.

It was shown that ligols have various commercial applications. For example, application of ligols in methylcellulose-based drilling fluids enables regulation of the viscosity of the gel and the time of gel formation (Figure 3), to enhance the stability of drilling fluids in mineralized water (Figure 4) and to increase thermostability of drilling fluid.

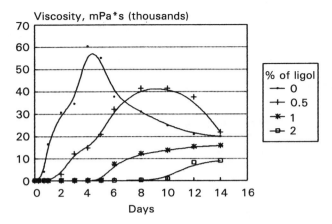

Figure 3. Viscosity changes of gel-forming systems, depending on ligol content.

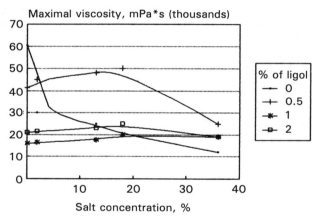

Figure 4. Hydrogel stability in mineralized water, depending on ligol content.

It was estimated, that ligols are better surfactants, compared with initial lignosulphonates (Figure 5). It makes possible to use ligols as an additive to flotation systems. For example, addition of ligols into flotation compositions increases the degree of extraction by 13% (relatively to phosphorous pentoxide) and by 5 up to 17,5% of KCl.

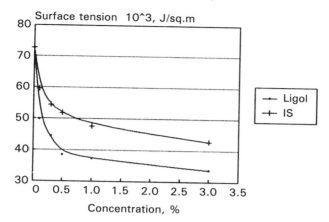

Figure 5. The surfactant properties of ligol and lignosulphonates in aqueous solutions.

CONCLUSIONS

Ligols are suitable for polyurethane production as a replacement of traditional ether-polyols.

Ligols may be used as a component for drilling fluids and as industrial surfactants.

Ligols have been successfully tested as flotation reagents.

REFERENCES

1. Mozheiko, L. N. , Gromova, M. F. , Sergeeva, V. N, and Bakalo, L. A. // Khimiya drevesiny (Wood chemistry), 1977, N2, p. 24-28; 29-32.

2. Mozheiko, L. N. , Gromova, M. F. , Bakalo, L. A. , and Sergeeva, V. N. // Vysokomol. Soyed. , A 23 (1), p. 126 (Polymer Science USSR, vol. 23, N1, p. 149.

3. Leo C. -F. Wu and Wolfgang G. Glasser // J. Appl. Polymer Sci. , vol. 29, p. 1111-1123.

81

Some new applications of modified lignin

Henryk Struszczyk, Karolina Grzebieniak, Krystyna Wrzésniewska-Tosik and Arkadiusz Wilczek - Institute of Chemical Fibres, 90-570 Lodz, 19 C. Sklodowska Str., Poland.

ABSTRACT

The manufacture of polymeric materials from renewable biomass will become more important in the near future. Lignin is one of the most abundant renewable natural products specially for production of chemicals. The aim of this report was to present a new application of modified lignin as the additive for modification of poly(ethylene terephthalate). The properties of modified polyester have also been discussed.

INTRODUCTION

The manufacture of polymeric materials from renewable biomass will become more important in the future. Of special interest is the conversion of agricultural residues and wastes into fuels as well as into chemicals, also high molecular weight.

Lignin, derived mainly from wood as a by product in the pulping process, is one of the most abundant renewable natural products. Use of lignin for production of chemicals seems to be a future possibility espe-

cially as a raw material for polymers (1-2).
The aim of this report is to present a new application
of modified lignin as an additive for modification of
poly(ethylene terephthalate) (PET).

EXPERIMENTAL
MATERIALS AND METHODS

Lignosulfonates of Ultrazine NAS (NAS) and Borresperse
NA(NA) (Borregaard Inc., Ltd, Sapsborg, Norway) as
well as experimental steam explosion lignin under a
code-name of L-7 (Cultor Oy, Kantvik, Finland) were
used for investigations.
Lignin samples were subjected to modification by terephthaloyl chloride (TC) and eventually additionally purified by ethylene glycol (EG) according to the original
methods (3-4). Poly(ethylene terephthalate) was synthesized from dimenthylterephthalate (DMT) and ethylene
glycol using a manganese acetate and antimony trioxide
as the catalysts by standard methods. Modified lignin
additives of 0.05-24.0 wt% of DMT were introduced in
different stages of polycondensation process (5).
Determination of properties of lignin-like additives
was carried out using standard methods.
Intrinsic viscosity of PET was realized by a viscosi-
metric method at 25°C in a mixture of phenol and tetra-
chloroethane with a weight ratio of 1:1. Melting point
of PET was determined using the Towsend-Crowther apparatus whereas its melt flow index was determined using
a plastometer with 0.5 mm spinneret and basic load of
2.16 kg at a temperature of 290°C after 6 min. an d 60
min. for the estimation of melt thermostability. The
carboxyl end group content was estimated by titration
with an ethyl alcohol solution of sodium hydroxide
with a PET sample dissolved in a benzyl alcohol. The
colour of PET samples was determined according to the

Hunter standard using ND-101D photocolorymeter of
Nippon Denshoku Kogo Co. Ltd.

RESULTS AND DISCUSSION

The general properties of lignin-terephthaloyl
chloride products are presented in Table 1.

Table 1. Some general properties of lignin modified
by terephthaloyl chloride

Property	Parameter
Form	powder
Colour	cream-white to beige
Melting point, °C	280 - 350
Solubility in water	non soluble
Solubility in organic solvents	hot ethylene glycol

Some behaviour of modified PET was investigated and
results obtained are presented as follows:
- effect of lignin-like additive introduction method
  (Table 2),
- effect of lignin-like additive type (Table 3),
- effect of amount of lignin-like additive introduced
  (Table 4),
- effect of lignin-like additive modification
  (Table 5).

Based on the results presented in Table 1-5 it can be
concluded that the lignin modified by terephthaloyl
chloride exists in a solid products form with melting
points in the range of 280-350°C and seems to be suitable for PET modification mainly on their introduc-

Table 2. Effect of lignin-like additive* introduction method for poly(ethylene terephthalate) synthesis on its properties

No	Introduction method	PET properties				Remarks
		$[\eta]$	m.p. °C	MFI ,g/10 min	COOH mval/kg	
1	before transesterification	-	-	-	-	no polymer was obtained
2	after transesterification	0.679	261.3	1.41	10.0	-
3	after polycondensation	0.359	260.0	-	250.0	rapid dropping of PET melt viscosity
	standard PET	0.710	261.8	1.03	32.2	-

* - 2.0 wt% type of lignin-like additive

MFI - melt flow index (290°C)

Table 3. Effect of lignin-like additive* type on modified poly(ethylene terephthalate) properties

| Symbol of sample | Lignin type | m.p. °C | Content, % | | | $[\eta]$ | m.p. °C | MFI g/10 min | MFI % | COOH mval/kg |
			C	H	S					
L-4	NAS	360	58.42	3.74	0.80	0.617	261.9	1.60	105.0	13.2
L-7	NAS	333-336	57.18	3.81	0.25	0.661	262.7	1.00	38.0	7.6
L-13	NAS	345-348	57.86	3.66	0.21	0.679	262.3	1.48	114.2	-
L-9	NA/EG[a]	333-336	57.93	4.01	0.20	0.679	262.3	1.41	97.2	10.0
L-10	NA	336-339	57.65	3.30	0.36	0.675	260.8	1.97	74.1	13.5
L-8	NA	336-340	57.56	3.84	0.25	0.639	262.0	1.50	46.0	15.0
L-12	NA	278-282	57.06	3.88	0.24	0.689	261.6	1.50	98.0	-
L-14	NA/EG[a]	338-341	57.18	3.79	0.26	0.614	262.6	1.49	101.7	-
L-6	L-7	350-370	66.99	4.93	-	-	262.9	1.65	69.7	10.0
Standard PET	-	-	-	-	-	0.710	261.8	1.03	182.5	32.2
Standard PET[b]	-	-	-	-	-	0.636	260.3	1.72	33.7	8.3

a - additional modification by ethylene glycol,   b - with thermal stabilizer

* - 2 wt% on DMT   MFI - augmentation after 1 h of heating (290°C)

Table 4. Effect of amount of lignin-like additive used on modified poly(ethylene terephthalate) properties

Symbol of sample	Lignin type	Additive amount ,wt%	$[\eta]$	m.p. ,°C	Modified PET properties MFI ,g/10 min.	MFI ,%	COOH ,mval/kg
L-9	NAS/EG	0.2	0.610	261.7	1.12	129.5	22.5
		2.0	0.679	262.3	1.41	97.9	10.0
		6.0	0.645	262.3	1.42	124.6	12.1
L-10	NA/EG	0.05	0.674	261.3	1.31	161.8	37.1
		0.10	0.686	261.6	1.12	195.5	27.9
		0.20	0.697	262.7	1.13	98.2	11.5
		1.00	0.634	262.5	1.70	67.1	9.5
		2.00	0.635	260.8	1.97	74.1	13.5
L-14	NA/EG	2.0	0.664	262.6	1.49	101.7	–
		24.0	0.666	262.5	1.48	117.6	–
Standard PET	–	–	0.710	261.8	1.03	182.6	32.2

Table 5. Effect of lignin-like additive on poly(ethylene terephthalate) colour

| | Lignin-like additive | | Modified lignin properties | | | | | |
| | | | colour index** | | | | | |
Symbol of sample	Lignin type	Colour of additive	L, %	a	b	$[\eta]$	m.p. ,°C	COOH ,mval/kg
L-7	NAS	l.beige	67.8	+ 0.4	+ 7.5	0.661	262.7	7.6
L-7*	NAS	white	71.1	- 0.5	+ 5.4	0.686	263.0	13.0
L-8	NA	l.beige	64.5	+ 0.7	+ 8.2	0.639	262.0	15.0
L-7*	NA	white	66.7	- 0.3	+ 6.4	0.686	261.2	12.3
Standard PET	-	-	68.7	- 1.0	+ 3.1	0.710	261.8	32.2
Standard PET^y	-	-	70.0	- 0.6	+ 1.0	0.636	260.3	8.3

* - purified

** - according to a Hunter scale where:

L - total reflection of light

a - grade od reddish (+) or greenish (-) colour

b - grade of yellowish (+) or bluish (-) colour

y - with thermal stabilizer

tion after the transesterfication stage (Table 2).
This method was used for further studies.

The lignin-like additives advantageously affected the
thermostability of modified PET melt at a temperature
of 290°C. These results correlate with suitable lower
content of carboxyl acid end groups (Table 3).

The lignin-like additives were introduced at levels
of 0.01 to 24.0% on DMT weight. No negative effect
was observed in this range (Table 4). Purification of
lignin-like additives by ethylene glycol was substantially affected on the colour of the modified PET
(Table 5).

The larger scope of selected modified PET synthesis
were also confirmed by the above results as well as
by the spinnability of the modified polyester being
similar to standard poly(ethylene terephthalate). The
fiber-forming properties as well as the modified PET
fibre behaviour such as dyeability, electrostacity,
pilling tendency and elasticity will be presented in
another report.

CONCLUSIONS

1. Lignin modified by terephthaloyl chloride can be
   succesfully used as a modifying additive for poly
   (ethylene terephthalate) synthesis especially when
   introduced directly after a transesterification
   stage.

2. Some properties of modified poly(ethylene terenph-
   thalate) such as average molecular weight, melting
   point or melt flow index were not sensitively
   affected by the lignin-like additive.

3. Modified poly(ethylene terephthalate) colour was
   dependent upon the lignin-like additive type
   whereas a purified additive was able to create a

modified polyester similar in colour to a standard
polymer.

REFERENCES

1. Glasser W.F., Sarkanen S., <u>Lignin: Properties and Materials</u>, ACS Symposium series No 397, Washington DC, 1989.   -

2. Struszczyk H., J. Macromol. Sci-chem., A 23(8), 1193 (1982).

3. Polish Pat. 134 256 (1981)

4. Polish Pat. Appl. P-289041 (1990)

5. Polish Pat. Appl. P-289767 (1991)

82

Utilization of lignocellulose for microbial protein production

D. Witkowska and J. Sobieszczański - Department of Biotechnology and
Microbiology, Academy of Agriculture, 50-375 Wroclaw, Norwida str. 25, Poland.

ABSTRACT

In our experiments the post liquid culture of Tri-
choderma reesei M7-1 containing cellulases, xylanases and polygalacturonases was used for degradation
of modified wheat straw. Reducing sugar (30g/L) were
obtained at 45°C after 72h of hydrolysis. Those hydrolysates enriched with the remaining components of
yeasts medium were utilized in biomass production.
The following kinds of yeasts were used. Candida utilis, Candida tropicalis and Trichosporon cutaneum.
Yeasts biomass 14-22.4g per 100g of wheat straw was
obtained as a result of this experiment. Therefore
production of high protein concentrates from lignocellulose substrates for animals is possible.

INTRODUCTION

Lignocellulose is a cheap and renewable raw material
which after bioconversion can be used for production
of chemicals or other products. The purpose of our
investigation was to use the straw - lignocellulose
after enzymatic hydrolysis to simple sugars, for
yeasts biomass production - as protein concentrated
feeding stuff.

EXPERIMENTAL

In the research, our UV mutant of <u>Trichoderma reesei</u>
M7-1 producing the extracellular cellulases, xylanases and polygalacturonases was used. A culture was
performed in modified Saunders medium with an addition of 2% cellulose Avicel and 2% sugar beet pulps
as the carbon source (6). This process was going on
at 28^0C, pH=5.0 in the Braun-Melsungen biostat, 1L
capacity. After 120 hours a post-culture liquid was
whirled in the Janetzki centrifuge at 3000xg and
then it was filtered through a Shotta filter G-2.
The following enzyme activities were determined in
the post-culture liquid: endoglucanases (CMC-ases)
(4,5), exoglucanases (FP-ases) (4,5), β-D-glucosida-
ses (1), xylanases (2) and polygalacturonases (3),
(Table 1).

Table 1

Enzymatic characterisic of post liquid
culture of <u>Trichoderma reesei</u> M7-1

CMCase	FPase	β-D-glucosi-dase	Xylanase	Polygalacturona-se
		Enzymes activity μM/ml/min.		
7.5	1.4	0.38	148.3	2.6

The wheat straw dried and powdered was treated with
NaOH solutions(various concentrations: 0.25M, 0.5M,
0.75M and 1.0M) 30 min. at 121^0C, 60 min. at 121^0C,
and 24 hours at 20^0C.

Samples of the modified wheat straw (3g) were placed in 200ml Erlenmayer's flasks then the post-
culture liquid (40ml) with an activity 1.4 FPA/ml
was added. The process was maintained in a shaking
water bath at 45^0C for 72 hours. After finishing
the hydrolysis, the samples were filtered. In the
next step assays were conducted for reducing sugars,
total sugars, glucose, pentoses.

The enzymatic hydrolysates including the simple sugars were used as a source of carbon for producing

the yeasts biomass. They were supplied with a source
of nitrogen and other components of the medium for
yeasts. The media were sterilised and inoculated
after chilling. A process of yeasts biomass production (Candida utilis, Candida tropicalis and Trichosporon cutaneum) was performed at 28°C for 48 hours in the shaker. After finishing the culture, the
amount of yeast biomass was determined by the weight
method and the amount of proteins (Nx 6.25) by Kjeldahl method. The sugar residues in the hydrolysate
filtrates were determined.

RESULTS

The most effective manner of wheat straw modification was the treatment with 1M NaOH at 121°C for 30
min. In this way a loss of the dry matter content
of the wheat straw was 45.6% (Figure 1). In general,
treatment of the wheat straw with NaOH concentrations
ranging from 0.25M to 1M at 121°C for 30 or 60 min.
resulted in 39.4-45.6% loss of lignin compounds in
the straw. On the other hand, treatment of the straw
with NaOH at the room temperature even for 24 hours
resulted in 26.4-36% loss of lignin compounds.

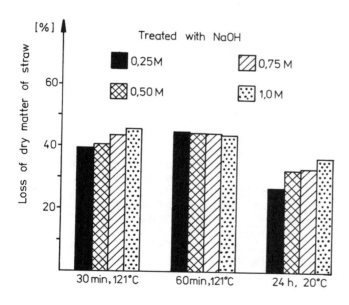

Figure 1. Effect of wheat straw modification on loss
          of lignin substances.

In a process of enzymatic hydrolysis of the modified
wheat straw (75g/L), various sugars were released:
total sugars - 48g/L; reducing sugars - 30g/L; glucose - 6.7g/L; pentoses - 10g/L.

Table 2

Production of yeast biomass on enzymatic
hydrolysates from wheat straw

Kind of yeast's strain	Biomass		Protein		
	g/L	g/100g of straw	g/L	g/100g of straw	g/g of utilized of red. sugars
Candida utilis Z-2	10.5	14.0	5.9	7.9	0.197
Candida tropi-calis E-7	14.4	19.2	5.3	7.1	0.175
Tricho-sporon cuta-neum	16.9	22.5	7.8	10.4	0.260

The yeasts Candida utilis Z-2. Candida tropicalis
C-7 and Trichosporon cutaneum cultivated in the
straw hydrolysate, produced from 10.5 to 16.9g/L
of biomass. An amount of the determined protein
ranged between 5.3 and 7.8g/L. Our calculations show
that 14.0 to 22.5g of biomass and 7.1 to 10.4g of
protein can be obtained from 100g of wheat straw
(Table 2). The most effective producer of biomass
and protein was Trichosporon cutaneum strain. During cultivation, the yeasts proved to utilize the
reducing sugars in 96-98%, glucose in 90-99.4% and
pentoses in 60-67% (Table 3). Thanks to microorganisms, straw and other lignocellulose wastes in the
farm can be converted to a high quality vitamin-
proteins concentrated feeding stuff.

Table 3

Utilization of sugars from enzymatic
hydrolysates in biomass production
by yeasts

Kind of yeast's strain	Total sugars	Reducing sugars (%)	Glucose	Pentoses
Candida utilis Z-2	94.4	98.0	99.4	67.0
Candida tropicalis C-7	93.3	98.0	99.4	65.0
Trichosporon cutaneum	87.5	96.0	98.9	60.0

REFERENCES

1. Deschamps,F.; Huet,M.C. Biotechnol.Lett.**1984**,6,
   55.
2. Deschamps,F.; Huet,M.C. Appl.Microbiol.Biotechnol.
   1985,22,177.
3. Hancock,J.G.; Millar,R.L. Phytopatology **1965**,55,
   346.
4. Mandels,M.; Andreotti,R.; Roche C. Biotechnol.
   Bioeng. Symp. **1976**,6,17.
5. Miller,G.L.; Anal. Chem. **1959**,31,426.
6. Witkowska,D.; Bień,M.; Sobieszczański,J. Micro-
   biologia SEM. **1989**,5,113.

Index